# STUDENT SOLUTIONS MANUAL TO ACCOMPANY

## ZILL/CULLEN'S
# DIFFERENTIAL EQUATIONS WITH BOUNDARY-VALUE PROBLEMS
### THIRD EDITION

**Warren S. Wright**
*Loyola Marymount University*

**Carol D. Wright**

## PWS Publishing Company
*Boston*

**PWS**
Publishing Company

20 Park Plaza
Boston, Massachusetts 02116

**International Thomson Publishing**
The trademark ITP is used under license.

**Printed in the United States of America**

2  3  4  5  6  7  8  9  10 - - 98  97  96  95  94

ISBN 0-534-93159-6

# Table of Contents

# 1 Introduction to Differential Equations

## Exercises 1.1

**3.** First-order; nonlinear because of $yy'$.

**6.** Second-order; nonlinear because of $\sin y$.

**9.** Third-order; linear.

**12.** From $y = 8$ we obtain $y' = 0$, so that $y' + 4y = 0 + 4(8) = 32$.

**15.** From $y = 5 \tan 5x$ we obtain $y' = 25 \sec^2 5x$. Then

$$y' = 25 \sec^2 5x = 25 \left(1 + \tan^2 5x\right) = 25 + (5 \tan 5x)^2 = 25 + y^2.$$

**18.** First write the differential equation in the form $2xy + \left(x^2 + 2y\right) y' = 0$. Implicitly differentiating $x^2 y + y^2 = c_1$ we obtain $2xy + \left(x^2 + 2y\right) y' = 0$.

**21.** Implicitly differentiating $y^2 = c_1 \left(x + \dfrac{1}{4} c_1\right)$ we obtain $y' = c_1/2y$. Then

$$2xy' + y(y')^2 = \frac{c_1 x}{y} + \frac{c_1^2}{4y} = \frac{y^2}{y} = y.$$

**24.** Differentiating $P = ac_1 e^{at}/\left(1 + bc_1 e^{at}\right)$ we obtain

$$\frac{dP}{dt} = \frac{\left(1 + bc_1 e^{at}\right) a^2 c_1 e^{at} - ac_1 e^{at} \cdot abc_1 e^{at}}{\left(1 + bc_1 e^{at}\right)^2}$$

$$= \frac{ac_1 e^{at}}{1 + bc_1 e^{at}} \cdot \frac{\left[a \left(1 + bc_1 e^{at}\right) - abc_1 e^{at}\right]}{1 + bc_1 e^{at}} = P(a - bP).$$

**27.** First write the differential equation in the form $y' = \dfrac{-x^2 - y^2}{x^2 - xy}$. Then $c_1(x + y)^2 = xe^{y/x}$ implies

$c_1 = \dfrac{xe^{y/x}}{(x + y)^2}$ and implicit differentiation gives $2c_1(x + y)(1 + y') = xe^{y/x} \dfrac{xy' - y}{x^2} + e^{y/x}$. Solving for $y'$ we obtain

$$y' = \frac{e^{y/x} - \dfrac{y}{x} e^{y/x} - 2c_1(x + y)}{2c_1(x + y) - e^{y/x}} = \frac{1 - \dfrac{y}{x} - \dfrac{2x}{x + y}}{\dfrac{2x}{x + y} - 1} = \frac{-x^2 - y^2}{x^2 - xy}.$$

**30.** From $y = e^{2x} + xe^{2x}$ we obtain $\dfrac{dy}{dx} = 3e^{2x} + 2xe^{3x}$ and $\dfrac{d^2y}{dx^2} = 8e^{2x} + 4xe^{2x}$ so that $\dfrac{d^2y}{dx^2} - 4\dfrac{dy}{dx} + 4y = 0$.

**33.** From $y = \ln|x + c_1| + c_2$ we obtain $y' = \dfrac{1}{x + c_1}$ and $y'' = \dfrac{-1}{(x + c_1)^2}$, so that $y'' + (y')^2 = 0$.

**36.** From $y = x\cos(\ln x)$ we obtain $y' = -\sin(\ln x) + \cos(\ln x)$ and $y'' = \dfrac{-1}{x}\cos(\ln x) - \dfrac{1}{x}\sin(\ln x)$, so that $x^2y'' - xy' + 2y = 0$.

**39.** From $y = x^2 e^x$ we obtain $y' = x^2 e^x + 2xe^x$, $y'' = x^2 e^x + 4xe^x 2e^x$, and $y''' = x^2 e^x + 6xe^x + 6e^x$, so that $y''' - 3y'' + 3y' - y = 0$.

**42.** From $y = \begin{cases} 0, & x < 0 \\ x^3, & x \geq 0 \end{cases}$ we obtain $y' = \begin{cases} 0, & x < 0 \\ 3x^2, & x \geq 0 \end{cases}$ so that $(y')^2 = \begin{cases} 0, & x < 0 \\ 9x^4, & x \geq 0. \end{cases}$

**45.** By inspection, $y = -1$ is a singular solution. Note that this is the "solution" obtained by computing the limit as $c$ approaches infinity of the one-parameter family of solutions.

**48.** From $y = e^{mx}$ we obtain $y' = me^{mx}$ and $y'' = m^2 e^{mx}$. Then $y'' + 10y' + 25y = 0$ implies

$$m^2 e^{mx} + 10me^{mx} + 25e^{mx} = (m+5)^2 e^{mx} = 0.$$

Since $e^{mx} > 0$ for all $x$, $m = -5$. Thus, $y = e^{-5x}$ is a solution.

**51.** It is easily shown that $y_1 = x^2$ and $y_2 = x^3$ are solutions. If $y_3 = c_1 y_1 + c_2 y_2 = c_1 x^2 + c_2 x^3$ then $y_3' = 2c_1 x + 3c_2 x^2$ and $y_3'' = 2c_1 + 6c_2 x$ so that $x^2 y_3'' - 4xy_3' + 6y_3 = 0$. Hence $c_1 y_1$, $c_2 y_2$, and $y_1 + y_2$ are solutions.

## ——— Exercises 1.2 ———————

**3. (a)** From $g = k/R^2$ we find $k = gR^2$.

**(b)** Using $a = \dfrac{d^2r}{dt^2}$ and part (a) we obtain $\dfrac{d^2r}{dt^2} = a = \dfrac{k}{r^2} = \dfrac{gR^2}{r^2}$ or $\dfrac{d^2r}{dt^2} - \dfrac{gR^2}{r^2} = 0$.

**(c)** Part (b) becomes $\dfrac{dv}{dr}\dfrac{dr}{dt} - \dfrac{gR^2}{r^2} = 0$ or $v\dfrac{dv}{dr} - \dfrac{gR^2}{r^2} = 0$.

**6.** By Kirchoff's second law we obtain $R\dfrac{dq}{dt} + \dfrac{1}{C}q = E(t)$.

**9.** The differential equation is $\dfrac{dh}{dt} = -\dfrac{0.6A_0}{A_w}\sqrt{2gh}$. We have $A_0 = \pi\left(\dfrac{1}{12}\right)^2 = \dfrac{\pi}{144}$ and $g = 32$. To find $A_w$ we solve $x^2 + (5 - h)^2 = 25$ where $x$ represents the radius of the circular area of the surface of the water whose depth is $h$. From $x = \sqrt{10h - h^2}$ we obtain $A_w = \pi(10h - h^2)$. Thus

$$\dfrac{dh}{dt} = -\dfrac{0.6\pi/144}{\pi(10h - h^2)}\sqrt{64h} = -\dfrac{1}{30h(10 - h)}\sqrt{h} = -\dfrac{1}{30\sqrt{h}\,(10 - h)}.$$

**12.** Equating Newton's law with the net forces in the $x$- and $y$-directions gives $m\dfrac{d^2x}{dt^2} = 0$ and $m\dfrac{d^2y}{dt^2} = -mg$, respectively.

**15.** To better understand the problem extend the line $L$ down to the x-axis. Then we see from the figure that $\phi = 2\theta$, $\tan\phi = \dfrac{x}{y}$, and $\dfrac{dy}{dx} = \tan\left(\dfrac{\pi}{2} - \theta\right) = \cot\theta$. Now

$$\tan\phi = \tan 2\theta = \frac{2\tan\theta}{1 - \tan^2\theta} = \frac{x}{y}, \text{ so } \frac{x}{y} = \frac{2(dx/dy)}{1 - (dx/dy)^2} \text{ and } x\left(\frac{dx}{dy}\right)^2 + 2y\left(\frac{dx}{dy}\right) = x.$$

**18.** Substituting into the differential equation we obtain $-(m_0 - at)g = (m_0 - at)\dfrac{dv}{dt} + b(-a)$ or $(m_0 - at)\dfrac{dv}{dt} = ab - m_0 g + agt.$

**21.** The differential equation is $\dfrac{dA}{dt} = k(M - A)$.

─────── **Chapter 1 Review Exercises** ───────

**3.** Second-order; partial.

**6.** From $y = c_1 \cos(\ln x) + c_2 \sin(\ln x)$ we obtain $y' = \dfrac{1}{x}\left[c_2 \cos(\ln x) - c_1 \sin(\ln x)\right]$ and

$$y'' = \frac{-1}{x^2}\left[c_1 \cos(\ln x) + c_2 \sin(\ln x) + c_2 \cos(\ln x) - c_1 \sin(\ln x)\right]$$

so that $x^2 y'' + xy' + y = 0$.

**9.** $y = x^2$

**12.** $y = 2$

**15.** $y = \sin x$, $y = \cos x$, $y = 0$

**18.** If $|x| < 2$ and $|y| > 2$, then $(dy/dx)^2 < 0$ and the differential equation has no real solutions. This is also true for $|x| > 2$ and $|y| < 2$.

# 2 First-Order Differential Equations

─────── **Exercises 2.1** ────────────────────

**3.** For $f(x, y) = \dfrac{y}{x}$ we have $\dfrac{\partial f}{\partial y} = \dfrac{1}{x}$. Thus the differential equation will have a unique solution in any region where $x \neq 0$.

**6.** For $f(x, y) = \dfrac{x^2}{1 + y^3}$ we have $\dfrac{\partial f}{\partial y} = \dfrac{-3x^2 y^2}{(1 + y^3)^2}$. Thus the differential equation will have a unique solution in any region where $y \neq -1$.

**9.** For $f(x, y) = x^3 \cos y$ we have $\dfrac{\partial f}{\partial y} = -x^2 \sin y$. Thus the differential equation will have a unique solution in the entire plane.

**12.** Two solutions are $y = 0$ and $y = x^2$. (Also, any constant multiple of $x^2$ is a solution.)

**15.** For $y = cx$ we have $y' = c$, from which we see that $y = cx$ is a solution of $xy' = y$ for all values of $c$. All of these solutions satisfy the initial condition $y(0) = 0$. The piecewise defined function is not a solution since it is not differentiable at $x = 0$.

**18.** We identify $f(x, y) = \sqrt{y^2 - 9}$ and $\partial f / \partial y = y^2 / \sqrt{y^2 - 9}$. We then note that $f(x, y)$ is discontinuous for $|y| < 3$ and that $\partial f / \partial y$ is discontinuous for $|y| < 3$. Applying Theorem 2.1 we see that the differential equation is not guaranteed to have a unique solution at $(5, 3)$.

─────── **Exercises 2.2** ────────────────────

In many of the following problems we will encounter an expression of the form $\ln |g(y)| = f(x) + c$. To solve for $g(y)$ we exponentiate both sides of the equation. This yields $|g(y)| = e^{f(x)+c} = e^c e^{f(x)}$ which implies $g(y) = \pm e^c e^{f(x)}$. Letting $c_1 = \pm e^c$ we obtain $g(y) = c_1 e^{f(x)}$.

**3.** From $dy = -e^{-3x}\, dx$ we obtain

$$y = \frac{1}{3} e^{-3x} + c.$$

**6.** From $dy = 2xe^{-x} dx$ we obtain

$$y = -2xe^{-x} - 2e^{-x} + c.$$

**9.** From $\dfrac{1}{y^3}\, dy = \dfrac{1}{x^2}\, dx$ we obtain

$$y^{-2} = \frac{2}{x} + c.$$

**4**

**12.** From $\left(\dfrac{1}{y} + 2y\right) dy = \sin x \, dx$ we obtain

$$\ln |y| + y^2 = -\cos x + c.$$

**15.** From $\dfrac{y}{2 + y^2} \, dy = \dfrac{x}{4 + x^2} \, dx$ we obtain

$$\ln |2 + y^2| = \ln |4 + x^2| + c \quad \text{or} \quad 2 + y^2 = c_1 \left(4 + x^2\right).$$

**18.** From $\dfrac{y^2}{y + 1} \, dy = \dfrac{1}{x^2} \, dx$ we obtain

$$\tfrac{1}{2}y^2 - y + \ln |y + 1| = -\dfrac{1}{x} + c \quad \text{or} \quad \tfrac{1}{2}y^2 - y + \ln |y + 1| = -\dfrac{1}{x} + c_1.$$

**21.** From $\dfrac{1}{S} \, dS = k \, dr$ we obtain $S = ce^{kr}$.

**24.** From $\dfrac{1}{N} \, dN = \left(te^{t+2} - 1\right) dt$ we obtain

$$\ln |N| = te^{t+2} - e^{t+2} - t + c.$$

**27.** From $\dfrac{e^{2y} - y}{e^y} \, dy = -\dfrac{\sin 2x}{\cos x} \, dx = -\dfrac{2 \sin x \cos x}{\cos x} \, dx$ or $\left(e^y - ye^{-y}\right) dy = -2 \sin x \, dx$ we obtain

$$e^y + ye^{-y} + e^{-y} = 2 \cos x + c.$$

**30.** From $\dfrac{y}{(1 + y^2)^{1/2}} \, dy = \dfrac{x}{(1 + x^2)^{1/2}} \, dx$ we obtain

$$\left(1 + y^2\right)^{1/2} = \left(1 + x^2\right)^{1/2} + c.$$

**33.** From $\dfrac{y - 2}{y + 3} \, dy = \dfrac{x - 1}{x + 4} \, dx$ or $\left(1 - \dfrac{5}{y - 3}\right) dy = \left(1 - \dfrac{5}{x + 4}\right) dx$ we obtain

$$y - 5 \ln |y - 3| = x - 5 \ln |x + 4| + c \quad \text{or} \quad \left(\dfrac{x + 4}{y - 3}\right)^5 = c_1 e^{x-y}.$$

**36.** From $\sec y \dfrac{dy}{dx} + \sin x \cos y - \cos x \sin y = \sin x \cos y + \cos x \sin y$ we find $\sec y \, dy = 2 \sin y \cos x \, dx$ or

$\dfrac{1}{2 \sin y \cos y} \, dy = \csc 2y \, dy = \cos x \, dx$. Then

$$\tfrac{1}{2} \ln |\csc 2y - \cot 2y| = \sin x + c.$$

**39.** From $\dfrac{1}{y^2} \, dy = \dfrac{1}{e^x + e^{-x}} \, dx = \dfrac{e^x}{(e^x)^2 + 1} \, dx$ we obtain

$$-\dfrac{1}{y} = \tan^{-1} e^x + c.$$

**42.** From $\dfrac{1}{1+(2y)^2}\,dy = \dfrac{-x}{1+(x^2)^2}\,dx$ we obtain

$$\frac{1}{2}\tan^{-1}2y = -\frac{1}{2}\tan^{-1}x^2 + c \quad \text{or} \quad \tan^{-1}2y + \tan^{-1}x^2 = c_1.$$

Using $y(1) = 0$ we find $c_1 = \pi/4$. The solution of the initial-value problem is

$$\tan^{-1}2y + \tan^{-1}x^2 = \frac{\pi}{4}.$$

**45.** From $\dfrac{1}{x^2+1}\,dx = 4\,dy$ we obtain $\tan^{-1}x = 4y + c$. Using $x(\pi/4) = 1$ we find $c = -3\pi/4$. The solution of the initial-value problem is

$$\tan^{-1}x = 4y - \frac{3\pi}{4} \quad \text{or} \quad x = \tan\left(4y - \frac{3\pi}{4}\right).$$

**48.** From $\dfrac{1}{1-2y}\,dy = dx$ we obtain $-\dfrac{1}{2}\ln|1-2y| = x + c$ or $1-2y = c_1e^{-2x}$. Using $y(0) = 5/2$ we find $c_1 = -4$. The solution of the initial-value problem is

$$1 - 2y = -4e^{-2x} \quad \text{or} \quad y = 2e^{-2x} + \frac{1}{2}.$$

**51.** By inspection a singular solution is $y = 1$.

**54.** Separating variables we obtain $\dfrac{dy}{(y-1)^2} = dx$. Then $-\dfrac{1}{y-1} = x + c$ and $y = \dfrac{x+c-1}{x+c}$. Setting $x = 0$ and $y = 1.01$ we obtain $c = -100$. The solution is

$$y = \frac{x-101}{x-100}.$$

**57.** Let $u = x + y + 1$ so that $du/dx = 1 + dy/dx$. Then $\dfrac{du}{dx} - 1 = u^2$ or $\dfrac{1}{1+u^2}\,du = dx$. Thus $\tan^{-1}u = x + c$ or $u = \tan(x+c)$, and

$$x + y + 1 = \tan(x+c) \quad \text{or} \quad y = \tan(x+c) - x - 1.$$

**60.** Let $u = x + y$ so that $du/dx = 1 + dy/dx$. Then $\dfrac{du}{dx} - 1 = \sin u$ or $\dfrac{1}{1+\sin u}\,du = dx$. Multiplying by $(1 - \sin u)/(1 - \sin u)$ we have $\dfrac{1-\sin u}{\cos^2 u}\,du = dx$ or $\left(\sec^2 u - \tan u \sec u\right)du = dx$. Thus

$$\tan u - \sec u = x + c \quad \text{or} \quad \tan(x+y) - \sec(x+y) = x + c.$$

**——— Exercises 2.3 ———**

**3.** Since $f(tx, ty) = \dfrac{(tx)^3(ty) - (tx)^2(ty)^2}{(tx + 8ty)^2} = t^2 f(x, y)$, the function is homogeneous of degree 2.

**6.** Since $f(tx, ty) = \sin \dfrac{x}{x + y} = f(x, y)$, the function is homogeneous of degree 0.

**9.** Since $f(tx, ty) = \left(\dfrac{1}{tx} + \dfrac{1}{ty}\right)^2 = \dfrac{1}{t^2} f(x, y)$, the function is homogeneous of degree $-2$.

**12.** Letting $y = ux$ we have

$$(x + ux)\, dx + x(u\, dx + x\, du) = 0$$

$$(1 + 2u)\, dx + x\, du = 0$$

$$\frac{dx}{x} + \frac{du}{1 + 2u} = 0$$

$$\ln|x| + \frac{1}{2}\ln|1 + 2u| = c$$

$$x^2 \left(1 + 2\frac{y}{x}\right) = c_1$$

$$x^2 + 2xy = c_1.$$

**15.** Letting $y = ux$ we have

$$\left(u^2 x^2 + ux^2\right) dx - x^2(u\, dx + x\, du) = 0$$

$$u^2\, dx - x\, du = 0$$

$$\frac{dx}{x} - \frac{du}{u^2} = 0$$

$$\ln|x| + \frac{1}{u} = c$$

$$\ln|x| + \frac{x}{y} = c$$

$$y \ln|x| + x = cy.$$

**18.** Letting $y = ux$ we have

$$(x + 3ux)\, dx - (3x + ux)(u\, dx + x\, du) = 0$$

$$\left(u^2 - 1\right) dx + x(u + 3)\, du = 0$$

$$\frac{dx}{x} + \frac{u + 3}{(u - 1)(u + 1)}\, du = 0$$

$$\ln|x| + 2\ln|u - 1| - \ln|u + 1| = c$$

$$\frac{x(u - 1)^2}{u + 1} = c_1$$

$$x\left(\frac{y}{x} - 1\right)^2 = c_1\left(\frac{y}{x} + 1\right)$$

$$(y - x)^2 = c_1(y + x).$$

**21.** Letting $x = vy$ we have

$$2v^2 y^3 (v\, dy + y\, dv) - \left(3v^3 y^3 + y^3\right) dy = 0$$

$$2v^2 y\, dv - \left(v^3 + 1\right) dy = 0$$

$$\frac{2v^2}{v^3 + 1}\, dv - \frac{dy}{y} = 0$$

$$\frac{2}{3} \ln\left|v^3 + 1\right| - \ln|y| = c$$

$$\left(v^3 + 1\right)^{2/3} = c_1 y$$

$$\left(\frac{x^3}{y^3} + 1\right)^2 = c_2 y^3$$

$$\left(x^3 + y^3\right)^2 = c_2 y^9.$$

**24.** Letting $y = ux$ we have

$$\left(u^3 x^3 + x^3 + u^2 x^3\right) dx - u^2 x^3 (u \, dx + x \, du) = 0$$

$$\left(1 + u^2\right) dx - u^2 x \, du = 0$$

$$\frac{dx}{x} - \frac{u^2}{u^2 + 1} du = 0$$

$$\ln|x| - u + \tan^{-1} u = c$$

$$\ln|x| - \frac{y}{x} + \tan^{-1} \frac{y}{x} = c.$$

**27.** Letting $y = ux$ we have

$$(ux + x \cot u) \, dx - x(u \, dx + x \, du) = 0$$

$$\cot u \, dx - x \, du = 0$$

$$\frac{dx}{x} - \tan u \, du = 0$$

$$\ln|x| + \ln|\cos u| = c$$

$$x \cos \frac{y}{x} = c.$$

**30.** Letting $y = ux$ we have

$$\left(x^2 + ux^2 + 3u^2 x^2\right) dx - \left(x^2 + 2ux^2\right)(u \, dx + x \, du) = 0$$

$$\left(1 + u^2\right) dx - x(1 + 2u) \, du = 0$$

$$\frac{dx}{x} - \frac{1 + 2u}{1 + u^2} du = 0$$

$$\ln|x| - \tan^{-1} u - \ln\left(1 + u^2\right) = c$$

$$\frac{x}{1 + u^2} = c_1 e^{\tan^{-1} u}$$

$$x^3 = \left(y^2 + x^2\right) c_1 e^{\tan^{-1} y/x}.$$

9

**33.** Letting $y = ux$ we have

$$\left(3ux^2 + u^2x^2\right) dx - 2x^2(u\, dx + x\, du) = 0$$

$$\left(u^2 + u\right) dx - 2x\, du = 0$$

$$\frac{dx}{x} - \frac{2\, du}{u(u+1)} = 0$$

$$\ln|x| - 2\ln|u| + 2\ln|u+1| = c$$

$$\frac{x(u+1)^2}{u^2} = c_1$$

$$x\left(\frac{y}{x}+1\right)^2 = c_1 \left(\frac{y}{x}\right)^2$$

$$x(y+x)^2 = c_1 y^2.$$

Using $y(1) = -2$ we find $c_1 = 1/4$. The solution of the initial-value problem is $4x(y+x)^2 = y^2$.

**36.** Letting $x = vy$ we have

$$y(v\, dy + y\, dv) + (y\cos v - vy)\, dy = 0$$

$$y\, dv + \cos v\, dy = 0$$

$$\sec v\, dv + \frac{dy}{y} = 0$$

$$\ln|\sec v + \tan v| + \ln|y| = c$$

$$y\left(\sec\frac{x}{y} + \tan\frac{x}{y}\right) = c_1.$$

Using $y(0) = 2$ we find $c_1 = 2$. The solution of the initial-value problem is $y\left(\sec\frac{x}{y} + \tan\frac{x}{y}\right) = 2$.

**39.** Letting $y = ux$ we have

$$\left(x - ux - u^{3/2}x\right) dx + \left(x + \sqrt{u}\, x\right)(u\, dx + x\, du) = 0$$

$$dx + x\left(1 + \sqrt{u}\right) du = 0$$

$$\frac{dx}{x} + \left(1 + \sqrt{u}\right) du = 0$$

$$\ln x + u + \frac{2}{3}u^{3/2} = c$$

$$3x^{3/2}\ln x + 3x^{1/2}y + 2y^{3/2} = c_1 x^{3/2}.$$

Using $y(1) = 1$ we find $c_1 = 5$. The solution of the initial-value problem is

$$3x^{3/2} \ln x + 3x^{1/2}y + 2y^{3/2} = 5x^{3/2}.$$

(Note: Since the solution involves $\sqrt{x}$ , $x \geq 0$ and we do not need an absolute value sign in $\ln x$.)

**42.** Letting $y = ux$ we have

$$\left(\sqrt{x} + \sqrt{ux}\right)^2 dx - x(u\, dx + x\, du) = 0$$

$$\left(1 + 2\sqrt{u}\right) dx - x\, du = 0$$

$$\frac{dx}{x} - \frac{du}{1 + 2\sqrt{u}} = 0$$

$$\ln|x| = \int \frac{du}{1 + 2\sqrt{u}} \qquad \boxed{u = t^2,\ \ du = 2t\, dt}$$

$$= \int \frac{2t}{1 + 2t}\, dt = t - \frac{1}{2}\ln|1 + 2t| + c$$

$$= \sqrt{\frac{y}{x}} - \frac{1}{2}\ln\left|1 + 2\sqrt{\frac{y}{x}}\right| + c$$

$$x^2\left(1 + 2\sqrt{\frac{y}{x}}\right) = c_1 e^{2\sqrt{y/x}}$$

$$x^{3/2}\left(\sqrt{x} + 2\sqrt{y}\right) = c_1 e^{2\sqrt{y/x}}.$$

Using $y(1) = 0$ we find $c_1 = 1$. The solution of the initial-value problem is

$$x^{3/2}\left(\sqrt{x} + 2\sqrt{y}\right) = e^{2\sqrt{y/x}}.$$

**45.** From $x = vy$ we obtain $dx = v\, dy + y\, dv$ and the differential equation becomes

$$M(vy, y)(v\, dy + y\, dv) + N(vy, y)\, dy = 0.$$

Using $M(vy, y) = y^n M(v, 1)$ and $N(vy, y) = y^2 N(v, 1)$ and simplifying we have

$$y^n M(v, 1)(v\, dy + y\, dv) + y^n N(v, 1)\, dy = 0$$

$$[vM(v, 1) + N(v, 1)]\, dy + yM(v, 1)\, dv = 0$$

$$\frac{dy}{y} + \frac{M(v, 1)\, dv}{vM(v, 1) + N(v, 1)} = 0.$$

**48.** If we let $u = y/x$, then by homogeneity $f(x, y) = x^n f\left(1, \frac{y}{x}\right) = x^n f(1, u)$. Using the chain rule for

**11**

partial derivatives, we obtain

$$\frac{\partial f(x,y)}{\partial x} = x^n \frac{\partial f(1,u)}{\partial u} \frac{\partial u}{\partial x} + nx^{n-1} f(1,u) = x^n \frac{\partial f(1,u)}{\partial u} \left(-\frac{y}{x^2}\right) + nx^{n-1} f(1,u)$$

$$= -yx^{n-2} \frac{\partial f(1,u)}{\partial u} + nx^{n-1} f(1,u)$$

and

$$\frac{\partial f(x,y)}{\partial y} = x^n \frac{\partial f(1,u)}{\partial u} \frac{\partial u}{\partial y} = x^n \frac{\partial f(1,u)}{\partial u} \left(\frac{1}{x}\right) = x^{n-1} \frac{\partial f(1,u)}{\partial u}.$$

Then

$$x \frac{\partial f}{\partial x} + y \frac{\partial f}{\partial y} = -yx^{n-1} \frac{\partial f(1,u)}{\partial u} + nx^n f(1,u) + yx^{n-1} \frac{\partial f(1,u)}{\partial u}$$

$$= nx^n f(1,u) = nx^n f\left(1, \frac{y}{x}\right) = nf(x,y).$$

## Exercises 2.4

**3.** Let $M = 5x + 4y$ and $N = 4x - 8y^3$ so that $M_y = 4 = N_x$. From $f_x = 5x + 4y$ we obtain $f = \frac{5}{2}x^2 + 4xy + h(y)$, $h'(y) = -8y^3$, and $h(y) = -2y^4$. The solution is

$$\frac{5}{2}x^2 + 4xy - 2y^4 = c.$$

**6.** Let $M = 4x^3 - 3y \sin 3x - y/x^2$ and $N = 2y - 1/x + \cos 3x$ so that $M_y = -3 \sin 3x - 1/x^2$ and $N_x = 1/x^2 - 3 \sin 3x$. The equation is not exact.

**9.** Let $M = y^3 - y^2 \sin x - x$ and $N = 3xy^2 + 2y \cos x$ so that $M_y = 3y^2 - 2y \sin x = N_x$. From $f_x = y^3 - y^2 \sin x - x$ we obtain $f = xy^3 + y^2 \cos x - \frac{1}{2}x^2 + h(y)$, $h'(y) = 0$, and $h(y) = 0$. The solution is

$$xy^3 + y^2 \cos x - \frac{1}{2}x^2 = c.$$

**12.** Let $M = 2x/y$ and $N = -x^2/y^2$ so that $M_y = -2x/y^2 = N_x$. From $f_x = 2x/y$ we obtain $f = \frac{x^2}{y} + h(y)$, $h'(y) = 0$, and $h(y) = 0$. The solution is $x^2 = cy$.

**15.** Let $M = 1 - 3/x + y$ and $N = 1 - 3/y + x$ so that $M_y = 1 = N_x$. From $f_x = 1 - 3/x + y$ we obtain $f = x - 3 \ln|x| + xy + h(y)$, $h'(y) = 1 - \frac{3}{y}$, and $h(y) = y - 3 \ln|y|$. The solution is

$$x + y + xy - 3 \ln|xy| = c.$$

**18.** Let $M = -2y$ and $N = 5y - 2x$ so that $M_y = -2 = N_x$. From $f_x = -2y$ we obtain $f = -2xy + h(y)$, $h'(y) = 5y$, and $h(y) = \dfrac{5}{2}y^2$. The solution is

$$-2xy + \frac{5}{2}y^2 = c.$$

**21.** Let $M = 4x^3 + 4xy$ and $N = 2x^2 + 2y - 1$ so that $M_y = 4x = N_x$. From $f_x = 4x^3 + 4xy$ we obtain $f = x^4 + 2x^2y + h(y)$, $h'(y) = 2y - 1$, and $h(y) = y^2 - y$. The solution is

$$x^4 + 2x^2y + y^2 - y = c.$$

**24.** Let $M = 1/x + 1/x^2 - y/\left(x^2 + y^2\right)$ and $N = ye^y + x/\left(x^2 + y^2\right)$ so that
$M_y = \left(y^2 - x^2\right)/\left(x^2 + y^2\right)^2 = N_x$. From $f_x = 1/x + 1/x^2 - y/\left(x^2 + y^2\right)$ we obtain
$f = \ln|x| - \dfrac{1}{x} - \arctan\left(\dfrac{x}{y}\right) + h(y)$, $h'(y) = ye^y$, and $h(y) = ye^y - e^y$. The solution is

$$\ln|x| - \frac{1}{x} - \arctan\left(\frac{x}{y}\right) + ye^y - e^y = c.$$

**27.** Let $M = 4y + 2x - 5$ and $N = 6y + 4x - 1$ so that $M_y = 4 = N_x$. From $f_x = 4y + 2x - 5$ we obtain $f = 4xy + x^2 - 5x + h(y)$, $h'(y) = 6y - 1$, and $h(y) = 3y^2 - y$. The general solution is $4xy + x^2 - 5x + 3y^2 - y = c$. If $y(-1) = 2$ then $c = 8$ and the solution of the initial-value problem is

$$4xy + x^2 - 5x + 3y^2 - y = 8.$$

**30.** Let $M = y^2 + y \sin x$ and $N = 2xy - \cos x - 1/\left(1 + y^2\right)$ so that $M_y = 2y + \sin x = N_x$. From $f_x = y^2 + y \sin x$ we obtain $f = xy^2 - y \cos x + h(y)$, $h'(y) = \dfrac{-1}{1 + y^2}$, and $h(y) = -\tan^{-1} y$. The general solution is $xy^2 - y \cos x - \tan^{-1} y = c$. If $y(0) = 1$ then $c = -1 - \pi/4$ and the solution of the initial-value problem is

$$xy^2 - y \cos x - \tan^{-1} y = -1 - \frac{\pi}{4}.$$

**33.** Equating $M_y = 4xy + e^x$ and $N_x = 4xy + ke^x$ we obtain $k = 1$.

**36.** Since $f_x = M(x,y) = y^{1/2}x^{-1/2} + x\left(x^2 + y\right)^{-1}$ we obtain $f = 2y^{1/2}x^{1/2} + \dfrac{1}{2}\ln\left|x^2 + y\right| + h(x)$ so that $f_y = y^{-1/2}x^{1/2} + \dfrac{1}{2}\left(x^2 + y\right)^{-1} + h'(x)$. Let

$$N(x,y) = y^{-1/2}x^{1/2} + \frac{1}{2}\left(x^2 + y\right)^{-1}.$$

**39.** Let $M = -x^2y^2 \sin x + 2xy^2 \cos x$ and $N = 2x^2y \cos x$ so that $M_y = -2x^2y \sin x + 4xy \cos x = N_x$. From $f_y = 2x^2y \cos x$ we obtain $f = x^2y^2 \cos x + h(y)$, $h'(y) = 0$, and $h(y) = 0$. The solution of the differential equation is

$$x^2y^2 \cos x = c.$$

**42.** Let $M = \left(x^2 + 2xy - y^2\right) / \left(x^2 + 2xy + y^2\right)$ and $N = \left(y^2 + 2xy - x^2\right) / \left(y^2 + 2xy + x^2\right)$ so that $M_y = -4xy/(x + y)^3 = N_x$. From $f_x = \left(x^2 + 2xy + y^2 - 2y^2\right) /(x + y)^2$ we obtain

$$f = x + \frac{2y^2}{x + y} + h(y), \quad h'(y) = -1, \text{ and } h(y) = -y. \text{ The solution of the differential equation is}$$

$$x^2 + y^2 = c(x + y).$$

## Exercises 2.5

**3.** For $y' + 4y = \frac{4}{3}$ an integrating factor is $e^{\int 4dx} = e^{4x}$ so that

$$\frac{d}{dx}\left[e^{4x}y\right] = \frac{4}{3}e^{4x} \quad \text{and} \quad y = \frac{1}{3} + ce^{-4x}$$

for $-\infty < x < \infty$.

**6.** For $y' - y = e^x$ an integrating factor is $e^{-\int dx} = e^{-x}$ so that

$$\frac{d}{dx}\left[e^{-x}y\right] = 1 \quad \text{and} \quad y = xe^x + ce^x$$

for $-\infty < x < \infty$.

**9.** For $y' + \frac{1}{x}y = \frac{1}{x^2}$ an integrating factor is $e^{\int(1/x)dx} = x$ so that

$$\frac{d}{dx}[xy] = \frac{1}{x} \quad \text{and} \quad y = \frac{1}{x}\ln x + \frac{c}{x}$$

for $0 < x < \infty$.

**12.** For $\frac{dx}{dy} - x = y$ an integrating factor is $e^{-\int dy} = e^{-y}$ so that

$$\frac{d}{dy}\left[e^{-y}x\right] = ye^{-y} \quad \text{and} \quad x = -y - 1 + ce^y$$

for $-\infty < y < \infty$.

**15.** For $y' + \frac{e^x}{1 + e^x}y = 0$ an integrating factor is $e^{\int[e^x/(1+e^x)]dx} = 1 + e^x$ so that

$$\frac{d}{dx}[1 + e^x y] = 0 \quad \text{and} \quad y = \frac{c}{1 + e^x}$$

for $-\infty < x < \infty$.

**18.** For $y' + (\cot x)y = 2\cos x$ an integrating factor is $e^{\int \cot x\, dx} = \sin x$ so that

$$\frac{d}{dx}[(\sin x)\,y] = 2\sin x \cos x \quad \text{and} \quad y = \sin x + c\csc x$$

for $0 < x < \pi$.

**21.** For $y' + \left(1 + \dfrac{2}{x}\right)y = \dfrac{e^x}{x^2}$ an integrating factor is $e^{\int [1+(2/x)]dx} = x^2 e^x$ so that

$$\frac{d}{dx}\left[x^2 e^x y\right] = e^{2x} \quad \text{and} \quad y = \frac{1}{2}\frac{e^x}{x^2} + \frac{ce^{-x}}{x^2}$$

for $0 < x < \infty$.

**24.** For $y' + \dfrac{2\sin x}{(1 - \cos x)}y = \tan x(1 - \cos x)$ an integrating factor is $e^{\int [2\sin x/(1-\cos x)]dx} = (1 - \cos x)^2$

so that

$$\frac{d}{dx}\left[(1 - \cos x)^2 y\right] = \tan x - \sin x \quad \text{and} \quad y(1 - \cos x)^2 = \ln|\sec x| + \cos x + c$$

for $0 < x < \pi/2$.

**27.** For $y' + \left(3 + \dfrac{1}{x}\right)y = \dfrac{e^{-3x}}{x}$ an integrating factor is $e^{\int [3+(1/x)]dx} = xe^{3x}$ so that

$$\frac{d}{dx}\left[xe^{3x} y\right] = 1 \quad \text{and} \quad y = e^{-3x} + \frac{ce^{-3x}}{x}$$

for $0 < x < \infty$.

**30.** For $y' + \dfrac{2}{x}y = \dfrac{1}{x}(e^x + \ln x)$ an integrating factor is $e^{\int (2/x)dx} = x^2$ so that

$$\frac{d}{dx}\left[x^2 y\right] = xe^x + x\ln x \quad \text{and} \quad x^2 y = xe^x - e^x + \frac{x^2}{2}\ln x - \frac{1}{4}x^2 + c$$

for $0 < x < \infty$.

**33.** For $\dfrac{dx}{dy} + \left(2y + \dfrac{1}{y}\right)x = 2$ an integrating factor is $e^{\int [2y+(1/y)]dy} = ye^{y^2}$ so that

$$\frac{d}{dy}\left[ye^{y^2} x\right] = 2ye^{y^2} \quad \text{and} \quad x = \frac{1}{y} + \frac{1}{y}ce^{-y^2}$$

for $0 < y < \infty$.

**36.** For $\dfrac{dP}{dt} + (2t - 1)P = 4t - 2$ an integrating factor is $e^{\int (2t-1)dt} = e^{t^2-t}$ so that

$$\frac{d}{dt}\left[Pe^{t^2-t}\right] = (4t - 2)e^{t^2-t} \quad \text{and} \quad P = 2 + ce^{t-t^2}$$

for $-\infty < t < \infty$.

**39.** For $y' + (\cosh x)y = 10\cosh x$ an integrating factor is $e^{\int \cosh x dx} = e^{\sinh x}$ so that

$$\frac{d}{dx}\left[e^{\sinh x} y\right] = 10(\cosh x)e^{\sinh x} \quad \text{and} \quad y = 10 + ce^{-\sinh x}$$

for $-\infty < x < \infty$.

**15**

**42.** For $y' - 2y = x\left(e^{3x} - e^{2x}\right)$ an integrating factor is $e^{-\int 2dx} = e^{-2x}$ so that

$$\frac{d}{dx}\left[e^{-2x}y\right] = xe^x - x \quad\text{and}\quad y = xe^{3x} - e^{3x} - \frac{1}{2}x^2e^{2x} + ce^{2x}$$

for $-\infty < x < \infty$. If $y(0) = 2$ then $c = 3$ and

$$y = xe^{3x} - e^{3x} - \frac{1}{2}x^2e^{2x} + 3e^{2x}.$$

**45.** For $y' + (\tan x)y = \cos^2 x$ an integrating factor is $e^{\int \tan x\, dx} = \sec x$ so that

$$\frac{d}{dx}\left[(\sec x)\,y\right] = \cos x \quad\text{and}\quad y = \sin x \cos x + c\cos x$$

for $-\pi/2 < x < \pi/2$. If $y(0) = -1$ then $c = -1$ and

$$y = \sin x \cos x - \cos x.$$

**48.** For $y' + \left(1 + \frac{2}{x}\right)y = \frac{2}{x}e^{-x}$ an integrating factor is $e^{\int(1+2/x)dx} = x^2e^x$ so that

$$\frac{d}{dx}\left[x^2e^x y\right] = 2x \quad\text{and}\quad y = e^{-x} + \frac{c}{x^2}e^{-x}$$

for $0 < x < \infty$. If $y(1) = 0$ then $c = -1$ and

$$y = e^{-x} - \frac{1}{x^2}e^{-x}.$$

**51.** For $y' + \frac{2}{x(x-2)}y = 0$ an integrating factor is $e^{\int[2/x(x-2)]dx} = \frac{x-2}{x}$ so that

$$\frac{d}{dx}\left[\frac{x-2}{x}y\right] = 0 \quad\text{and}\quad (x-2)y = cx$$

for $2 < x < \infty$. If $y(3) = 6$ then $c = 2$ and

$$y = \frac{2x}{x-2}.$$

**54.** For $y' + \left(\sec^2 x\right)y = \sec^2 x$ an integrating factor is $e^{\int(\sec^2 x)dx} = e^{\tan x}$ so that

$$\frac{d}{dx}\left[e^{\tan x}y\right] = \sec^2 x\, e^{\tan x} \quad\text{and}\quad y = 1 + ce^{-\tan x}$$

for $-\pi/2 < x < \pi/2$. If $y(0) = -3$ then $c = -4$ and

$$y = 1 - 4e^{-\tan x}.$$

**57.** For $y' + 2xy = f(x)$ an integrating factor is $e^{x^2}$ so that

$$ye^{x^2} = \begin{cases} \frac{1}{2}e^{x^2} + c_1, & 0 \le x < 1; \\ c_2, & x \ge 1. \end{cases}$$

If $y(0) = 2$ then $c_1 = 3/2$ and for continuity we must have $c_2 = \frac{1}{2}e + \frac{3}{2}$ so that

$$y = \begin{cases} \frac{1}{2} + \frac{3}{2}e^{-x^2}, & 0 \le x < 1; \\ \left(\frac{1}{2}e + \frac{3}{2}\right)e^{-x^2}, & x \ge 1. \end{cases}$$

## Exercises 2.6

**3.** From $y' + y = xy^4$ and $w = y^{-3}$ we obtain $\dfrac{dw}{dx} - 3w = -3x$. An integrating factor is $e^{-3x}$ so that

$$e^{-3x} = xe^{-3x} + \frac{1}{3}e^{-3x} + c \quad \text{or} \quad y^{-3} = x + \frac{1}{3} + ce^{3x}.$$

**6.** From $y' + \dfrac{2}{3\left(1+x^2\right)}y = \dfrac{2x}{3\left(1+x^2\right)}y^4$ and $w = y^{-3}$ we obtain $\dfrac{dw}{dx} - \dfrac{2x}{1+x^2}w = \dfrac{-2x}{1+x^2}$. An integrating factor is $\dfrac{1}{1+x^2}$ so that

$$\frac{w}{1+x^2} = \frac{1}{1+x^2} + c \quad \text{or} \quad y^{-3} = 1 + c\left(1+x^2\right).$$

**9.** From $\dfrac{dx}{dy} - yx = y^3x^2$ and $w = x^{-1}$ we obtain $\dfrac{dw}{dy} + yx = -y^3$. An integrating factor is $e^{y^2/2}$ so that

$$e^{y^2/2}w = -ye^{y^2/2} + 2e^{y^2/2} + c \quad \text{or} \quad x^{-1} = 2 - y^2 + ce^{-y^2/2}.$$

If $y(1) = 0$ then $c = -1$ and

$$x^{-1} = 2 - y^2 - e^{-y^2/2}.$$

**12.** Identify $P(x) = 1 - x$, $Q(x) = -1$, and $R(x) = x$. Then $\dfrac{dw}{dx} + (-1 + 2x)w = -x$. An integrating factor is $e^{x^2-x}$ so that

$$e^{x^2-x}w = -\int xe^{x^2-x}dx + c \quad \text{or} \quad u = \frac{-e^{x^2-x}}{\int xe^{x^2-x}dx + c}.$$

Thus, $y = 1 + u$.

**15.** Identify $P(x) = e^{2x}$, $Q(x) = 1 + 2e^x$, and $R(x) = 1$. Then $\dfrac{dw}{dx} + (1 + 2e^x - 2e^x)w = -1$. An integrating factor is $e^x$ so that

$$e^xw = -e^x + c \quad \text{or} \quad u = \frac{1}{ce^{-x} - 1}.$$

Thus, $y = -e^x + u$.

**18.** Identify $P(x) = 9$, $Q(x) = 6$, $R(x) = 1$, and $y_1 = -3$. An integrating factor for $\dfrac{dw}{dx} + (6-6)w = -1$ is 1 so that

$$w = -x + c \quad \text{or} \quad u = \frac{1}{-x+c}.$$

Thus, $y = -3 + u$.

**21.** Let $y = xy' + f(y')$ where $f(t) = -t^3$. A family of solutions is $y = cx - c^3$. The singular solution is given by

$$x = 3t^2 \quad \text{and} \quad y = 2t^3 \quad \text{or} \quad 27y^2 = 4x^3.$$

**24.** Let $y = xy' + f(y')$ where $f(t) = \ln t$. A family of solutions is $y = cx + \ln c$. The singular solution is given by

$$x = -\frac{1}{t} \quad \text{and} \quad y = \ln t - 1 \quad \text{or} \quad y = \ln\left(-\frac{1}{x}\right) - 1.$$

**27.** If $y' + y^2 - Q(x)y - P(x) = 0$ and $y = \dfrac{w'}{w}$ then $\dfrac{dy}{dx} = \dfrac{ww'' - w'w'}{w^2}$ and $w'' - Q(x)w' - P(x)w = 0$.

**30.** From $x = -f'(t)$ and $y = f(t) - tf'(t)$ we obtain

$$\frac{dy}{dx} = \frac{dy/dt}{dx/dt} = \frac{-tf''(t)}{-f''(t)} = t$$

for $f''(t) \neq 0$. Substituting into $y = xy' + f(y')$ we find $f(t) - tf'(t) = xt + f(t)$. Since $x = -f'(t)$, this becomes $f(t) - tf'(t) = -tf'(t) + f(t)$, which is an identity. Thus, the parametric equations form a solution of $y = xy' + f(y')$.

## Exercises 2.7

**3.** Let $u = ye^x$. Then $y = ue^{-x}$ and $dy = -ue^{-x}\,dx + e^{-x}\,du$, and the equation becomes

$$ue^{-x}\,dx + (1+u)(-ue^{-x}\,dx + e^{-x}\,du) = 0 \quad \text{or} \quad (1+u)\,du = u^2\,dx.$$

Separating variables and integrating we find

$$-\frac{1}{u} + \ln|u| = x + c \implies -\frac{1}{ye^x} + \ln|y| + x = x + c \implies y\ln|y| = e^{-x} + cy.$$

**6.** Let $u = x + y$ so that $\dfrac{du}{dx} = 1 + \dfrac{dy}{dx}$. The equation becomes

$$\left(\frac{du}{dx} - 1\right) + u + 1 = u^2 e^{3x} \quad \text{or} \quad \frac{du}{dx} + u = u^2 e^{3x}.$$

This is a Bernoulli equation and we use the substitution $w = u^{-1}$ to obtain $\dfrac{dw}{dx} - w = e^{-3x}$. An integrating factor is $e^{-x}$, so

$$\frac{d}{dx}[e^{-x}w] = e^{-2x} \implies w = -\frac{1}{2}e^{3x} + ce^x \implies u = \frac{1}{-\frac{1}{2}e^{3x} + ce^x} \implies y = \frac{2}{-e^{3x} + c_1 e^x} - x.$$

**9.** Let $u = \ln(\tan y)$ so that $\dfrac{du}{dx} = \dfrac{\sec^2 y}{\tan y}\dfrac{dy}{dx} = 2\csc 2y \dfrac{dy}{dx}$. The equation becomes $x\dfrac{du}{dx} = 2x - u$ or $\dfrac{du}{dx} + \dfrac{1}{x}u = 2$. An integrating factor is $x$, so

$$\frac{d}{dx}[xu] = 2x \implies u = x + \frac{c}{x} \implies \ln(\tan y) = x + \frac{c}{x}.$$

**12.** Let $u = e^y$ so that $u' = e^y y'$. The equation becomes $xu' - 2u = x^2$ or $u' - \dfrac{2}{x}u = x$. An integrating factor is $x^{-2}$, so

$$\frac{d}{dx}\left[x^{-2}u\right] = \frac{1}{x} \implies u = x^2 \ln|x| + cx^2 \implies e^y = x^2 \ln|x| + cx^2.$$

**15.** Let $u = y^2 \ln x$ so that $\dfrac{du}{dy} = \dfrac{y^2}{x}\dfrac{dx}{dy} + 2y\ln x$ or $\dfrac{x}{y}\dfrac{du}{dy} = y\dfrac{dy}{dx} + 2x\ln x$. The equation becomes $\dfrac{x}{y}\dfrac{du}{dy} = xe^y$ or $\dfrac{1}{y}\dfrac{du}{dy} = e^y$. Separating variables we have

$$du = ye^y \implies u = ye^y - e^y + c \implies y^2 \ln x = ye^y - e^y + c.$$

**18.** Let $u = y'$ so that $u' = y''$. The equation becomes $u' - \dfrac{1}{x}u = u^2$, which is Bernoulli. Using the substitution $w = u^{-1}$ we obtain $\dfrac{dw}{dx} + \dfrac{1}{x}w = -1$. An integrating factor is $x$, so

$$\frac{d}{dx}[xw] = -x \implies w = -\frac{1}{2}x + \frac{1}{x}c \implies \frac{1}{u} = \frac{c_1 - x^2}{2x} \implies u = \frac{2x}{c_1 - x^2} \implies y = -\ln\left|c_1 - x^2\right| + c_2.$$

**21.** Let $u = y'$ so that $u' = y''$. The equation becomes $u = xu' + (u')^3 + 1$. This is a Clairaut equation with $f(t) = 1 + t^3$. A family of solutions is

$$u = c_1 x + \left(1 + c_1^3\right) \quad \text{and} \quad y = \frac{1}{2}c_1 x^2 + \left(1 + c_1^3\right)x + c_2.$$

A singular solution is given by $x = -3t^2$ and $u = 1 + t^3 - t\left(-3t^2\right) = 1 + 4t^3$. Eliminating the parameter we obtain

$$u = 1 + 4\left(-\frac{x}{3}\right)^{3/2} \quad \text{and} \quad y = x - \frac{24}{5}\left(-\frac{x}{3}\right)^{5/2}.$$

**24.** Let $u = y'$ so that $u' = y''$. The equation becomes $u' + u \tan x = 0$. Separating variables we obtain

$$\frac{du}{u} = -(\tan x)\,dx \implies \ln|u| = \ln|\cos x| + c \implies u = c_1 \cos x \implies y = c_1 \sin x + c_2.$$

**27.** We need to solve $\left[1 + (y')^2\right]^{3/2} = y''$. Let $u = y'$ so that $u' = y''$. The equation becomes $\left(1 + u^2\right)^{3/2} = u'$ or $\left(1 + u^2\right)^{3/2} = \dfrac{du}{dx}$. Separating variables and using the substitution $u = \tan\theta$ we have

$$\frac{du}{(1+u^2)^{3/2}} = dx \implies \int \frac{\sec^2\theta}{\left(1 + \tan^2\theta\right)^{3/2}}\,d\theta = x \implies \int \frac{\sec^2\theta}{\sec^3\theta}\,d\theta = x$$

$$\implies \int \cos\theta\,d\theta = x \implies \sin\theta = x \implies \frac{u}{\sqrt{1+u^2}} = x$$

$$\implies \frac{y'}{\sqrt{1+(y')^2}} = x \implies (y')^2 = x^2\left[1 + (y')^2\right] = \frac{x^2}{1 - x^2}$$

$$\implies y' = \frac{x}{\sqrt{1 - x^2}} \quad (\text{for } x > 0) \implies y = -\sqrt{1 - x^2}.$$

---

## Exercises 2.8

**3.** Identify $x_0 = 0$, $y_0 = 1$, and $f(t, y_{n-1}(t)) = 2ty_{n-1}(t)$. Picard's formula is

$$y_n(x) = 1 + 2 \int_0^x t y_{n-1}(t)\,dt$$

for $n = 1, 2, 3, \ldots$. Iterating we find

$$y_1(x) = 1 + x^2 \qquad\qquad y_3(x) = 1 + x^2 + \frac{1}{2}x^4 + \frac{1}{6}x^6$$

$$y_2(x) = 1 + x^2 + \frac{1}{2}x^4 \qquad\qquad y_4(x) = 1 + x^2 + \frac{1}{2}x^4 + \frac{1}{6}x^6 + \frac{1}{24}x^8.$$

As $n \to \infty$, $y_n(x) \to e^{x^2}$.

**6.** Identify $x_0 = 0$, $y_0 = 1$, and $f(t, y_{n-1}(t)) = 2e^t - y_{n-1}(t)$. Picard's formula is

$$y_n(x) = 2e^x - 1 - \int_0^x y_{n-1}(t)\,dt$$

for $n = 1, 2, 3, \ldots$. Iterating we find

$$y_1(x) = 2e^x - 1 - x \qquad\qquad y_3(x) = 2e^x - 1 - x - \frac{1}{2}x^2 - \frac{1}{6}x^3$$

$$y_2(x) = 1 + x + \frac{1}{2}x^2 \qquad\qquad y_4(x) = 1 + x + \frac{1}{2}x^2 + \frac{1}{6}x^3 + \frac{1}{24}x^4.$$

As $n \to \infty$, $y_n(x) \to e^x$.

# Chapter 2 Review Exercises

**3.** False; since $y = 0$ is a solution.

**6.** Separating variables we obtain

$$\cos^2 x \, dx = \frac{y}{y^2 + 1} \, dy \implies \frac{1}{2}x + \frac{1}{4}\sin 2x = \frac{1}{2}\ln\left(y^2 + 1\right) + c \implies 2x + \sin 2x = 2\ln\left(y^2 + 1\right) + c.$$

**9.** The equation is homogeneous, so let $y = ux$. Then $dy = u \, dx + x \, du$ and the differential equation becomes $ux^2(u \, dx + x \, du) = \left(3u^2x^2 + x^2\right) dx$ or $ux \, du = \left(2u^2 + 1\right) dx$. Separating variables we obtain

$$\frac{u}{2u^2 + 1} \, du = \frac{dx}{x} \implies \frac{1}{4}\ln\left(2u^2 + 1\right) = \ln x + c \implies 2u^2 + 1 = c_1 x^4$$

$$\implies 2\frac{y^2}{x^2} + 1 = c_1 x^4 \implies 2y^2 + x^2 = c_1 x^6.$$

If $y(-1) = 2$ then $c_1 = 9$ and the solution of the initial-value problem is $2y^2 + x^2 = 9x^6$.

**12.** Let $u = xy$ so that $du = x \, dy + y \, dx$. The differential equation becomes

$$du - y \, dx + \left(u + y - x^2 - 2x\right) dx = 0 \quad \text{or} \quad \frac{du}{dx} + u = x^2 + 2x.$$

An integrating factor is $e^x$, so

$$\frac{d}{dx}[e^x u] = \left(x^2 + 2x\right) e^x \implies e^x u = x^2 e^x + c \implies y = x + \frac{c}{x}e^{-x}.$$

**15.** The differential equation is Bernoulli. Using $w = y^{-1}$ we obtain $-xy^2 \dfrac{dw}{dx} + 4y = x^4 y^2$ or

$\dfrac{dw}{dx} - \dfrac{4}{x}w = -x^3$. An integrating factor is $x^{-4}$, so

$$\frac{d}{dx}\left[x^{-4}w\right] = -\frac{1}{x} \implies x^{-4}w = -\ln x + c \implies w = -x^4 \ln x + cx^4 \implies y = \left(cx^4 - x^4 \ln x\right)^{-1}.$$

If $y(1) = 1$ then $c = 1$ and $y = \left(x^4 - x^4 \ln x\right)^{-1}$.

**18.** Let $u = y'$ so that $u' = y''$. The equation becomes $u' = x - u$ or $u' + u = x$. An integrating factor is $e^x$, so

$$\frac{d}{dx}[e^x u] = xe^x \implies e^x u = xe^x - e^x + c_1 \implies y' = x - 1 + c_1 e^{-x} \implies y = \frac{1}{2}x^2 - x - c_1 e^{-x} + c_2.$$

# 3 Applications of First-Order Differential Equations

————— **Exercises 3.1** —————

**3.** From $y = c_1 x^2$ we obtain $y' = \dfrac{2y}{x}$ so that the differential equation of the orthogonal family is $y' = -\dfrac{x}{2y}$. Then

$$2y\,dy = -x\,dx \quad \text{and} \quad 2y^2 + x^2 = c_2.$$

**6.** From $2x^2 + y^2 = c_1^2$ we obtain $y' = -\dfrac{2x}{y}$ so that the differential equation of the orthogonal family is $y' = \dfrac{y}{2x}$. Then

$$\frac{1}{y}\,dy = \frac{1}{2x}\,dx \quad \text{and} \quad y^2 = c_2 x.$$

**9.** From $y^2 = c_1 x^3$ we obtain $y' = \dfrac{3y}{2x}$ so that the differential equation of the orthogonal family is $y' = -\dfrac{2x}{3y}$. Then

$$3y\,dy = -2x\,dx \quad \text{and} \quad 3y^2 + 2x^2 = c_2.$$

**12.** From $y = \dfrac{1 + c_1 x}{1 - c_1 x}$ we obtain $y' = \dfrac{y^2 - 1}{2x}$ so that the differential equation of the orthogonal family is $y' = \dfrac{2x}{1 - y^2}$. Then

$$\left(1 - y^2\right) dy = 2x\,dx \quad \text{and} \quad 3y - 3x^2 - y^3 = c_2.$$

**15.** From $y^3 + 3x^2 y = c_1$ we obtain $y' = -\dfrac{2xy}{x^2 + y^2}$ so that the differential equation of the orthogonal family is $y' = \dfrac{x^2 + y^2}{2xy}$. This is a homogeneous differential equation. Let $y = ux$ so that $y' = u + xu'$. Then

$$\frac{2u}{1 - u^2}\,du = \frac{dx}{x} \implies -\ln\left|1 - u^2\right| = \ln|x| + c \implies x\left(1 - \frac{y^2}{x^2}\right) = c_1 \implies x^2 - y^2 = c_1 x.$$

**18.** From $y = \dfrac{1}{c_1 + x}$ we obtain $y' = -y^2$ so that the differential equation of the orthogonal family is $y' = \dfrac{1}{y^2}$. Then

$$y^2\, dy = dx \quad \text{and} \quad y^3 = 3x + c.$$

**21.** From $y = \dfrac{1}{\ln c_1 x}$ we obtain $y' = -\dfrac{y^2}{x}$ so that the differential equation of the orthogonal family is $y' = \dfrac{x}{y^2}$. Then

$$y^2\, dy = x\, dx \quad \text{and} \quad 2y^3 = 3x^2 + c.$$

**24.** From $y = c_1 \sin x$ we obtain $y' = y \cot x$ so that the differential equation of the orthogonal family is $y' = -\dfrac{\tan x}{y}$. Then

$$y\, dy = -\tan x\, dx \quad \text{and} \quad y^2 = 2 \ln |\cos x| + c_2.$$

**27.** From $x + y = c_1 e^y$ we obtain $y' = \dfrac{1}{x + y - 1}$ so that the differential equation of the orthogonal family is $y' = 1 - x - y$. Then $y' + y = 1 - x$. An integrating factor is $e^x$, so

$$\frac{d}{dx}[e^x y] = e^x - xe^x \implies e^x y = 2e^x - xe^x + c \implies y = 2 - x + ce^{-x}.$$

If $y(0) = 5$ then $c = 3$ and $y = 2 - x + 3e^{-x}$.

**30.** From $r = c_1(1 + \cos \theta)$ we obtain $r\dfrac{d\theta}{dr} = -\dfrac{1 + \cos \theta}{\sin \theta}$ so that the differential equation of the orthogonal family is $r\dfrac{d\theta}{dr} = \dfrac{\sin \theta}{1 + \cos \theta}$. Then

$$\frac{1 + \cos \theta}{\sin \theta}\, d\theta = \frac{dr}{r} \implies \frac{\sin \theta}{1 - \cos \theta}\, d\theta = \frac{dr}{r} \implies \ln |1 - \cos \theta| = \ln |r| + c \implies r = c_1(1 - \cos \theta).$$

**33.** From $r = c_1 \sec \theta$ we obtain $r\dfrac{d\theta}{dr} = \cot \theta$ so that the differential equation of the orthogonal family is $r\dfrac{d\theta}{dr} = -\tan \theta$. Then

$$-\cot \theta = \frac{dr}{r} \implies -\ln |\sin \theta| = \ln |r| + c \implies r = c_1 \csc \theta.$$

**36.** Since the differential equation of the original family is $f(x, y) = \dfrac{y}{x}$, the differential equation of the isogonal family is $y' = \dfrac{y/x \pm 1}{1 \mp y/x} = \dfrac{y \pm x}{x \mp y}$. This is homogeneous so let $y = ux$. Then $y' = u + xu'$

**23**

and

$$xu' = \frac{\pm 1 \pm u^2}{1 \mp u} \implies \pm \frac{1 \mp u}{1 + u^2}\, du = \frac{dx}{x} \implies \pm \tan^{-1} u - \frac{1}{2} \ln\left(1 + u^2\right) = \ln|x| + c$$

$$\implies \pm 2 \tan^{-1} \frac{y}{x} - \ln\left(1 + \frac{y^2}{x^2}\right) = 2\ln|x| + c_1 \implies \pm 2 \tan \frac{y}{x} - \ln\left(x^2 + y^2\right) = c_1.$$

**39.** From $y^2 = c_1(2x + c_1)$ we obtain $c_1 = -x \pm \sqrt{x^2 + y^2}$ and

$$y' = -\frac{x}{y} + \sqrt{\left(\frac{x}{y}\right)^2 + 1} \quad \text{or} \quad y' = -\frac{x}{y} - \sqrt{\left(\frac{x}{y}\right)^2 + 1}.$$

Self–orthogonality follows from the fact that the product of these derivatives is $-1$.

**42.** We have $\psi_1 - \psi_2 = \dfrac{\pi}{2}$ so that $\tan \psi_1 = \tan\left(\psi_2 + \dfrac{\pi}{2}\right) = -\cot \psi_2 = -\dfrac{1}{\tan \psi_2}$.

# Exercises 3.2

**3.** Let $P = P(t)$ be the population at time $t$. From $dP/dt = kt$ and $P(0) = P_0 = 500$ we obtain $P = 500e^{kt}$. Using $P(10) = 575$ we find $k = \frac{1}{10} \ln 1.15$. Then $P(30) = 500e^{3\ln 1.15} \approx 760$ years.

**6.** Let $N = N(t)$ be the amount at time $t$. From $dN/dt = kt$ and $N(0) = 100$ we obtain $N = 100e^{kt}$. Using $N(6) = 97$ we find $k = \frac{1}{6} \ln 0.97$. Then $N(24) = 100e^{(1/6)(\ln 0.97)24} = 100(0.97)^4 \approx 88.5$ mg.

**9.** Let $I = I(t)$ be the intensity, $t$ the thickness, and $I(0) = I_0$. If $dI/dt = kI$ and $I(3) = .25I_0$ then $I = I_0 e^{kt}$, $k = \frac{1}{3} \ln .25$, and $I(15) = .00098I_0$.

**12.** Assume that $dT/dt = k(T - 5)$ so that $T = 5 + ce^{kt}$. If $T(1) = 55°$ and $T(5) = 30°$ then $k = -\frac{1}{4} \ln 2$ and $c = 59.4611$ so that $T(0) = 64.4611°$.

**15.** Assume $L\, di/dt + Ri = E(t)$, $L = .1$, $R = 50$, and $E(t) = 50$ so that $i = \frac{3}{5} + ce^{-500t}$. If $i(0) = 0$ then $c = -3/5$ and $\lim_{t \to \infty} i(t) = 3/5$.

**18.** Assume $R\, dq/dt + (1/c)q = E(t)$, $R = 1000$, $C = 5 \times 10^{-6}$, and $E(t) = 200$ so that $q = 1/1000 + ce^{-200t}$ and $i = -200ce^{-200t}$. If $i(0) = .4$ then $c = -1/500$, $q(.005) = .003$ coulombs, and $i(.005) = .1472$ amps. As $t \to \infty$ we have $q \to 1/1000$.

**21.** From $dA/dt = 4 - A/50$ we obtain $A = 200 + ce^{-t/50}$. If $A(0) = 30$ then $c = -170$ and $A = 200 - 170e^{-t/50}$.

**24.** From $\dfrac{dA}{dt} = 10 - \dfrac{10A}{500 - (10 - 5)t} = 10 - \dfrac{2A}{100 - t}$ we obtain $A = 1000 - 10t + c(100 - t)^2$. If $A(0) = 0$ then $c = -\dfrac{1}{10}$. The tank is empty in 100 minutes.

**27. (a)** From $m\,dv/dt = mg - kv$ we obtain $v = gm/k + ce^{-kt/m}$. If $v(0) = v_0$ then $c = v_0 - gm/k$ and the solution of the initial-value problem is

$$v = \frac{gm}{k} + \left(v_0 - \frac{gm}{k}\right)e^{-kt/m}.$$

**(b)** As $t \to \infty$ the limiting velocity is $gm/k$.

**(c)** From $ds/dt = v$ and $s(0) = s_0$ we obtain

$$s = \frac{gm}{k}t - \frac{m}{k}\left(v_0 - \frac{gm}{k}\right)e^{-kt/m} + s_0 + \frac{m}{k}\left(v_0 - \frac{gm}{k}\right).$$

**30.** From $V\,dC/dt = kA(C_s - C)$ and $C(0) = C_0$ we obtain $C = C_s + (C_0 - C_s)e^{-kAt/V}$.

**33.** From $r^2 d\theta = (L/m)\,dt$ we obtain $A = \dfrac{1}{2}\displaystyle\int_{\theta_1}^{\theta_2} r^2 d\theta = \dfrac{1}{2}\dfrac{L}{m}\displaystyle\int_a^b dt = \dfrac{1}{2}\dfrac{L}{m}(b - a)$.

## ——— Exercises 3.3 ———

**3.** From $\dfrac{dP}{dt} = P\left(10^{-1} - 10^{-7}P\right)$ and $P(0) = 5000$ we obtain $P = \dfrac{500}{.0005 + .0995e^{-.1t}}$ so that

$P \to 1{,}000{,}000$ as $t \to \infty$. If $P(t) = 500{,}000$ then $t = 52.9$ months.

**6.** From Problem 5 we have $P = e^{a/b}e^{-ce^{-bt}}$ so that

$$\frac{dP}{dt} = bce^{a/b - bt}e^{-ce^{-bt}} \quad \text{and} \quad \frac{d^2 P}{dt^2} = b^2 ce^{a/b - bt}e^{-ce^{-bt}}\left(ce^{-bt} - 1\right).$$

Setting $d^2 P/dt^2 = 0$ and using $c = a/b - \ln P_0$ we obtain $t = (1/b)\ln(a/b - \ln P_0)$ and $P = e^{a/b - 1}$.

**9.** If $\alpha \neq \beta$, $\dfrac{dX}{dt} = k(\alpha - X)(\beta - X)$, and $X(0) = 0$ then $\left(\dfrac{1/(\beta - \alpha)}{\alpha - X} + \dfrac{1/(\alpha - \beta)}{\beta - X}\right)dX = k\,dt$ so that

$X = \dfrac{\alpha\beta - \alpha\beta e^{(\alpha - \beta)kt}}{\beta - \alpha e^{(\alpha - \beta)kt}}$. If $\alpha = \beta$ then $\dfrac{1}{(\alpha - X)^2}dX = k\,dt$ and $X = \alpha - \dfrac{1}{kt + c}$.

**12.** From $\dfrac{d^2 y}{dx^2} = \dfrac{w}{T_1}\sqrt{1 + \left(\dfrac{dy}{dx}\right)^2}$, $p = \dfrac{dy}{dx}$, and $y'(0) = 0$ we obtain $p + \sqrt{1 + p^2} = e^{wx/T_1}$ so that

$p = \sinh\dfrac{w}{T_1}x$. From $y(0) = 1$ it follows that $y = \dfrac{T_1}{w}\cosh\dfrac{w}{T_1}x + 1 - \dfrac{T_1}{w}$.

**15.** From $\dfrac{dh}{dt} = -\dfrac{\sqrt{h}}{25}$ and $h(0) = 20$ we obtain $h = \left(\sqrt{20} - \dfrac{t}{50}\right)^2$. If $h(t) = 0$ then $t = 50\sqrt{20}$ seconds.

**18.** From $m\dfrac{dv}{dt} = mg - kv^2$ and $v(0) = v_0$ we obtain

$$\left[\frac{1/2g}{1 - \sqrt{k/mg}\,v} + \frac{1/2g}{1 + \sqrt{k/mg}\,v}\right]dv = dt$$

so that
$$\frac{v + \sqrt{mg/k}}{v - \sqrt{mg/k}} = \frac{v_0 + \sqrt{mg/k}}{v_0 - \sqrt{mg/k}} e^{2\sqrt{gk/m}\, t}.$$

Divide this equation by $e^{2\sqrt{gk/m}\, t}$ and multiply by $v - \sqrt{mg/k}$ to see that $v \to \sqrt{mg/k}$ as $t \to \infty$.

21. Using $\dfrac{dy}{dx} = \dfrac{dy}{dt} \Big/ \dfrac{dx}{dt}$ we obtain $\left(\dfrac{-\gamma + \delta y}{y}\right) dy = \left(\dfrac{\alpha - \beta x}{x}\right) dx$. Using $x \geq 0$ and $y \geq 0$ we have
$-\gamma \ln y + \delta y = \alpha \ln x - \beta x + c$.

# ———— Chapter 3 Review Exercises ————

3. From $y - 2 = c_1(x - 1)^2$ we obtain $y' = \dfrac{2(y - 2)}{x - 1}$ so that the differential equation of the orthogonal

family is $y' = \dfrac{1 - x}{2(y - 2)}$. The orthogonal trajectories are $(y - 2)^2 = x - \dfrac{1}{2}x^2 + c_2$.

6. Let $A = A(t)$ be the volume of $CO_2$ at time $t$. From $\dfrac{dA}{dt} = 1.2 - \dfrac{A}{4}$ and $A(0) = 16\,\text{ft}^3$ we obtain
$A = 4.8 + 11.2e^{-t/4}$. Since $A(10) = 5.7\,\text{ft}^3$, the concentration is $0.017\%$. As $t \to \infty$ we have
$A \to 4.8\,\text{ft}^3$ or $0.06\%$.

9. (a) The differential equation is
$$\frac{dT}{dt} = k[T - T_2 - B(T_1 - T)] = k[(1 + B)T - (BT_1 + T_2)].$$

Separating variables we obtain $\dfrac{dT}{(1 + B)T - (BT_1 + T_2)} = k\, dt$. Then

$\dfrac{1}{1 + B} \ln |(1 + B)T - (BT_1 + T_2)| = kt + c$ and $T(t) = \dfrac{BT_1 + T_2}{1 + B} + c_3 e^{k(1+B)t}$.

Since $T(0) = T_1$ we must have $c_3 = \dfrac{T_1 - T_2}{1 + B}$ and so
$$T(t) = \frac{BT_1 + T_2}{1 + B} + \frac{T_1 - T_2}{1 + B} e^{k(1+B)t}.$$

(b) Since $k < 0$, $\lim\limits_{t \to \infty} e^{k(1+B)t} = 0$ and $\lim\limits_{t \to \infty} T(t) = \dfrac{BT_1 + T_2}{1 + B}$.

(c) Since $T_s = T_2 + B(T_1 - T)$, $\lim\limits_{t \to \infty} T_s = T_2 + BT_1 - B\left(\dfrac{BT_1 + T_2}{1 + B}\right) = \dfrac{BT_1 + T_2}{1 + B}$.

# 4 Linear Differential Equations of Higher Order

─────── **Exercises 4.1** ───────────────────

**3.** From $y = c_1 e^{4x} + c_2 e^{-x}$ we find $y' = 4c_1 e^{4x} - c_2 e^{-x}$. Then $y(0) = c_1 + c_2 = 1$, $y'(0) = 4c_1 - c_2 = 2$ so that $c_1 = 3/5$ and $c_2 = 2/5$. The solution is

$$y = \frac{3}{5}e^{4x} + \frac{2}{5}e^{-x}.$$

**6.** From $y = c_1 + c_2 x^2$ we find $y' = 2c_2 x$. Then $y(0) = c_1 = 0$, $y'(0) = 2c_2 \cdot 0 = 0$ and $y'(0) = 1$ is not possible. Since $a_2(x) = x$ is 0 at $x = 0$, Theorem 4.1 is not violated.

**9.** From $y = c_1 e^x \cos x + c_2 e^x \sin x$ we find

$$y' = c_1 e^x(-\sin x + \cos x) + c_2 e^x(\cos x + \sin x).$$

  **(a)** We have $y(0) = c_1 = 1$, $y'(0) = c_1 + c_2 = 0$ so that $c_1 = 1$ and $c_2 = -1$. The solution is
    $y = e^x \cos x - e^x \sin x$.

  **(b)** We have $y(0) = c_1 = 1$, $y(\pi) = -c_1 e^\pi = -1$, which is not possible.

  **(c)** We have $y(0) = c_1 = 1$, $y(\pi/2) = c_2 e^{\pi/2} = 1$ so that $c_1 = 1$ and $c_2 = e^{-\pi/2}$. The solution is
    $y = e^x \cos x + e^{-\pi/2} e^x \sin x$.

  **(d)** We have $y(0) = c_1 = 0$, $y(\pi) = -c_1 e^\pi = 0$ so that $c_1 = 0$ and $c_2$ is arbitrary. Solutions are
    $y = c_2 e^x \sin x$, for any real numbers $c_2$.

**12.** Since $a_0(x) = \tan x$ and $x_0 = 0$ the problem has a unique solution for $-\pi/2 < x < \pi/2$.

**15.** Since $(-4)x + (3)x^2 + (1)(4x - 3x^2) = 0$ the functions are linearly dependent.

**18.** Since $(1)\cos 2x + (1)1 + (-2)\cos^2 x = 0$ the functions are linearly dependent.

**21.** The functions are linearly independent since

$$W\left(1 + x, x, x^2\right) = \begin{vmatrix} 1+x & x & x^2 \\ 1 & 1 & 2x \\ 0 & 0 & 2 \end{vmatrix} = 2 \neq 0.$$

**24.** $W\left(1 + x, x^3\right) = \begin{vmatrix} 1+x & x^3 \\ 1 & 3x^2 \end{vmatrix} = x^2(3 + 2x) \neq 0$ for $-\infty < x < \infty$.

**27.** $W\left(e^x, e^{-x}, e^{4x}\right) = \begin{vmatrix} e^x & e^{-x} & e^{4x} \\ e^x & -e^{-x} & 4e^{4x} \\ e^x & e^{-x} & 16e^{4x} \end{vmatrix} = -30e^{4x} \neq 0$ for $-\infty < x < \infty$.

**30. (a)** The graphs of $f_1$ and $f_2$ are as shown. Obviously, neither function is a constant multiple of the other on $-\infty < x < \infty$. Hence, $f_1$ and $f_2$ are linearly independent on $(-\infty, \infty)$.

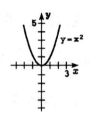 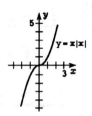

**(b)** For $x \geq 0$, $f_2 = x^2$ and so

$$W(f_1, f_2) = \begin{vmatrix} x^2 & x^2 \\ 2x & 2x \end{vmatrix} = 2x^3 - 2x^3 = 0.$$

For $x < 0$, $f_2 = -x^2$ and

$$W(f_1, f_2) = \begin{vmatrix} x^2 & -x^2 \\ 2x & -2x \end{vmatrix} = -2x^3 + 2x^3 = 0.$$

We conclude that $W(f_1, f_2) = 0$ for all real values of $x$.

**33.** The functions satisfy the differential equation and are linearly independent since

$$W\left(e^{-3x}, e^{4x}\right) = 7e^x \neq 0$$

for $-\infty < x < \infty$. The general solution is

$$y = c_1 e^{-3x} + c_2 e^{4x}.$$

**36.** The functions satisfy the differential equation and are linearly independent since

$$W\left(e^{x/2}, xe^{x/2}\right) = e^x \neq 0$$

for $-\infty < x < \infty$. The general solution is

$$y = c_1 e^{x/2} + c_2 x e^{x/2}.$$

**39.** The functions satisfy the differential equation and are linearly independent since

$$W\left(x, x^{-2}, x^{-2} \ln x\right) = 9x^{-6} \neq 0$$

for $0 < x < \infty$. The general solution is

$$y = c_1 x + c_2 x^{-2} + c_3 x^{-2} \ln x.$$

**42.** The functions $y_1 = \cos x$ and $y_2 = \sin x$ form a fundamental set of solutions of the homogeneous equation, and $y_p = x \sin x + (\cos x) \ln(\cos x)$ is a particular solution of the nonhomogeneous equation.

**45. (a)** From the graphs of $y_1 = x^3$ and $y_2 = |x|^3$ we see that the functions are linearly independent since they cannot be multiples of each other. It is easily shown that $y_1 = x^3$ solves $x^2 y'' - 4xy' + 6y = 0$. To show that $y_2 = |x|^3$ is a solution let $y_2 = x^3$ for $x \geq 0$ and let $y_2 = -x^3$ for $x < 0$.

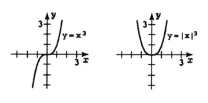

**(b)** If $x \geq 0$ then $y_2 = x^3$ and $W(y_1, y_2) = \begin{vmatrix} x^3 & x^3 \\ 3x^2 & 3x^2 \end{vmatrix} = 0$. If $x < 0$ then $y_2 = -x^3$ and

$$W(y_1, y_2) = \begin{vmatrix} x^3 & -x^3 \\ 3x^2 & -3x^2 \end{vmatrix} = 0.$$

**(c)** Part (b) does not violate Theorem 4.4 since $a_2(x) = x^2$ is zero at $x = 0$.

**(d)** The functions $Y_1 = x^3$ and $Y_2 = x^2$ are solutions of $x^2 y'' - 4xy' + 6y = 0$. They are linearly independent since $W\left(x^3, x^2\right) = x^4 \neq 0$ for $-\infty < x < \infty$.

**(e)** The function $y = x^3$ satisfies $y(0) = 0$ and $y'(0) = 0$.

**(f)** Neither is the general solution since we form a general solution on an interval for which $a_2(x) \neq 0$ for every $x$ in the interval.

**48.** We identify $a_2(x) = 1 - x^2$ and $a_1(x) = -2x$. Then from Abel's formula in Problem 47 we have

$$W = ce^{-\int [a_1(x)/a_2(x)]\,dx} = ce^{-\int [-2x/(1-x^2)]\,dx} = ce^{-\ln(1-x^2)} = \frac{c}{1 - x^2}.$$

## ——— Exercises 4.2 ———

In Problems 3-9 we use reduction of order to find a second solution. In Problems 12-30 we use formula (4) from the text.

**3.** Define $y = u(x)e^{2x}$ so

$$y' = 2ue^{2x} + u'e^{2x}, \quad y'' = e^{2x}u'' + 4e^{2x}u' + 4e^{2x}u, \quad \text{and} \quad y'' - 4y' + 4y = 4e^{2x}u'' = 0.$$

Therefore $u'' = 0$ and $u = c_1 x + c_2$. Taking $c_1 = 1$ and $c_2 = 0$ we see that a second solution is $y_2 = xe^{2x}$.

**6.** Define $y = u(x)\sin 3x$ so

$$y' = 3u\cos 3x + u'\sin 3x, \quad y'' = u''\sin 3x + 6u'\cos 3x - 9u\sin 3x,$$

and

$$y'' + 9y = (\sin 3x)u'' + 6(\cos 3x)u' = 0 \quad \text{or} \quad u'' + 9(\cot 3x)u' = 0.$$

## Exercises 4.2

If $w = u'$ we obtain the first-order equation $w' + 6(\cot 3x)w = 0$ which has the integrating factor $e^{6\int \cot 3x\, dx} = \sin^2 3x$. Now

$$\frac{d}{dx}[(\sin^2 3x)w] = 0 \quad \text{gives} \quad (\sin^2 3x)w = c.$$

Therefore $w = u' = c\csc^2 3x$ and $u = c_1 \cot 3x$. A second solution is $y_2 = \cot 3x \sin 3x = \cos 3x$.

**9.** Define $y = u(x)e^{2x/3}$ so

$$y' = \frac{2}{3}e^{2x/3}u + e^{2x/3}u', \quad y'' = e^{2x/3}u'' + \frac{4}{3}e^{2x/3}u' + \frac{4}{9}e^{2x/3}u$$

and

$$9y'' - 12y' + 4y = 9e^{2x/3}u'' = 0.$$

Therefore $u'' = 0$ and $u = c_1 x + c_2$. Taking $c_1 = 1$ and $c_2 = 0$ we see that a second solution is $y_2 = xe^{2x/3}$.

**12.** Identifying $P(x) = 2/x$ we have

$$y_2 = x^2 \int \frac{e^{-\int(2/x)\,dx}}{x^4}\, dx = x^2 \int x^{-6}\, dx = -\frac{1}{5}x^{-3}.$$

A second solution is $y_2 = x^{-3}$.

**15.** Identifying $P(x) = 2(1+x)/\left(1 - 2x - x^2\right)$ we have

$$y_2 = (x+1) \int \frac{e^{-\int 2(1+x)dx/(1-2x-x^2)}}{(x+1)^2}\, dx = (x+1) \int \frac{e^{\ln(1-2x-x^2)}}{(x+1)^2}\, dx$$

$$= (x+1) \int \frac{1 - 2x - x^2}{(x+1)^2}\, dx = (x+1) \int \left[\frac{2}{(x+1)^2} - 1\right] dx$$

$$= (x+1) \left[-\frac{2}{x+1} - x\right] = -2 - x^2 - x.$$

A second solution is $y_2 = x^2 + x + 2$.

**18.** Identifying $P(x) = -3/x$ we have

$$y_2 = x^2 \cos(\ln x) \int \frac{e^{-\int -3\,dx/x}}{x^4 \cos^2(\ln x)}\, dx = x^2 \cos(\ln x) \int \frac{x^3}{x^4 \cos^2(\ln x)}\, dx$$

$$= x^2 \cos(\ln x)\tan(\ln x) = x^2 \sin(\ln x).$$

A second solution is $y_2 = x^2 \sin(\ln x)$.

**21.** Identifying $P(x) = -1/x$ we have

$$y_2 = x \int \frac{e^{-\int -dx/x}}{x^2}\, dx = x \int \frac{dx}{x} = x \ln|x|.$$

A second solution is $y_2 = x \ln|x|$.

**24.** Identifying $P(x) = 1/x$ we have

$$y_2 = \cos(\ln x) \int \frac{e^{-\int dx/x}}{\cos^2(\ln x)} \, dx = \cos(\ln x) \int \frac{1/x}{\cos^2(\ln x)} \, dx = \cos(\ln x)\tan(\ln x) = \sin(\ln x).$$

A second solution is $y_2 = \sin(\ln x)$.

**27.** Identifying $P(x) = -(9x + 6)/(3x + 1)$ we have

$$y_2 = e^{3x} \int \frac{e^{-\int -(9x+6)dx/(3x+1)}}{e^{6x}} \, dx = e^{3x} \int \frac{e^{\int[3+3/(3x+1)]dx}}{e^{6x}} \, dx = e^{3x} \int \frac{e^{3x+\ln(3x+1)}}{e^{6x}} \, dx$$

$$= e^{3x} \int \frac{(3x+1)e^{3x}}{e^{6x}} \, dx = e^{3x} \int (3x+1)e^{-3x} \, dx = e^{3x}\left(-xe^{-3x} - \frac{2}{3}e^{-3x}\right) = -x - \frac{2}{3}.$$

A second solution is $y_2 = 3x + 2$.

**30.** Identifying $P(x) = -(2 + x)/x$ we have

$$y_2 = \int e^{-\int -(2+x)dx/x}\,dx = \int e^{2\ln x+x}\,dx = \int x^2 e^x \, dx = \left(x^2 - 2x + 2\right)e^x.$$

A second solution is $y_2 = \left(x^2 - 2x + 2\right)e^x$.

**33.** Identifying $P(x) = -3$ we have

$$y_2 = e^x \int \frac{e^{-\int -3\,dx}}{e^{2x}} \, dx = e^x \int e^x \, dx = e^{2x}.$$

To find a particular solution we try $y_p = Ae^{3x}$. Then $y' = 3Ae^{3x}$, $y'' = 9Ae^{3x}$, and $9Ae^{3x} - 3\left(3Ae^{3x}\right) + 2Ae^{3x} = 5e^{3x}$. Thus $A = 5/2$ and $y_p = \frac{5}{2}e^{3x}$. The general solution is

$$y = c_1 e^x + c_2 e^{2x} + \frac{5}{2}e^{3x}.$$

# Exercises 4.3

**3.** From $m^2 - 36 = 0$ we obtain $m = 6$ and $m = -6$ so that

$$y = c_1 e^{6x} + c_2 e^{-6x}.$$

**6.** From $3m^2 + 1 = 0$ we obtain $m = i/\sqrt{3}$ and $m = -i/\sqrt{3}$ so that

$$y = c_1 \cos x/\sqrt{3} + c_2 \sin x/\sqrt{3}.$$

**9.** From $m^2 + 8m + 16 = 0$ we obtain $m = -4$ and $m = -4$ so that

$$y = c_1 e^{-4x} + c_2 x e^{-4x}.$$

**12.** From $m^2 + 4m - 1 = 0$ we obtain $m = -2 \pm \sqrt{5}$ so that
$$y = c_1 e^{(-2+\sqrt{5})x} + c_2 e^{(-2-\sqrt{5})x}.$$

**15.** From $m^2 - 4m + 5 = 0$ we obtain $m = 2 \pm i$ so that
$$y = e^{2x}(c_1 \cos x + c_2 \sin x).$$

**18.** From $2m^2 + 2m + 1 = 0$ we obtain $m = -1/2 \pm i/2$ so that
$$y = e^{-x/2}(c_1 \cos x/2 + c_2 \sin x/2).$$

**21.** From $m^3 - 1 = 0$ we obtain $m = 1$ and $m = -1/2 \pm \sqrt{3}\,i/2$ so that
$$y = c_1 e^x + e^{-x/2}\left(c_2 \cos \sqrt{3}\,x/2 + c_3 \sin \sqrt{3}\,x/2\right).$$

**24.** From $m^3 + 3m^2 - 4m - 12 = 0$ we obtain $m = -2$, $m = 2$, and $m = -3$ so that
$$y = c_1 e^{-2x} + c_2 e^{2x} + c_3 e^{-3x}.$$

**27.** From $m^3 + 3m^2 + 3m + 1 = 0$ we obtain $m = -1$, $m = -1$, and $m = -1$ so that
$$y = c_1 e^{-x} + c_2 x e^{-x} + c_3 x^2 e^{-x}.$$

**30.** From $m^4 - 2m^2 + 1 = 0$ we obtain $m = 1$, $m = 1$, $m = -1$, and $m = -1$ so that
$$y = c_1 e^x + c_2 x e^x + c_3 e^{-x} + c_4 x e^{-x}.$$

**33.** From $m^5 - 16m = 0$ we obtain $m = 0$, $m = 2$, $m = -2$, and $m = \pm 2i$ so that
$$y = c_1 + c_2 e^{2x} + c_3 e^{-2x} + c_4 \cos 2x + c_5 \sin 2x.$$

**36.** From $2m^5 - 7m^4 + 12m^3 + 8m^2 = 0$ we obtain $m = 0$, $m = 0$, $m = -1/2$, and $m = 2 \pm 2i$ so that
$$y = c_1 + c_2 x + c_3 e^{-x/2} + e^{2x}(c_4 \cos 2x + c_5 \sin 2x).$$

**39.** From $m^2 + 6m + 5 = 0$ we obtain $m = -1$ and $m = -5$ so that $y = c_1 e^{-x} + c_2 e^{-5x}$. If $y(0) = 0$ and $y'(0) = 3$ then $c_1 + c_2 = 0$, $-c_1 - 5c_2 = 3$, so $c_1 = 3/4$, $c_2 = -3/4$, and
$$y = \frac{3}{4}e^{-x} - \frac{3}{4}e^{-5x}.$$

**42.** From $m^2 - 2m + 1 = 0$ we obtain $m = 1$ and $m = 1$ so that $y = c_1 e^x + c_2 x e^x$. If $y(0) = 5$ and $y'(0) = 10$ then $c_1 = 5$, $c_1 + c_2 = 10$ so $c_1 = 5$, $c_2 = 5$, and
$$y = 5e^x + 5x e^x.$$

**45.** From $m^2 - 3m + 2 = 0$ we obtain $m = 1$ and $m = 2$ so that $y = c_1 e^x + c_2 e^{2x}$. If $y(1) = 0$ and $y'(1) = 1$ then $c_1 e + c_2 e^2 = 0$, $c_1 e + 2c_2 e^2 = 0$ so $c_1 = -e^{-1}$, $c_2 = e^{-2}$, and
$$y = -e^{x-1} + e^{2x-2}.$$

**48.** From $m^3 + 2m^2 - 5m - 6 = 0$ we obtain $m = -1$, $m = 2$, and $m = -3$ so that

$$y = c_1 e^{-x} + c_2 e^{2x} + c_3 e^{-3x}.$$

If $y(0) = 0$, $y'(0) = 0$, and $y''(0) = 1$ then

$$c_1 + c_2 + c_3 = 0, \quad -c_1 + 2c_2 - 3c_3 = 0, \quad c_1 + 4c_2 + 9c_3 = 1,$$

so $c_1 = -1/6$, $c_2 = 1/15$, $c_3 = 1/10$, and

$$y = -\frac{1}{6}e^{-x} + \frac{1}{15}e^{2x} + \frac{1}{10}e^{-3x}.$$

**51.** From $m^4 - 3m^3 + 3m^2 - m = 0$ we obtain $m = 0$, $m = 1$, $m = 1$, and $m = 1$ so that
$y = c_1 + c_2 e^x + c_3 x e^x + c_4 x^2 e^x$. If $y(0) = 0$, $y'(0) = 0$, $y''(0) = 1$, and $y'''(0) = 1$ then

$$c_1 + c_2 = 0, \quad c_2 + c_3 = 0, \quad c_2 + 2c_3 + 2c_4 = 1, \quad c_2 + 3c_3 + 6c_4 = 1,$$

so $c_1 = 2$, $c_2 = -2$, $c_3 = 2$, $c_4 = -1/2$, and

$$y = 2 - 2e^x + 2xe^x - \frac{1}{2}x^2 e^x.$$

**54.** From $m^2 + 4 = 0$ we obtain $m = \pm 2i$ so that $y = c_1 \cos 2x + c_2 \sin 2x$. If $y(0) = 0$ and $y(\pi) = 0$
then $c_1 = 0$ and $y = c_2 \sin 2x$.

**57.** Since $(m - 4)(m + 5)^2 = m^3 + 6m^2 - 15m - 100$ the differential equation is

$$y''' + 6y'' - 15y' - 100y = 0.$$

**60.** From the solution $y_1 = e^{-4x} \cos x$ we conclude that $m_1 = -4 + i$ and $m_2 = -4 - i$ are roots of the
auxiliary equation. Hence another solution must be $y_2 = e^{-4x} \sin x$. Now dividing the polynomial
$m^3 + 6m^2 + m - 34$ by $[m - (-4 + i)][m - (-4 - i)] = m^2 + 8m + 17$ gives $m - 2$. Therefore $m_3 = 2$
is the third root of the auxiliary equation, and the general solution of the differential equation is

$$y = c_1 e^{-4x} \cos x + c_2 e^{-4x} \sin x + c_3 e^{2x}.$$

**63.** Since $m^2(m - 7) = m^3 - 7m^2$, a differential equation is

$$y''' - 7y'' = 0.$$

**3.** From $m^2 - 10m + 25 = 0$ we find $m_1 = m_2 = 5$. Then $y_c = c_1 e^{5x} + c_2 x e^{5x}$ and we assume $y_p = Ax + B$. Substituting into the differential equation we obtain $25A = 30$ and $-10A + 25B = 3$. Then $A = \frac{6}{5}$, $B = \frac{6}{5}$, $y_p = \frac{6}{5}x + \frac{6}{5}$, and

$$y = c_1 e^{5x} + c_2 x e^{5x} + \frac{6}{5}x + \frac{6}{5}.$$

**6.** From $m^2 - 8m + 20 = 0$ we find $m_1 = 2 + 4i$ and $m_2 = 2 - 4i$. Then $y_c = e^{2x}(c_1 \cos 4x + c_2 \sin 4x)$ and we assume $y_p = Ax^2 + Bx + C + (Dx + E)e^x$. Substituting into the differential equation we obtain

$$2A - 8B + 20C = 0$$

$$-6D + 13E = 0$$

$$-16A + 20B = 0$$

$$13D = -26$$

$$20A = 100.$$

Then $A = 5$, $B = 4$, $C = \frac{11}{10}$, $D = -2$, $E = -\frac{12}{13}$, $y_p = 5x^2 + 4x + \frac{11}{10} + \left(-2x - \frac{12}{13}\right)e^x$ and

$$y = e^{2x}(c_1 \cos 4x + c_2 \sin 4x) + 5x^2 + 4x + \frac{11}{10} + \left(-2x - \frac{12}{13}\right)e^x.$$

**9.** From $m^2 - m = 0$ we find $m_1 = 1$ and $m_2 = 0$. Then $y_c = c_1 e^x + c_2$ and we assume $y_p = Ax$. Substituting into the differential equation we obtain $-A = -3$. Then $A = 3$, $y_p = 3x$ and $y = c_1 e^x + c_2 + 3x$.

**12.** From $m^2 - 16 = 0$ we find $m_1 = 4$ and $m_2 = -4$. Then $y_c = c_1 e^{4x} + c_2 e^{-4x}$ and we assume $y_p = Axe^{4x}$. Substituting into the differential equation we obtain $8A = 2$. Then $A = \frac{1}{4}$, $y_p = \frac{1}{4}xe^{4x}$ and

$$y = c_1 e^{4x} + c_2 e^{-4x} + \frac{1}{4}xe^{4x}.$$

**15.** From $m^2 + 1 = 0$ we find $m_1 = i$ and $m_2 = -i$. Then $y_c = c_1 \cos x + c_2 \sin x$ and we assume $y_p = (Ax^2 + Bx)\cos x + (Cx^2 + Dx)\sin x$. Substituting into the differential equation we obtain $4C = 0$, $2A + 2D = 0$, $-4A = 2$, and $-2B + 2C = 0$. Then $A = -\frac{1}{2}$, $B = 0$, $C = 0$, $D = \frac{1}{2}$, $y_p = -\frac{1}{2}x^2 \cos x + \frac{1}{2}x \sin x$, and

$$y = c_1 \cos x + c_2 \sin x - \frac{1}{2}x^2 \cos x + \frac{1}{2}x \sin x.$$

**18.** From $m^2 - 2m + 2 = 0$ we find $m_1 = 1 + i$ and $m_2 = 1 - i$. Then $y_c = e^x(c_1 \cos x + c_2 \sin x)$ and we assume $y_p = Ae^{2x} \cos x + Be^{2x} \sin x$. Substituting into the differential equation we obtain $A + 2B = 1$ and $-2A + B = -3$. Then $A = \frac{7}{5}$, $B = -\frac{1}{5}$, $y_p = \frac{7}{5}e^{2x} \cos x - \frac{1}{5}e^{2x} \sin x$ and

$$y = e^x(c_1 \cos x + c_2 \sin x) + \frac{7}{5}e^{2x} \cos x - \frac{1}{5}e^{2x} \sin x.$$

**21.** From $m^3 - 6m^2 = 0$ we find $m_1 = m_2 = 0$ and $m_3 = 6$. Then $y_c = c_1 + c_2 x + c_3 e^{6x}$ and we assume $y_p = Ax^2 + B \cos x + C \sin x$. Substituting into the differential equation we obtain $-12A = 3$, $6B - C = -1$, and $B + 6C = 0$. Then $A = -\frac{1}{4}$, $B = -\frac{6}{37}$, $C = \frac{1}{37}$, $y_p = -\frac{1}{4}x^2 - \frac{6}{37} \cos x + \frac{1}{37} \sin x$, and

$$y = c_1 + c_2 x + c_3 e^{6x} - \frac{1}{4}x^2 - \frac{6}{37} \cos x + \frac{1}{37} \sin x.$$

**24.** From $m^3 - m^2 - 4m + 4 = 0$ we find $m_1 = 1$, $m_2 = 2$, and $m_3 = -2$. Then $y_c = c_1 e^x + c_2 e^{2x} + c_3 e^{-2x}$ and we assume $y_p = A + Bxe^x + Cxe^{2x}$. Substituting into the differential equation we obtain $4A = 5$, $-3B = -1$, and $4C = 1$. Then $A = \frac{5}{4}$, $B = \frac{1}{3}$, $C = \frac{1}{4}$, $y_p = \frac{5}{4} + \frac{1}{3}xe^x + \frac{1}{4}xe^{2x}$, and

$$y = c_1 e^x + c_2 e^{2x} + c_3 e^{-2x} + \frac{5}{4} + \frac{1}{3}xe^x + \frac{1}{4}xe^{2x}.$$

**27.** We write $8 \sin^2 x = 4 - 4 \cos 2x$. From $m^2 + 4 = 0$ we find $m_1 = 2i$ and $m_2 = -2i$. Then $y_c = c_1 \cos 2x + c_2 \sin 2x$ and we assume $y_p = A + Bx \cos 2x + Cx \sin 2x$. Substituting into the differential equation we obtain $4A = 4$, $-4B = 0$, and $4C = -4$. Then $A = 1$, $B = 0$, $C = -1$, and

$$y_p = 1 - x \sin 2x.$$

**30.** We have $y_c = c_1 e^{-2x} + c_2 e^{x/2}$ and we assume $y_p = Ax^2 + Bx + C$. Substituting into the differential equation we find $A = -7$, $B = -19$, and $C = -37$. Thus $y = c_1 e^{-2x} + c_2 e^{x/2} - 7x^2 - 19x - 37$. From the initial conditions we obtain $c_1 = -\frac{1}{5}$ and $c_2 = \frac{186}{5}$, so

$$y = -\frac{1}{5}e^{-2x} + \frac{186}{5}e^{x/2} - 7x^2 - 19x - 37.$$

**33.** We have $y_c = e^{-2x}(c_1 \cos x + c_2 \sin x)$ and we assume $y_p = Ae^{-4x}$. Substituting into the differential equation we find $A = 5$. Thus $y = e^{-2x}(c_1 \cos x + c_2 \sin x) + 7e^{-4x}$. From the initial conditions we obtain $c_1 = -10$ and $c_2 = 9$, so

$$y = e^{-2x}(-10 \cos x + 9 \sin x + 7e^{-4x}).$$

**36.** We have $x_c = c_1 \cos \omega t + c_2 \sin \omega t$ and we assume $x_p = A \cos \gamma t + B \sin \gamma t, \gamma \neq \omega$. Substituting into the differential equation we find $A = F_0/(\omega^2 - \gamma^2)$ and $B = 0$. Thus

$$x = c_1 \cos \omega t + c_2 \sin \omega t + \frac{F_0}{(\omega^2 - \gamma^2)} \cos \gamma t.$$

From the initial conditions we obtain $c_1 = -F_0/(\omega^2 - \gamma^2)$ and $c_2 = 0$, so

$$x = -\frac{F_0}{(\omega^2 - \gamma^2)}\cos \omega t + \frac{F_0}{(\omega^2 - \gamma^2)}\cos \gamma t.$$

**39.** We have $y_c = c_1 + c_2 e^x + c_3 x e^x$ and we assume $y_p = Ax + Bx^2 e^x + Ce^{5x}$. Substituting into the differential equation we find $A = 2$, $B = -12$, and $C = \frac{1}{2}$. Thus

$$y = c_1 + c_2 e^x + c_3 x e^x + 2x - 12x^2 e^x + \frac{1}{2}e^{5x}.$$

From the initial conditions we obtain $c_1 = 11$, $c_2 = -11$, and $c_3 = 9$, so

$$y = 11 - 11e^x + 9xe^x + 2x - 12x^2 e^x + \frac{1}{2}e^{5x}.$$

**42.** We have $y_c = e^x(c_1 \cos x + c_2 \sin x)$ and we assume $y_p = Ax + B$. Substituting into the differential equation we find $A = 1$ and $B = 0$. Thus $y = e^x(c_1 \cos x + c_2 \sin x) + x$. From $y(0) = 0$ and $y(\pi) = \pi$ we obtain

$$c_1 = 0$$

$$\pi - e^\pi c_1 = \pi.$$

Solving this system we find $c_1 = 0$ and $c_2$ is any real number. The solution of the boundary-value problem is

$$y = c_2 e^x \sin x + x.$$

## Exercises 4.5

**3.** $(3D^2 - 5D + 1)y = e^x$

**6.** $(D^4 - 2D^2 + D) = e^{-3x} + e^{2x}$

**9.** $D^2 - 4D - 12 = (D - 6)(D + 2)$

**12.** $D^3 + 4D = D(D^2 + 4)$

**15.** $D^4 + 8D = D(D + 2)(D^2 - 2D + 4)$

**18.** $(2D - 1)y = (2D - 1)4e^{x/2} = 8De^{x/2} - 4e^{x/2} = 4e^{x/2} - 4e^{x/2} = 0$

**21.** $D^4$ because of $x^3$

**24.** $D^2(D - 6)^2$ because of $x$ and $xe^{6x}$

**27.** $D^3(D^2 + 16)$ because of $x^2$ and $\sin 4x$

**30.** $D(D - 1)(D - 2)$ because of $1$, $e^x$, and $e^{2x}$

**33.** $1$, $x$, $x^2$, $x^3$, $x^4$

**36.** $D^2 - 9D - 36 = (D - 12)(D + 3)$; $e^{12x}$, $e^{-3x}$

**39.** $D^3 - 10D^2 + 25D = D(D - 5)^2$; $1$, $e^{5x}$, $xe^{5x}$

## —————— Exercises 4.6 ——————

**3.** Applying $D$ to the differential equation we obtain

$$D(D^2 + D)y = D^2(D + 1)y = 0.$$

Then

$$y = \underbrace{c_1 + c_2 e^{-x}}_{y_c} + c_3 x$$

and $y_p = Ax$. Substituting $y_p$ into the differential equation yields $A = 3$. The general solution is

$$y = c_1 + c_2 e^{-3x} + 3x.$$

**6.** Applying $D^2$ to the differential equation we obtain

$$D^2(D^2 + 3D)y = D^3(D + 3)y = 0.$$

Then

$$y = \underbrace{c_1 + c_2 e^{-3x}}_{y_c} + c_3 x^2 + c_4 x$$

and $y_p = Ax^2 + Bx$. Substituting $y_p$ into the differential equation yields $6Ax + (2A + 3B) = 4x - 5$. Equating coefficients gives

$$6A = 4$$

$$2A + 3B = -5.$$

Then $A = 2/3$, $B = -19/9$, and the general solution is

$$y = c_1 + c_2 e^{-3x} + \frac{2}{3}x^2 - \frac{19}{9}x.$$

**9.** Applying $D - 4$ to the differential equation we obtain

$$(D - 4)(D^2 - D - 12)y = (D - 4)^2(D + 3)y = 0.$$

Then

$$y = \underbrace{c_1 e^{4x} + c_2 e^{-3x}}_{y_c} + c_3 x e^{4x}$$

and $y_p = Axe^{4x}$. Substituting $y_p$ into the differential equation yields $7Ae^{4x} = e^{4x}$. Equating coefficients gives $A = 1/7$. The general solution is

$$y = c_1 e^{4x} + c_2 e^{-3x} + \frac{1}{7}xe^{4x}.$$

**37**

## Exercises 4.6

**12.** Applying $D^2(D+2)$ to the differential equation we obtain
$$D^2(D+2)(D^2+6D+8)y = D^2(D+2)^2(D+4)y = 0.$$

Then
$$y = \underbrace{c_1e^{-2x} + c_2e^{-4x}}_{y_c} + c_3xe^{-2x} + c_4x + c_5$$

and $y_p = Axe^{-2x} + Bx + C$. Substituting $y_p$ into the differential equation yields $2Ae^{-2x} + 8Bx + (6B+8C) = 3e^{-2x} + 2x$. Equating coefficients gives

$$2A = 3$$

$$8B = 2$$

$$6B + 8C = 0.$$

Then $A = 3/2$, $B = 1/4$, $C = -3/16$ , and the general solution is

$$y = c_1e^{-2x} + c_2e^{-4x} + \frac{3}{2}xe^{-2x} + \frac{1}{4}x - \frac{3}{16}.$$

**15.** Applying $(D-4)^2$ to the differential equation we obtain
$$(D-4)^2(D^2+6D+9)y = (D-4)^2(D+3)^2y = 0.$$

Then
$$y = \underbrace{c_1e^{-3x} + c_2xe^{-3x}}_{y_c} + c_3xe^{4x} + c_4e^{4x}$$

and $y_p = Axe^{4x} + Be^{4x}$. Substituting $y_p$ into the differential equation yields $49Axe^{4x} + (14A + 49B)e^{4x} = -xe^{4x}$. Equating coefficients gives

$$49A = -1$$

$$14A + 49B = 0.$$

Then $A = -1/49$, $B = 2/343$, and the general solution is

$$y = c_1e^{-3x} + c_2xe^{-3x} - \frac{1}{49}xe^{4x} + \frac{2}{343}e^{4x}.$$

**18.** Applying $(D+1)^3$ to the differential equation we obtain
$$(D+1)^3(D^2+2D+1)y = (D+1)^5y = 0.$$

Then
$$y = \underbrace{c_1e^{-x} + c_2xe^{-x}}_{y_c} + c_3x^4e^{-x} + c_4x^3e^{-x} + c_5x^2e^{-x}$$

and $y_p = Ax^4e^{-x} + Bx^3e^{-x} + Cx^2e^{-x}$. Substituting $y_p$ into the differential equation yields $12Ax^2e^{-x} + 6Bxe^{-x} + 2Ce^{-x} = x^2e^{-x}$. Equating coefficients gives $A = 1/12$, $B = 0$, and $C = 0$. The general solution is

$$y = c_1e^{-x} + c_2xe^{-x} + \frac{1}{2}x^4e^{-x}.$$

**21.** Applying $D^2 + 25$ to the differential equation we obtain

$$(D^2 + 25)(D^2 + 25) = (D^2 + 25)^2 = 0.$$

Then

$$y = \underbrace{c_1 \cos 5x + c_2 \sin 5x}_{y_c} + c_3 x \cos 5x + c_4 x \cos 5x$$

and $y_p = Ax \cos 5x + Bx \sin 5x$. Substituting $y_p$ into the differential equation yields $10B \cos 5x - 10A \sin 5x = 20 \sin 5x$. Equating coefficients gives $A = -2$ and $B = 0$. The general solution is

$$y = c_1 \cos 5x + c_2 \sin 5x - 2x \cos 5x.$$

**24.** Writing $\cos^2 x = \frac{1}{2}(1 + \cos 2x)$ and applying $D(D^2 + 4)$ to the differential equation we obtain

$$D(D^2 + 4)(D^2 + 4) = D(D^2 + 4)^2 = 0.$$

Then

$$y = \underbrace{c_1 \cos 2x + c_2 \sin 2x}_{y_c} + c_3 x \cos 2x + c_4 x \sin 2x + c_5$$

and $y_p = Ax \cos 2x + Bx \sin 2x + C$. Substituting $y_p$ into the differential equation yields $-4A \sin 2x + 4B \cos 2x + 4C = \frac{1}{2} + \frac{1}{2} \cos 2x$. Equating coefficients gives $A = 0$, $B = 1/8$, and $C = 1/8$. The general solution is

$$y = c_1 \cos 2x + c_2 \sin 2x + \frac{1}{8}x \sin 2x + \frac{1}{8}.$$

**27.** Applying $D^2(D - 1)$ to the differential equation we obtain

$$D^2(D - 1)(D^3 - 3D^2 + 3D - 1) = D^2(D - 1)^4 = 0.$$

Then

$$y = \underbrace{c_1e^x + c_2xe^x + c_3x^2e^x}_{y_c} + c_4 + c_5x + c_6x^3e^x$$

and $y_p = A + Bx + Cx^3e^x$. Substituting $y_p$ into the differential equation yields $(-A + 3B) - Bx + 6Ce^x = 16 - x + e^x$. Equating coefficients gives

$$-A + 3B = 16$$

$$-B = -1$$

$$6C = 1.$$

**39**

Then $A = -13$, $B = 1$, and $C = 1/6$, and the general solution is

$$y = c_1 e^x + c_2 x e^x + c_3 x^2 e^x - 13 + x + \frac{1}{6} x^3 e^x.$$

**30.** Applying $D^3(D - 2)$ to the differential equation we obtain

$$D^3(D - 2)(D^4 - 4D^2) = D^5(D - 2)^2(D + 2) = 0.$$

Then

$$y = \underbrace{c_1 + c_2 x + c_3 e^{2x} + c_4 e^{-2x}}_{y_c} + c_5 x^2 + c_6 x^3 + c_7 x^4 + c_8 x e^{2x}$$

and $y_p = Ax^2 + Bx^3 + Cx^4 + Dxe^{2x}$. Substituting $y_p$ into the differential equation yields
$(-8A + 24C) - 24Bx - 48Cx^2 + 16De^{2x} = 5x^2 - e^{2x}$. Equating coefficients gives

$$-8A + 24C = 0$$

$$-24B = 0$$

$$-48C = 5$$

$$16D = -1.$$

Then $A = -5/16$, $B = 0$, $C = -5/48$, and $D = -1/16$, and the general solution is

$$y = c_1 + c_2 x + c_3 e^{2x} + c_4 e^{-2x} - \frac{5}{16} x^2 - \frac{5}{48} x^4 - \frac{1}{16} x e^{2x}.$$

**33.** The complementary function is $y_c = c_1 e^{8x} + c_2 e^{-8x}$. Using $D$ to annihilate 16 we find $y_p = A$.
Substituting $y_p$ into the differential equation we obtain $-64A = 16$. Thus $A = -1/4$ and

$$y = c_1 e^{8x} + c_2 e^{-8x} - \frac{1}{4}$$

$$y' = 8c_1 e^{8x} - 8c_2 e^{-8x}.$$

The initial conditions imply

$$c_1 + c_2 = \frac{5}{4}$$

$$8c_1 - 8c_2 = 0.$$

Thus $c_1 = c_2 = 5/8$ and

$$y = \frac{5}{8} e^{8x} + \frac{5}{8} e^{-8x} - \frac{1}{4}.$$

**36.** The complementary function is $y_c = c_1 e^x + c_2 e^{-6x}$. Using $D - 2$ to annihilate $10e^{2x}$ we find
$y_p = Ae^{2x}$. Substituting $y_p$ into the differential equation we obtain $8Ae^{2x} = 10e^{2x}$. Thus $A = 5/4$

and

$$y = c_1 e^x + c_2 e^{-6x} + \frac{5}{4} e^{2x}$$

$$y' = c_1 e^x - 6c_2 e^{-6x} + \frac{5}{2} e^{2x}.$$

The initial conditions imply

$$c_1 + c_2 = -\frac{1}{4}$$

$$c_1 - 6c_2 = -\frac{3}{2}.$$

Thus $c_1 = -3/7$ and $c_2 = 5/28$, and

$$y = -\frac{3}{7} e^x + \frac{5}{28} e^{-6x} + \frac{5}{4} e^{2x}$$

**39.** The complementary function is $y_c = e^{2x}(c_1 \cos 2x + c_2 \sin 2x)$. Using $D^4$ to annihilate $x^3$ we find $y_p = A + Bx + Cx^2 + Dx^3$. Substituting $y_p$ into the differential equation we obtain $(8A - 4B + 2C) + (8B - 8C + 6D)x + (8C - 12D)x^2 + 8Dx^3 = x^3$. Thus $A = 0$, $B = 3/32$, $C = 3/16$, and $D = 1/8$, and

$$y = e^{2x}(c_1 \cos 2x + c_2 \sin 2x) + \frac{3}{32} x + \frac{3}{16} x^2 + \frac{1}{8} x^3$$

$$y' = e^{2x}\left[c_1(2\cos 2x - 2\sin 2x) + c_2(2\cos 2x + 2\sin 2x)\right] + \frac{3}{32} + \frac{3}{8} x + \frac{3}{8} x^2.$$

The initial conditions imply

$$c_1 = 2$$

$$2c_1 + 2c_2 + \frac{3}{32} = 4.$$

Thus $c_1 = 2$, $c_2 = -3/64$, and

$$y = e^{2x}\left(2\cos 2x - \frac{3}{64} \sin 2x\right) + \frac{3}{32} x + \frac{3}{16} x^2 + \frac{1}{8} x^3.$$

**42.** The complementary function is $y_c = c_1 + c_2 e^{-x}$. Using $D(D+1)(D^2+1)^3$ to annihilate $9 - e^{-x} + x^2 \sin x$ we obtain

$$y_p = Ax + Bxe^{-x} + C\cos x + D\sin x + Ex\cos x + Fx\sin x + Gx^2 \cos x + Hx^2 \sin x.$$

———— **Exercises 4.7** ————————————————

The particular solution, $y_p = u_1 y_1 + u_2 y_2$, in the following problems can take on a variety of forms, especially where trigonometric functions are involved. The validity of a particular form can best be checked by substituting it back into the differential equation.

**3.** The auxiliary equation is $m^2 + 1 = 0$, so $y_c = c_1 \cos x + c_2 \sin x$ and

$$W = \begin{vmatrix} \cos x & \sin x \\ -\sin x & \cos x \end{vmatrix} = 1.$$

Identifying $f(x) = \sin x$ we obtain

$$u_1' = -\sin^2 x$$

$$u_2' = \cos x \sin x.$$

Then

$$u_1 = \frac{1}{4}\sin 2x - \frac{1}{2}x = \frac{1}{2}\sin x \cos x - \frac{1}{2}x$$

$$u_2 = -\frac{1}{2}\cos^2 x.$$

and

$$y = c_1 \cos x + c_2 \sin x + \frac{1}{2}\sin x \cos^2 x - \frac{1}{2}x \cos x - \frac{1}{2}\cos^2 x \sin x$$

$$= c_1 \cos x + c_2 \sin x - \frac{1}{2}x \cos x$$

for $-\infty < x < \infty$.

**6.** The auxiliary equation is $m^2 + 1 = 0$, so $y_c = c_1 \cos x + c_2 \sin x$ and

$$W = \begin{vmatrix} \cos x & \sin x \\ -\sin x & \cos x \end{vmatrix} = 1.$$

Identifying $f(x) = \sec^2 x$ we obtain

$$u_1' = -\frac{\sin x}{\cos^2 x}$$

$$u_2' = \sec x.$$

Then

$$u_1 = -\frac{1}{\cos x} = -\sec x$$

$$u_2 = \ln|\sec x + \tan x|$$

and

$$y = c_1 \cos x + c_2 \sin x - \cos x \sec x + \sin x \ln|\sec x + \tan x|$$

$$= c_1 \cos x + c_2 \sin x - 1 + \sin x \ln|\sec x + \tan x|$$

for $-\pi/2 < x < \pi/2$.

**9.** The auxiliary equation is $m^2 - 4 = 0$, so $y_c = c_1 e^{2x} + c_2 e^{-2x}$ and

$$W = \begin{vmatrix} e^{2x} & e^{-2x} \\ 2e^{2x} & -2e^{-2x} \end{vmatrix} = -4.$$

Identifying $f(x) = e^{2x}/x$ we obtain $u_1' = 1/4x$ and $u_2' = -e^{4x}/4x$. Then

$$u_1 = \frac{1}{4}\ln|x|, \qquad u_2 = -\frac{1}{4}\int_{x_0}^{x} \frac{e^{4t}}{t}\, dt$$

and

$$y = c_1 e^{2x} + c_2 e^{-2x} + \frac{1}{4}\left(e^{2x}\ln|x| - e^{-2x}\int_{x_0}^{x} \frac{e^{4t}}{t}\, dt\right), \qquad x_0 > 0$$

for $x > 0$.

**12.** The auxiliary equation is $m^2 - 3m + 2 = (m-1)(m-2) = 0$, so $y_c = c_1 e^x + c_2 e^{2x}$ and

$$W = \begin{vmatrix} e^x & e^{2x} \\ e^x & 2e^{2x} \end{vmatrix} = e^{3x}.$$

Identifying $f(x) = e^{3x}/(1+e^x)$ we obtain

$$u_1' = -\frac{e^{2x}}{1+e^x} = \frac{e^x}{1+e^x} - e^x$$

$$u_2' = \frac{e^x}{1+e^x}.$$

Then $u_1 = \ln(1+e^x) - e^x$, $u_2 = \ln(1+e^x)$, and

$$y = c_1 e^x + c_2 e^{2x} + e^x \ln(1+e^x) - e^{2x} + e^{2x}\ln(1+e^x)$$

$$= c_1 e^x + c_3 e^{2x} + (1+e^x)e^x \ln(1+e^x)$$

for $-\infty < x < \infty$.

**15.** The auxiliary equation is $m^2 - 2m + 1 = (m-1)^2 = 0$, so $y_c = c_1 e^x + c_2 x e^x$ and

$$W = \begin{vmatrix} e^x & x e^x \\ e^x & x e^x + e^x \end{vmatrix} = e^{2x}.$$

**43**

Identifying $f(x) = e^x/\left(1+x^2\right)$ we obtain

$$u_1' = -\frac{xe^xe^x}{e^{2x}\left(1+x^2\right)} = -\frac{x}{1+x^2}$$

$$u_2' = \frac{e^xe^x}{e^{2x}\left(1+x^2\right)} = \frac{1}{1+x^2}.$$

Then $u_1 = -\frac{1}{2}\ln\left(1+x^2\right)$, $u_2 = \tan^{-1}x$, and

$$y = c_1e^x + c_2xe^x - \frac{1}{2}e^x\ln\left(1+x^2\right) + xe^x\tan^{-1}x$$

for $-\infty < x < \infty$.

18. The auxiliary equation is $m^2 + 10m + 25 = (m+5)^2 = 0$, so $y_c = c_1e^{-5x} + c_2xe^{-5x}$ and

$$W = \begin{vmatrix} e^{-5x} & xe^{-5x} \\ -5e^{-5x} & -5xe^{-5x} + e^{-5x} \end{vmatrix} = e^{-10x}.$$

Identifying $f(x) = e^{-10x}/x^2$ we obtain

$$u_1' = -\frac{xe^{-5x}e^{-10x}}{x^2e^{-10x}} = -\frac{e^{-5x}}{x}$$

$$u_2' = \frac{e^{-5x}e^{-10x}}{x^2e^{-10x}} = \frac{e^{-5x}}{x^2}.$$

Then

$$u_1 = -\int_{x_0}^x \frac{e^{-5t}}{t}\,dt, \quad x_0 > 0$$

$$u_2 = \int_{x_0}^x \frac{e^{-5t}}{t^2}\,dt, \quad x_0 > 0$$

and

$$y = c_1e^{-5x} + c_2xe^{-5x} - e^{-5x}\int_{x_0}^x \frac{e^{-5t}}{t}\,dt + xe^{-5x}\int_{x_0}^x \frac{e^{-5t}}{t^2}\,dt$$

for $x > 0$.

21. The auxiliary equation is $m^3 + m = m(m^2+1) = 0$, so $y_c = c_1 + c_2\cos x + c_3\sin x$ and

$$W = \begin{vmatrix} 1 & \cos x & \sin x \\ 0 & -\sin x & \cos x \\ 0 & -\cos x & -\sin x \end{vmatrix} = 1.$$

**44**

Identifying $f(x) = \tan x$ we obtain

$$u'_1 = W_1 = \begin{vmatrix} 0 & \cos x & \sin x \\ 0 & -\sin x & \cos x \\ \tan x & -\cos x & -\sin x \end{vmatrix} = \tan x$$

$$u'_2 = W_2 = \begin{vmatrix} 1 & 0 & \sin x \\ 0 & 0 & \cos x \\ 0 & \tan x & -\sin x \end{vmatrix} = -\sin x$$

$$u'_3 = W_3 = \begin{vmatrix} 1 & \cos x & 0 \\ 0 & -\sin x & 0 \\ 0 & -\cos x & \tan x \end{vmatrix} = -\sin x \tan x = \frac{\cos^2 x - 1}{\cos x} = \cos x - \sec x.$$

Then

$$u_1 = -\ln|\cos x|$$

$$u_2 = \cos x$$

$$u_3 = \sin x - \ln|\sec x + \tan x|$$

and

$$y = c_1 + c_2 \cos x + c_3 \sin x - \ln|\cos x| + \cos^2 x$$

$$+ \sin^2 x - \sin x \ln|\sec x + \tan x|$$

$$= c_4 + c_2 \cos x + c_3 \sin x - \ln|\cos x| - \sin x \ln|\sec x + \tan x|$$

for $-\infty < x < \infty$.

**24.** The auxiliary equation is $2m^3 - 6m^2 = 2m^2(m-3) = 0$, so $y_c = c_1 + c_2 x + c_3 e^{3x}$ and

$$W = \begin{vmatrix} 1 & x & e^{3x} \\ 0 & 1 & 3e^{3x} \\ 0 & 0 & 9e^{3x} \end{vmatrix} = 9e^{3x}.$$

Identifying $f(x) = x^2/2$ we obtain

$$u_1' = \frac{1}{9e^{3x}}W_1 = \frac{1}{9e^{3x}}\begin{vmatrix} 0 & x & e^{3x} \\ 0 & 1 & 3e^{3x} \\ x^2/2 & 0 & 9e^{3x} \end{vmatrix} = \frac{\frac{3}{2}x^3e^{3x} - \frac{1}{2}x^2e^{3x}}{9e^{3x}} = \frac{1}{6}x^3 - \frac{1}{18}x^2$$

$$u_2' = \frac{1}{9e^{3x}}W_2 = \frac{1}{9e^{3x}}\begin{vmatrix} 1 & 0 & e^{3x} \\ 0 & 0 & 3e^{3x} \\ 0 & x^2/2 & 9e^{3x} \end{vmatrix} = \frac{-\frac{3}{2}x^2e^{3x}}{9e^{3x}} = -\frac{1}{6}x^2$$

$$u_3' = \frac{1}{9e^{3x}}W_3 = \frac{1}{9e^{3x}}\begin{vmatrix} 1 & x & 0 \\ 0 & 1 & 0 \\ 0 & 0 & x^2/2 \end{vmatrix} = \frac{\frac{1}{2}x^2}{9e^{3x}} = \frac{1}{18}x^2e^{-3x}.$$

Then

$$u_1 = \frac{1}{24}x^4 - \frac{1}{54}x^3$$

$$u_2 = -\frac{1}{18}x^3$$

$$u_3 = -\frac{1}{54}x^2e^{-3x} - \frac{1}{81}xe^{-3x} - \frac{1}{243}e^{-3x}$$

and

$$y = c_1 + c_2x + c_3e^{3x} + \frac{1}{24}x^4 - \frac{1}{54}x^3 - \frac{1}{18}x^4 - \frac{1}{54}x^2 - \frac{1}{81}x - \frac{1}{243}$$

$$= c_4 + c_5x + c_3e^{3x} - \frac{1}{72}x^4 - \frac{1}{54}x^3 - \frac{1}{54}x^2$$

for $-\infty < x < \infty$.

**27.** The auxiliary equation is $m^2 + 2m - 8 = (m-2)(m+4) = 0$, so $y_c = c_1e^{2x} + c_2e^{-4x}$ and

$$W = \begin{vmatrix} e^{2x} & e^{-4x} \\ 2e^{2x} & -4e^{-4x} \end{vmatrix} = -6e^{-2x}.$$

Identifying $f(x) = 2e^{-2x} - e^{-x}$ we obtain

$$u_1' = \frac{1}{3}e^{-4x} - \frac{1}{6}e^{-3x}$$

$$u_2' = -\frac{1}{6}e^{3x} - \frac{1}{3}e^{2x}.$$

Then

$$u_1 = -\frac{1}{12}e^{-4x} + \frac{1}{18}e^{-3x}$$

$$u_2 = \frac{1}{18}e^{3x} - \frac{1}{6}e^{2x}.$$

Thus

$$y = c_1 e^{2x} + c_2 e^{-4x} - \frac{1}{12} e^{-2x} + \frac{1}{18} e^{-x} + \frac{1}{18} e^{-x} - \frac{1}{6} e^{-2x}$$

$$= c_1 e^{2x} + c_2 e^{-4x} - \frac{1}{4} e^{-2x} + \frac{1}{9} e^{-x}$$

and

$$y' = 2c_1 e^{2x} - 4c_2 e^{-4x} + \frac{1}{2} e^{-2x} - \frac{1}{9} e^{-x}.$$

The initial conditions imply

$$c_1 + c_2 - \frac{5}{36} = 1$$

$$2c_1 - 4c_2 + \frac{7}{18} = 0.$$

Thus $c_1 = 25/36$ and $c_2 = 4/9$, and

$$y = \frac{25}{36} e^{2x} + \frac{4}{9} e^{-4x} - \frac{1}{4} e^{-2x} + \frac{1}{9} e^{-x}.$$

**30.** Write the equation in the form

$$y'' - \frac{4}{x} y' + \frac{6}{x^2} y = \frac{1}{x^3}$$

and identify $f(x) = 1/x^3$. From $y_1 = x^2$ and $y_2 = x^3$ we compute

$$W(y_1, y_2) = \begin{vmatrix} x^2 & x^3 \\ 2x & 3x^2 \end{vmatrix} = 3x^4 - 2x^4 = x^4.$$

Now

$$u_1' = -\frac{x^3/x^3}{x^4} = -\frac{1}{x^4} \quad \text{so} \quad u_1 = \frac{1}{3x^3},$$

and

$$u_2' = \frac{x^2/x^3}{x^4} = \frac{1}{x^5} \quad \text{so} \quad u_2 = \frac{1}{4x^4}.$$

Thus

$$y_p = \frac{x^2}{3x^3} - \frac{x^3}{4x^4} = \frac{1}{12x}$$

and

$$y = y_c + y_p = c_1 x^2 + c_2 x^3 + \frac{1}{12x}.$$

**33. (a)** We have $y_c = c_1 e^{-x} + c_2 x e^{-x}$ and we assume $y_p = Ax^2 + Bx + C$. Substituting into the

**47**

differential equation we find

$$A = 4$$

$$4A + B = 0$$

$$2A + 2B + C = -3$$

so that $A = 4$, $B = -16$, and $C = 21$. A particular solutionis $y_p = 4x^2 - 16x + 21$.

**(b)** We have $y_c = c_1 e^{-x} + c_2 x e^{-x}$ and

$$W = \begin{vmatrix} e^{-x} & xe^{-x} \\ -e^{-x} & -xe^{-x} + e^{-x} \end{vmatrix} = e^{-2x}.$$

Identifying $f(x) = e^{-x}/x$ we obtain

$$u_1' = -\frac{xe^{-x}e^{-x}/x}{e^{-2x}} = -1$$

$$u_2' = \frac{e^{-x}e^{-x}/x}{e^{-2x}} = \frac{1}{x}.$$

Then $u_1 = -x$, $u_2 = \ln x$ and

$$y_p = -xe^{-x} + xe^{-x}\ln x.$$

Since $-xe^{-x}$ is a solution of the homogeneous differential equation, we take $y_p = xe^{-x}\ln x$.

**(c)** Adding the results of **(a)** and **(b)** we have

$$y_p = 4x^2 - 16x + 21 + xe^{-x}\ln x.$$

# Chapter 4 Review Exercises

**3.** False; consider $f_1(x) = 0$ and $f_2(x) = x$. These are linearly dependent even though $x$ is not a multiple of 0. The statement would be true if it read: "Two functions $f_1(x)$ and $f_2(x)$ are linearly independent on an interval if *neither* is a constant multiple of the other."

**6.** True

**9.** $A + Bxe^x$

**12.** Identifying $P(x) = -2 - 2/x$ we have $\int P\, dx = -2x - 2\ln x$ and

$$y_2 = e^x \int \frac{e^{2x+\ln x^2}}{e^{2x}}\, dx = e^x \int x^2\, dx = \frac{1}{3}x^3 e^x.$$

**15.** From $m^3 + 10m^2 + 25m = 0$ we obtain $m = 0$, $m = -5$, and $m = -5$ so that

$$y = c_1 + c_2 e^{-5x} + c_3 x e^{-5x}.$$

**18.** From $2m^4 + 3m^3 + 2m^2 + 6m - 4 = 0$ we obtain $m = 1/2$, $m = -2$, and $m = \pm\sqrt{2}\,i$ so that

$$y = c_1 e^{x/2} + c_2 e^{-2x} + c_3 \cos\sqrt{2}\,x + c_4 \sin\sqrt{2}\,x.$$

**21.** Applying $D(D^2 + 1)$ to the differential equation we obtain

$$D(D^2 + 1)(D^3 - 5D^2 + 6D) = D^2(D^2 + 1)(D - 2)(D - 3) = 0.$$

Then

$$y = \underbrace{c_1 + c_2 e^{2x} + c_3 e^{3x}}_{y_c} + c_4 x + c_5 \cos x + c_6 \sin x$$

and $y_p = Ax + B\cos x + C\sin x$. Substituting $y_p$ into the differential equation yields

$$6A + (5B + 5C)\cos x + (-5B + 5C)\sin x = 8 + 2\sin x.$$

Equating coefficients gives $A = 4/3$, $B = -1/5$, and $C = 1/5$. The general solution is

$$y = c_1 + c_2 e^{2x} + c_3 e^{3x} + \frac{4}{3}x - \frac{1}{5}\cos x + \frac{1}{5}\sin x.$$

**24.** The auxiliary equation is $m^2 - 1 = (m - 1)(m + 1) = 0$ so that $m = \pm 1$ and $y = c_1 e^x + c_2 e^{-x}$. Assuming $y_p = Ax + B + C\sin x$ and substituting into the differential equation we find $A = -1$, $B = 0$, and $C = -\frac{1}{2}$. Thus $y_p = -x - \frac{1}{2}\sin x$ and

$$y = c_1 e^x + c_2 e^{-x} - x - \frac{1}{2}\sin x.$$

Setting $y(0) = 2$ and $y'(0) = 3$ we obtain

$$c_1 + c_2 = 2$$

$$c_1 - c_2 - \frac{3}{2} = 3.$$

Solving this system we find $c_1 = \frac{13}{4}$ and $c_2 = -\frac{5}{4}$. The solution of the initial-value problem is

$$y = \frac{13}{4}e^x - \frac{5}{4}e^{-x} - x - \frac{1}{2}\sin x.$$

**27.** The auxiliary equation is $2m^3 - 13m^2 + 24m - 9 = (2m - 1)(m - 3)^2 = 0$ so that

$$y_c = c_1 e^{x/2} + c_2 e^{3x} + c_3 x e^{3x}.$$

A particular solution is $y_p = -4$ and the general solution is

$$y = c_1 e^{x/2} + c_2 e^{3x} + c_3 x e^{3x} - 4.$$

Setting $y(0) = -4$, $y'(0) = 0$, and $y''(0) = \frac{5}{2}$ we obtain

$$c_1 + c_2 - 4 = -4$$

$$\frac{1}{2}c_1 + 3c_2 + c_3 = 0$$

$$\frac{1}{4}c_1 + 9c_2 + 6c_3 = \frac{5}{2}.$$

Solving this system we find $c_1 = \frac{2}{5}$, $c_2 = -\frac{2}{5}$, and $c_3 = 1$. Thus

$$y = \frac{2}{5}e^{x/2} - \frac{2}{5}e^{3x} + xe^{3x} - 4.$$

# 5 Applications of Second-Order Differential Equations: Vibrational Models

————— **Exercises 5.1** —————

**3.** Applying the initial conditions to $x(t) = c_1 \cos 5t + c_2 \sin 5t$ and $x'(t) = -5c_1 \sin 5t + 5c_2 \cos 5t$ gives

$$x(0) = c_1 = -2 \quad \text{and} \quad x'(0) = 5c_2 = 10.$$

Then $c_1 = -2$, $c_2 = 2$, and

$$A = \sqrt{4+4} = 2\sqrt{2} \quad \text{and} \quad \tan \phi = \frac{-2}{2} = -1.$$

Since $\sin \phi < 0$ and $\cos \phi > 0$, $\phi$ is a fourth quadrant angle and $\phi = -\pi/4$. Thus $x(t) = 2\sqrt{2} \sin(5t - \pi/4)$.

**6.** Applying the initial conditions to $x(t) = c_1 \cos 8t + c_2 \sin 8t$ and $x'(t) = -8c_1 \sin 8t + 8c_2 \cos 8t$ gives

$$x(0) = c_1 = 4 \quad \text{and} \quad x'(0) = 8c_2 = 16.$$

Then $c_1 = 4$, $c_2 = 2$, and

$$A = \sqrt{16+4} = 2\sqrt{5} \quad \text{and} \quad \tan \phi = \frac{4}{2} = 2.$$

Since $\sin \phi > 0$ and $\cos \phi > 0$, $\phi$ is a first quadrant angle and $\phi = \tan^{-1} 2 \approx 1.107$. Thus $x(t) \approx 2\sqrt{5} \sin(8t + 1.107)$.

**9.** From $mx'' + 16x = 0$ we obtain

$$x = c_1 \cos \frac{4}{\sqrt{m}} t + c_2 \sin \frac{4}{\sqrt{m}} t$$

so that the period $\pi/4 = \pi\sqrt{m}/2$, $m = 1/4$ slug, and the weight is 8 lb.

**12.** From $20x'' + kx = 0$ we obtain

$$x = c_1 \cos \frac{1}{2}\sqrt{\frac{k}{5}} t + c_2 \sin \frac{1}{2}\sqrt{\frac{k}{5}} t$$

so that the frequency $2/\pi = \frac{1}{4}\sqrt{k/5}\,\pi$ and $k = 320$ N/m. If $80x'' + 320x = 0$ then $x = c_1 \cos 2t + c_2 \sin 2t$ so that the frequency is $2/2\pi = 1/\pi$ vibrations/second.

**15.** From $\frac{5}{8}x'' + 40x = 0$, $x(0) = 1/2$, and $x'(0) = 0$ we obtain $x = \frac{1}{2}\cos 8t$.

(a) $x(\pi/12) = -1/4$, $x(\pi/8) = -1/2$, $x(\pi/6) = -1/4$, $x(\pi/8) = 1/2$, $x(9\pi/32) = \sqrt{2}/4$.

(b) $x' = -4\sin 8t$ so that $x'(3\pi/16) = 4$ ft/s directed downward.

(c) If $x = \frac{1}{2}\cos 8t = 0$ then $t = (2n+1)\pi/16$ for $n = 0, 1, 2, \ldots$.

18. From $x'' + 16x = 0$, $x(0) = -1$, and $x'(0) = -2$ we obtain

$$x = -\cos 4t - \frac{1}{2}\sin 4t = \frac{\sqrt{5}}{2}\cos(4t - 3.6).$$

The period is $\pi/2$ seconds and the amplitude is $\sqrt{5}/2$ feet. In $4\pi$ seconds it will make 8 complete vibrations.

21. From $2x'' + 200x = 0$, $x(0) = -2/3$, and $x'(0) = 5$ we obtain

(a) $x = -\frac{2}{3}\cos 10t + \frac{1}{2}\sin 10t = \frac{5}{6}\sin(10t - 0.927)$.

(b) The amplitude is $5/6$ ft and the period is $2\pi/10 = \pi/5$

(c) $3\pi = \pi k/5$ and $k = 15$ cycles.

(d) If $x = 0$ and the weight is moving downward for the second time, then $10t - 0.927 = 2\pi$ or $t = 0.721$ s.

(e) If $x' = \frac{25}{3}\cos(10t - 0.927) = 0$ then $10t - 0.927 = \pi/2 + n\pi$ or $t = (2n+1)\pi/20 + 0.0927$ for $n = 0, 1, 2, \ldots$.

(f) $x(3) = -0.597$ ft

(g) $x'(3) = -5.814$ ft/s

(h) $x''(3) = 59.702$ ft/s$^2$

(i) If $x = 0$ then $t = \frac{1}{10}(0.927 + n\pi)$ for $n = 0, 1, 2, \ldots$ and $x'(t) = \pm\frac{25}{3}$ ft/s.

(j) If $x = 5/12$ then $t = \frac{1}{10}(\pi/6 + 0.927 + 2n\pi)$ and $t = \frac{1}{10}(5\pi/6 + 0.927 + 2n\pi)$ for $n = 0, 1, 2, \ldots$.

(k) If $x = 5/12$ and $x' < 0$ then $t = \frac{1}{10}(5\pi/6 + 0.927 + 2n\pi)$ for $n = 0, 1, 2, \ldots$.

24. Let $m$ denote the mass in slugs of the first weight. Let $k_1$ and $k_2$ be the spring constants and $k = 4k_1k_2/(k_1 + k_2)$ the effective spring constant of the system. Now, the numerical value of the first weight is $W = mg = 32m$, so

$$32m = k_1\left(\frac{1}{3}\right) \quad \text{and} \quad 32m = k_2\left(\frac{1}{2}\right).$$

From these equations we find $2k_1 = 3k_2$. The given period of the combined system is $2\pi/w = \pi/15$, so $w = 30$. Since the mas of an 8-pound weight is $1/4$ slug, we have from $w^2 = k/m$

$$30^2 = \frac{k}{1/4} = 4k \quad \text{or} \quad k = 225.$$

We now have the system of equations

$$\frac{4k_1k_2}{k_1 + k_2} = 225$$

$$2k_1 = 3k_2.$$

Solving the second equation for $k_1$ and substituting in the first equation, we obtain

$$\frac{4(3k_2/2)k_2}{3k_2/2 + k_2} = \frac{12k_2^2}{5k_2} = \frac{12k_2}{5} = 225.$$

Thus, $k_2 = 375/4$ and $k_1 = 1125/8$. Finally, the value of the first weight is

$$W = 32m = \frac{k_1}{3} = \frac{1125/8}{3} = \frac{375}{8} \approx 46.88 \text{ lb.}$$

**27.** $x = 2\sqrt{2}\left[\frac{-1}{\sqrt{2}}\cos 5t - \frac{-1}{\sqrt{2}}\sin 5t\right] = 2\sqrt{2}\cos(5t + 5\pi/4)$.

**30.** If $x = A\sin(\omega t + \phi)$ the extremes for $x$ occur when $x' = A\omega\cos(\omega t + \phi) = 0$, or $t = (\pi/2 - \phi + 2n\pi)\frac{1}{\omega}$ and $t = (-\pi/2 - \phi + 2n\pi)$ for $n = 0, 1, 2, \ldots$ . Thus, the time interval between successive maxima is $2\pi/\omega$.

# Exercises 5.2

**3. (a)** above                    **(b)** heading upward

**6. (a)** above                    **(b)** heading downward

**9. (a)** From $x'' + 10x' + 16x = 0$, $x(0) = 1$, and $x'(0) = 0$ we obtain $x = \frac{4}{3}e^{-2t} - \frac{1}{3}e^{-8t}$.

   **(b)** From $x'' + x' + 16x = 0$, $x(0) = 1$, and $x'(0) = -12$ then $x = -\frac{2}{3}e^{-2t} + \frac{5}{3}e^{-8t}$.

**12. (a)** From $\frac{1}{4}x'' + x' + 5x = 0$, $x(0) = 1/2$, and $x'(0) = 1$ we obtain $x = e^{-2t}\left(\frac{1}{2}\cos 4t + \frac{1}{2}\sin 4t\right)$.

   **(b)** $x = e^{-2t}\frac{1}{\sqrt{2}}\left(\frac{\sqrt{2}}{2}\cos 4t + \frac{\sqrt{2}}{2}\sin 4t\right) = \frac{1}{\sqrt{2}}e^{-2t}\sin\left(4t + \frac{\pi}{4}\right)$.

   **(c)** If $x = 0$ then $4t + \pi/4 = \pi, 2\pi, 3\pi, \ldots$ so that the times heading downward are $t = (7 + 8n)\pi/16$ for $n = 0, 1, 2, \ldots$ .

   **(d)**

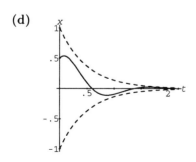

## Exercises 5.2

**15.** From $40x'' + 560x' + 3920x = 0$, $x(0) = 0$, and $x'(0) = 2$ we obtain

$$x = \frac{2}{7}e^{-7t}\sin 7t.$$

**18.** From $x'' + \beta x' + 25x = 0$ we see that the roots of the auxiliary equation are $m = -\frac{\beta}{2} \pm \frac{1}{2}\sqrt{100 - \beta^2}\,i$. The quasi-period is $\pi/2 = 2\pi / \frac{1}{2}\sqrt{100 - \beta^2}$ so that $\beta = 6$.

**21.** The time interval between successive values of $t$ for which (15) touches the graphs of $y = \pm Ae^{-\lambda t}$

is

$$t = \frac{(2n+3)\pi/2 - \phi}{\sqrt{\omega^2 - \lambda^2}} - \frac{(2n+1)\pi/2 - \phi}{\sqrt{\omega^2 - \lambda^2}} = \frac{\pi}{\sqrt{\omega^2 - \lambda^2}}.$$

**24. (a)** If $\delta > 0$ is very small then $x_n$ is slightly larger than $x_{n+2}$ and the rate of damping is slow.

**(b)** If $x = \frac{1}{\sqrt{2}}e^{-2t}\sin(4t + \pi/4)$ then $\delta = 2\pi\lambda/\sqrt{\omega^2 - \lambda^2} = 4\pi/4 = \pi$.

## ———— Exercises 5.3 ————

**3.** From $x'' + 8x' + 16x = 8\sin 4t$, $x(0) = 0$, and $x'(0) = 0$ we obtain $x_c = c_1e^{-4t} + c_2te^{-4t}$ and $x_p = -\frac{1}{4}\cos 4t$ so that the equation of motion is

$$x = \frac{1}{4}e^{-4t} + te^{-4t} - \frac{1}{4}\cos 4t.$$

**6.** Since $x = \frac{\sqrt{85}}{4}\sin(4t - 0.219) - \frac{\sqrt{17}}{2}e^{-2t}\sin(4t - 2.897)$, the amplitude approaches $\sqrt{85}/4$ as $t \to \infty$.

**9. (a)** From $100x'' + 1600x = 1600\sin 8t$, $x(0) = 0$, and $x'(0) = 0$ we obtain $x_c = c_1\cos 4t + c_2\sin 4t$ and $x_p = -\frac{1}{3}\sin 8t$ so that

$$x = \frac{2}{3}\sin 4t - \frac{1}{3}\sin 8t.$$

**(b)** If $x = \frac{1}{3}\sin 4t(2 - 2\cos 4t) = 0$ then $t = n\pi/4$ for $n = 0, 1, 2, \ldots$.

**(c)** If $x' = \frac{8}{3}\cos 4t - \frac{8}{3}\cos 8t = \frac{8}{3}(1 - \cos 4t)(1 + 2\cos 4t) = 0$ then $t = \pi/3 + n\pi/2$ and $t = \pi/6 + n\pi/2$ for $n = 0, 1, 2, \ldots$ at the extreme values. *Note*: There are many other values of $t$ for which $x' = 0$.

**(d)** $x(\pi/6 + n\pi/2) = \sqrt{3}/2$ cm. and $x(\pi/3 + n\pi/2) = -\sqrt{3}/2$ cm.

**(e)**

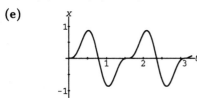

54

**12. (a)** If $x'' + \beta x' + 3x = 0$ and $0 < \beta < 2\sqrt{3}$ then the roots of the auxiliary equation are

$m = \frac{1}{2}\left(-\beta \pm \sqrt{\beta^2 - 12}\right)$; this is underdamped motion. The system is in resonance when

$\gamma = \sqrt{3 - \beta^2/2}$, where we require that $3 - \beta/2 > 0$, or $0 < \beta < \sqrt{6}$.

**(b)** When $F_0 = 3$, the resonance curve is given by

$$g(\gamma) = \frac{3}{\sqrt{(3 - \gamma^2)^2 + \beta^2 \gamma^2}},$$

and the family of graphs is shown for various values of $\beta$.

**15. (a)** From $x'' + \omega^2 x = F_0 \cos \gamma t$, $x(0) = 0$, and $x'(0) = 0$ we obtain $x_c = c_1 \cos \omega t + c_2 \sin \omega t$ and $x_p = (F_0 \cos \gamma t)/\left(\omega^2 - \gamma^2\right)$ so that

$$x = -\frac{F_0}{\omega^2 - \gamma^2}\cos \omega t + \frac{F_0}{\omega^2 - \gamma^2}\cos \gamma t.$$

**(b)** $\displaystyle \lim_{\gamma \to \omega} \frac{F_0}{\omega^2 - \gamma^2}(\cos \gamma t - \cos \omega t) = \lim_{\gamma \to \omega}\frac{-F_0 t \sin \gamma t}{-2\gamma} = \frac{F_0}{2\omega}t \sin \omega t.$

**18.** From $x'' + 9x = 5 \sin 3t$, $x(0) = 2$, and $x'(0) = 0$ we obtain $x_c = c_1 \cos 3t + c_2 \sin 3t$, $x_p = -\frac{5}{6}t \cos 3t$, and

$$x = 2 \cos 3t + \frac{5}{18}\sin 3t - \frac{5}{6}t \cos 3t.$$

## ────── Exercises 5.4 ──────

**3.** Since $R^2 - 4L/C = -20 < 0$, the circuit is underdamped.

**6.** Solving $\frac{1}{4}q'' + 20q' + 300q = 0$ we obtain $q(t) = c_1 e^{-20t} + c_2 e^{-60t}$. The initial conditions $q(0) = 4$ and $q'(0) = 0$ imply $c_1 = 6$ and $c_2 = -2$. Thus

$$q(t) = 6e^{-20t} - 2e^{-60t}.$$

Setting $q = 0$ we find $e^{40t} = 1/3$ which implies $t < 0$. Therefore the charge is never 0.

**9.** Solving $q'' + 2q' + 4q = 0$ we obtain $y_c = e^{-t}\left(\cos \sqrt{3}\,t + \sin \sqrt{3}\,t\right)$. The steady-state charge has the form $y_p = A \cos t + B \sin t$. Substituting into the differential equation we find

$$(3A + 2B)\cos t + (3B - 2A)\sin t = 50 \cos t.$$

Thus, $A = 150/13$ and $B = 100/13$. The steady-state charge is

$$q_p(t) = \frac{150}{13}\cos t + \frac{100}{13}\sin t$$

and the steady-state current is

$$i_p(t) = -\frac{150}{13}\sin t + \frac{100}{13}\cos t.$$

12. Solving $\frac{1}{2}q'' + 20q' + 1000q = 0$ we obtain $q_c(t) = (c_1\cos 40t + c_2\sin 40t)$. **The steady-state charge has the form** $q_p(t) = A\sin 60t + B\cos 60t + C\sin 40t + D\cos 40t$. **Substituting into the differential** equation we find

$$(-1600A - 2400B)\sin 60t + (2400A - 1600B)\cos 60t$$

$$+ (400C - 1600D)\sin 40t + (1600C + 400D)\cos 40t$$

$$= 200\sin 60t + 400\cos 40t.$$

Equating coefficients we obtain $A = -1/26$, $B = -3/52$, $C = 4/17$, and $D = 1/17$. **The steady-**state charge is

$$q_p(t) = -\frac{1}{26}\sin 60t - \frac{3}{52}\cos 60t + \frac{4}{17}\sin 40t + \frac{1}{17}\cos 40t$$

and the steady-state current is

$$i_p(t) = -\frac{30}{13}\cos 60t + \frac{45}{13}\sin 60t + \frac{160}{17}\cos 40t - \frac{40}{17}\sin 40t.$$

15. By Problem 10 the amplitude of the steady-state current is $E_0/Z$, where $Z = \sqrt{X^2 + R^2}$ and $X = L\gamma - 1/C\gamma$. Since $E_0$ is constant the amplitude will be a maximum when $Z$ is a minimum. Since $R$ is constant, $Z$ will be a minimum when $X = 0$. Solving $L\gamma - 1/C\gamma = 0$ for $C$ we obtain $C = 1/L\gamma^2$.

18. When the circuit is in resonance the form of $q_p(t)$ is $q_p(t) = At\cos kt + Bt\sin kt$ where $k = 1/\sqrt{LC}$. Substituting $q_p(t)$ into the differential equation we find

$$q_p'' + k^2 q = -2kA\sin kt + 2kB\cos kt = \frac{E_0}{L}\cos kt.$$

Equating coefficients we obtain $A = 0$ and $B = E_0/2kL$. The charge is

$$q(t) = c_1\cos kt + c_2\sin kt + \frac{E_0}{2kL}t\sin kt.$$

The initial conditions $q(0) = q_0$ and $q'(0) = i_0$ imply $c_1 = q_0$ and $c_2 = i_0/k$. The current is

$$i(t) = -c_1 k\sin kt + c_2 k\cos kt + \frac{E_0}{2kL}(kt\cos kt + \sin kt)$$

$$= \left(\frac{E_0}{2kL} - q_0 k\right)\sin kt + i_0\cos kt + \frac{E_0}{2L}t\cos kt.$$

# Chapter 5 Review Exercises

**3.** $5/4$ m., since $x = -\cos 4t + \frac{3}{4}\sin 4t$.

**6.** False

**9.** $9/2$, since $x = c_1 \cos \sqrt{2k}\, t + c_2 \sin \sqrt{2k}\, t$.

**12.** From $x'' + \beta x' + 64x = 0$ we see that oscillatory motion results if $\beta^2 - 256 < 0$ or $0 \le |\beta| < 16$.

**15.** Writing $\frac{1}{8}x'' + \frac{8}{3}x = \cos \gamma t + \sin \gamma t$ in the form $x'' + \frac{64}{3}x = 8\cos \gamma t + 8\sin \gamma t$ we identify $\lambda = 0$ and $\omega^2 = 64/3$. From Example 4 in Section 5.3 we see that the system is in a state of pure resonance when $\gamma = \sqrt{64/3} = 8/\sqrt{3}$.

**18. (a)** Let $k$ be the effective spring constant and $x_1$ and $x_2$ the elongation of springs $k_1$ and $k_2$. The restoring forces satisfy $k_1 x_1 = k_2 x_2$ so $x_2 = (k_1/k_2)x_1$. From $k(x_1 + x_2) = k_1 x_1$ we have

$$k\left(x_1 + \frac{k_1}{k_2}x_2\right) = k_1 x_1$$

$$k\left(\frac{k_2 + k_1}{k_2}\right) = k_1$$

$$k = \frac{k_1 k_2}{k_1 + k_2}$$

$$\frac{1}{k} = \frac{1}{k_1} + \frac{1}{k_2}.$$

**(b)** From $k_1 = 2W$ and $k_2 = 4W$ we find $1/k = 1/2W + 1/4W = 3/4W$. Then $k = 4W/3 = 4mg/3$. The differential equation $mx'' + kx = 0$ then becomes $x'' + (4g/3)x = 0$. The solution is

$$x(t) = c_1 \cos 2\sqrt{\frac{g}{3}}\, t + c_2 \sin 2\sqrt{\frac{g}{3}}\, t.$$

The initial conditions $x(0) = 1$ and $x'(0) = 2/3$ imply $c_1 = 1$ and $c_2 = 1/\sqrt{3g}$.

**(c)** To compute the maximum speed of the weight we compute

$$x'(t) = 2\sqrt{\frac{g}{3}}\sin 2\sqrt{\frac{g}{3}}\, t + \frac{2}{3}\cos 2\sqrt{\frac{g}{3}}\, t \quad \text{and} \quad |x'(t)| = \sqrt{4\frac{g}{3} + \frac{4}{9}} = \frac{2}{3}\sqrt{3g + 1}.$$

**57**

# 6 Differential Equations with Variable Coefficients

_____ **Exercises 6.1** _____

**3.** The auxiliary equation is $m^2 = 0$ so that $y = c_1 + c_2 \ln x$.

**6.** The auxiliary equation is $m^2 + 4m + 3 = (m+1)(m+3) = 0$ so that $y = c_1 x^{-1} + c_2 x^{-3}$.

**9.** The auxiliary equation is $25m^2 + 1 = 0$ so that $y = c_1 \cos\left(\frac{1}{5}\ln x\right) + c_2\left(\frac{1}{5}\ln x\right)$.

**12.** The auxiliary equation is $m^2 + 7m + 6 = (m+1)(m+6) = 0$ so that $y = c_1 x^{-1} + c_2 x^{-6}$.

**15.** The auxiliary equation is $3m^2 + 3m + 1 = 0$ so that $y = x^{-1/2}\left[c_1 \cos\left(\frac{\sqrt{3}}{6}\ln x\right) + c_2 \sin\left(\frac{\sqrt{3}}{6}\ln x\right)\right]$.

**18.** Assuming that $y = x^m$ and substituting into the differential equation we obtain

$$m(m-1)(m-2) + m - 1 = m^3 - 3m^2 + 3m - 1 = (m-1)^3 = 0.$$

Thus

$$y = c_1 x + c_2 x \ln x + c_3 x(\ln x)^2.$$

**21.** Assuming that $y = x^m$ and substituting into the differential equation we obtain

$$m(m-1)(m-2)(m-3) + 6m(m-1)(m-2) = m^4 - 7m^2 + 6m = m(m-1)(m-2)(m+3) = 0.$$

Thus

$$y = c_1 + c_2 x + c_3 x^2 + c_4 x^{-3}.$$

**24.** The auxiliary equation is $m^2 - 6m + 8 = (m-2)(m-4) = 0$, so that

$$y = c_1 x^2 + c_2 x^4 \quad \text{and} \quad y' = 2c_1 x + 4c_2 x^3.$$

The initial conditions imply

$$4c_1 + 16c_2 = 32$$

$$4c_1 + 32c_2 = 0.$$

Thus, $c_1 = 16$, $c_2 = -2$, and $y = 16x^2 - 2x^4$.

**27.** In this problem we use the substitution $t = -x$ since the initial conditions are on the interval $(-\infty, 0)$. Then

$$\frac{dy}{dt} = \frac{dy}{dx}\frac{dx}{dt} = -\frac{dy}{dx}$$

and

$$\frac{d^2y}{dt^2} = \frac{d}{dt}\left(\frac{dy}{dt}\right) = \frac{d}{dt}\left(-\frac{dy}{dx}\right) = -\frac{d}{dt}(y') = -\frac{dy'}{dx}\frac{dx}{dt} = -\frac{d^2y}{dx^2}\frac{dx}{dt} = \frac{d^2y}{dx^2}.$$

The differential equation and initial conditions become

$$4t^2\frac{d^2y}{dt^2} + y = 0; \quad y(t)\Big|_{t=1} = 2, \quad y'(t)\Big|_{t=1} = -4.$$

The auxiliary equation is $4m^2 - 4m + 1 = (2m - 1)^2 = 0$, so that

$$y = c_1t^{1/2} + c_2t^{1/2}\ln t \quad \text{and} \quad y' = \frac{1}{2}c_1t^{-1/2} + c_2\left(t^{-1/2} + \frac{1}{2}t^{-1/2}\ln t\right).$$

The initial conditions imply $c_1 = 2$ and $1 + c_2 = -4$. Thus

$$y = 2t^{1/2} - 5t^{1/2}\ln t = 2(-x)^{1/2} - 5(-x)^{1/2}\ln(-x), \quad x < 0.$$

**30.** The auxiliary equation is $m^2 - 5m = m(m - 5) = 0$ so that $y_c = c_1 + c_2x^5$ and

$$W(1, x^5) = \begin{vmatrix} 1 & x^5 \\ 0 & 5x^4 \end{vmatrix} = 5x^4.$$

·Identifying $f(x) = x^3$ we obtain $u_1' = -\frac{1}{5}x^4$ and $u_2' = 1/5x$. Then $u_1 = -\frac{1}{25}x^5$, $u_2 = \frac{1}{5}\ln x$, and

$$y = c_1 + c_2x^5 - \frac{1}{25}x^5 + \frac{1}{5}x^5\ln x = c_1 + c_3x^5 + \frac{1}{5}x^5\ln x.$$

**33.** The auxiliary equation is $m^2 - 2m + 1 = (m - 1)^2 = 0$ so that $y_c = c_1x + c_2x\ln x$ and

$$W(x, x\ln x) = \begin{vmatrix} x & x\ln x \\ 1 & 1 + \ln x \end{vmatrix} = x.$$

Identifying $f(x) = 2/x$ we obtain $u_1' = -2\ln x/x$ and $u_2' = 2/x$. Then $u_1 = -(\ln x)^2$, $u_2 = 2\ln x$, and

$$y = c_1x + c_2x\ln x - x(\ln x)^2 + 2x(\ln x)^2 = c_1x + c_2x\ln x + x(\ln x)^2.$$

**36.** From Example 6 in the text: When $x = e^t$ or $t = \ln x$,

$$\frac{dy}{dx} = \frac{1}{x}\frac{dy}{dt} \quad \text{and} \quad \frac{d^2y}{dx^2} = \frac{1}{x^2}\left[\frac{d^2y}{dt^2} - \frac{dy}{dt}\right].$$

Substituting into the differential equation we obtain

$$\frac{d^2y}{dt^2} - 5\frac{dy}{dt} + 6y = 2t.$$

The auxiliary equation is $m^2 - 5m + 6 = (m - 2)(m - 3) = 0$ so that $y_c = c_1e^{2t} + c_2e^{3t}$. Using undetermined coefficients we try $y_p = At + B$. This leads to $(-5A + 6B) + 6At = 2t$, so that $A = 1/3$, $B = 5/18$, and

$$y = c_1e^{2t} + c_2e^{3t} + \frac{1}{3}t + \frac{5}{18} = c_1x^2 + c_2x^3 + \frac{1}{3}\ln x + \frac{5}{18}.$$

## Exercises 6.1

**39.** From Example 6 in the text: When $x = e^t$ or $t = \ln x$,

$$\frac{dy}{dx} = \frac{1}{x}\frac{dy}{dt} \quad \text{and} \quad \frac{d^2y}{dx^2} = \frac{1}{x^2}\left[\frac{d^2y}{dt^2} - \frac{dy}{dt}\right].$$

Substituting into the differential equation we obtain

$$\frac{d^2y}{dt^2} + 8\frac{dy}{dt} - 20y = 5e^{-3t}.$$

The auxiliary equation is $m^2 + 8m - 20 = (m+10)(m-2) = 0$ so that $y_c = c_1 e^{-10t} + c_2 e^{2t}$. Using undetermined coefficients we try $y_p = Ae^{-3t}$. This leads to $-35Ae^{-3t} = 5e^{-3t}$, so that $A = -1/7$ and

$$y = c_1 e^{-10t} + c_2 e^{2t} - \frac{1}{7}e^{-3t} = c_1 x^{-10} + c_2 x^2 - \frac{1}{7}x^{-3}.$$

**42.** The auxiliary equation is $m^2 = 0$ so that $u(r) = c_1 + c_2 \ln r$. The boundary conditions $u(a) = u_0$ and $u(b) = u_1$ yield the system $c_1 + c_2 \ln a = u_0$, $c_1 + c_2 \ln b = u_1$. Solving gives

$$c_1 = \frac{u_1 \ln a - u_0 \ln b}{\ln(a/b)} \quad \text{and} \quad c_2 = \frac{u_0 - u_1}{\ln(a/b)}.$$

Thus,

$$u(r) = \frac{u_1 \ln a - u_0 \ln b}{\ln(a/b)} + \frac{u_0 - u_1}{\ln(a/b)}\ln r = \frac{u_0 \ln(r/b) - u_1 \ln(r/a)}{\ln(a/b)}.$$

**45.** Letting $t = x + 2$ we obtain

$$\frac{dy}{dx} = \frac{dy}{dt}$$

and

$$\frac{d^2y}{dx^2} = \frac{d}{dx}\left(\frac{dy}{dt}\right) = \frac{d^2y}{dt^2}\frac{dt}{dx} = \frac{d^2y}{dt^2}.$$

Substituting into the differential equation we obtain

$$t^2\frac{d^2y}{dt^2} + t\frac{dy}{dt} + y = 0.$$

The auxiliary equation is $m^2 + 1 = 0$ so that

$$y = c_1 \cos(\ln t) + c_2 \sin(\ln t) = c_1 \cos\left[\ln(x+2)\right] + c_2 \sin\left[\ln(x+2)\right].$$

## Exercises 6.2

**3.** $\lim_{n\to\infty}\left|\dfrac{a_{n+1}}{a_n}\right| = \lim_{n\to\infty}\left|\dfrac{2^{n+1}x^{n+1}/(n+1)}{2^n x^n/n}\right| = \lim_{n\to\infty}\dfrac{2n}{n+1}|x| = 2|x|$

The series is absolutely convergent for $2|x| < 1$ or $|x| < 1/2$. At $x = -1/2$, the series $\displaystyle\sum_{k=1}^{\infty}\dfrac{(-1)^k}{k}$

converges by the alternating series test. At $x = 1/2$, the series $\displaystyle\sum_{k=1}^{\infty}\dfrac{1}{k}$ is the harmonic series which

diverges. Thus, the given series converges on $[-1/2, 1/2)$.

**6.** $\lim_{n\to\infty}\left|\dfrac{a_{n+1}}{a_n}\right| = \lim_{n\to\infty}\left|\dfrac{(x+7)^{n+1}/\sqrt{n+1}}{(x+7)^n\sqrt{n}}\right| = \lim_{n\to\infty}\sqrt{\dfrac{n}{n+1}}\,|x+7| = |x+7|$

The series is absolutely convergent for $|x+7| < 1$ or on $(-8, 6)$. At $x = -8$, the series $\displaystyle\sum_{n=1}^{\infty}\dfrac{(-1)^n}{\sqrt{n}}$

converges by the alternating series test. At $x = -6$, the series $\displaystyle\sum_{n=1}^{\infty}\dfrac{1}{\sqrt{n}}$ is a divergent $p$-series. Thus,

the given series converges on $[-8, -6)$.

**9.** $\lim_{n\to\infty}\left|\dfrac{a_{n+1}}{a_n}\right| = \lim_{n\to\infty}\left|\dfrac{(n+1)!2^{n+1}x^{n+1}}{n!2^n x^n}\right| = \lim_{n\to\infty}2(n+1)|x| = \infty, \quad x \neq 0$

The series converges only at $x = 0$.

**12.** $e^{-x}\cos x = \left(1 - x + \dfrac{x^2}{2} - \dfrac{x^3}{6} + \dfrac{x^4}{24} - \cdots\right)\left(1 - \dfrac{x^2}{2} + \dfrac{x^4}{24} - \cdots\right) = 1 - x + \dfrac{x^3}{3} - \dfrac{x^4}{6} + \cdots$

**15.** $\left(x - \dfrac{x^3}{3} + \dfrac{x^5}{5} - \dfrac{x^7}{7} + \cdots\right)^2 = x^2 - \dfrac{2x^4}{3} + \dfrac{23x^6}{45} - \dfrac{44x^8}{105} + \cdots$

**18.** $e^x + e^{-x} = \left(1 + x + \dfrac{x^2}{2} + \dfrac{x^3}{6} + \dfrac{x^4}{24} + \dfrac{x^5}{120} + \dfrac{x^6}{720} + \cdots\right)$

$$+ \left(1 - x + \dfrac{x^2}{2} - \dfrac{x^3}{6} + \dfrac{x^4}{24} - \dfrac{x^5}{120} + \dfrac{x^6}{720} - \cdots\right)$$

$$= 2 + x^2 + \dfrac{x^4}{12} + \dfrac{x^6}{360} + \cdots$$

$$\dfrac{1}{e^x + e^{-x}} = \dfrac{1}{2 + x^2 + \dfrac{x^4}{12} + \dfrac{x^6}{360} + \cdots} = \dfrac{1}{2} - \dfrac{x^2}{4} + \dfrac{5x^4}{48} - \dfrac{61x^6}{1440} + \cdots$$

**21.** Separating variables we obtain

$$\dfrac{dy}{y} = -dx \implies \ln|y| = -x + c \implies y = c_1 e^{-x}.$$

Substituting $y = \sum_{n=0}^{\infty} c_n x^n$ into the differential equation leads to

$$y' + y = \sum_{\substack{n=1 \\ k=n-1}}^{\infty} n c_n x^{n-1} + \sum_{\substack{n=0 \\ k=n}}^{\infty} c_n x^n = \sum_{k=0}^{\infty} (k+1)c_{k+1} x^k + \sum_{k=0}^{\infty} c_k x^k = \sum_{k=0}^{\infty} [(k+1)c_{k+1} + c_k] x^k = 0.$$

Thus

$$(k+1)c_{k+1} + c_k = 0$$

and

$$c_{k+1} = -\frac{1}{k+1} c_k, \quad k = 0, 1, 2, \ldots .$$

Iterating we find

$$c_1 = -c_0$$

$$c_2 = -\frac{1}{2} c_1 = \frac{1}{2} c_0$$

$$c_3 = -\frac{1}{3} c_2 = -\frac{1}{6} c_0$$

$$c_4 = -\frac{1}{4} c_3 = \frac{1}{24} c_0$$

and so on. Therefore

$$y = c_0 - c_0 x + \frac{1}{2} c_0 x^2 - \frac{1}{6} c_0 x^3 + \frac{1}{24} c_0 x^4 - \cdots = c_0 \left[ 1 - x + \frac{1}{2} x^2 - \frac{1}{6} x^3 + \frac{1}{24} x^4 - \cdots \right]$$

$$= c_0 \sum_{n=0}^{\infty} \frac{1}{n!} (-x)^n = c_0 e^{-x}.$$

**24.** Separating variables we obtain

$$\frac{dy}{y} = -x^3 dx \implies \ln|y| = -\frac{1}{4} x^4 + c \implies y = c_1 e^{-x^4/4}.$$

Substituting $y = \sum_{n=0}^{\infty} c_n x^n$ into the differential equation leads to

$$y' + x^3 2y = \sum_{\substack{n=1 \\ k=n-4}}^{\infty} n c_n x^{n-1} + \sum_{\substack{n=0 \\ k=n}}^{\infty} c_n x^{n+3} = \sum_{k=-3}^{\infty} (k+4)c_{k+4} x^{k+3} - \sum_{k=0}^{\infty} c_k x^{k+3}$$

$$= c_1 + 2c_2 x + 3c_3 x^2 + \sum_{k=0}^{\infty} [(k+4)c_{k+4} + c_k] x^{k+2} = 0.$$

Thus

$$c_1 = c_2 = c_3 = 0,$$

$$(k+4)c_{k+4} + c_k = 0$$

and

$$c_{k+4} = -\frac{1}{k+4}c_k, \quad k = 0,1,2,\ldots.$$

Iterating we find

$$c_4 = -\frac{1}{4}c_0$$

$$c_5 = c_6 = c_7 = 0$$

$$c_8 = -\frac{1}{8}c_4 = \frac{1}{2}\cdot\frac{1}{4^2}c_0$$

$$c_9 = c_{10} = c_{11} = 0$$

$$c_{12} = -\frac{1}{12}c_8 = -\frac{1}{2\cdot 3}\cdot\frac{1}{4^3}c_0$$

and so on. Therefore

$$y = c_0 - \frac{1}{4}c_0 x^4 + \frac{1}{2}\cdot\frac{1}{4^2}c_0 x^8 - \frac{1}{2\cdot 3}\cdot\frac{1}{4^3}c_0 x^{12} + \cdots$$

$$= c_0\left[1 - \frac{x^4}{4} + \frac{1}{2}\left(\frac{x^4}{4}\right)^2 - \frac{1}{2\cdot 3}\left(\frac{x^4}{4}\right)^3 + \cdots\right] = c_0\sum_{n=0}^{\infty}\frac{1}{n!}\left(\frac{-x^4}{4}\right)^n = c_0 e^{-x^4/4}.$$

**27.** The auxiliary equation is $m^2 + 1 = 0$, so $y = c_1\cos x + c_2\sin x$. Substituting $y = \displaystyle\sum_{n=0}^{\infty}c_n x^n$ into the differential equation leads to

$$y'' + y = \underbrace{\sum_{n=2}^{\infty}n(n-1)c_n x^{n-2}}_{k=n-2} + \underbrace{\sum_{n=0}^{\infty}c_n x^n}_{k=n} = \sum_{k=0}^{\infty}(k+2)(k+1)c_{k+2}x^k + \sum_{k=0}^{\infty}c_k x^k$$

$$= \sum_{k=0}^{\infty}[(k+2)(k+1)c_{k+2} + c_k]x^k = 0.$$

Thus

$$(k+2)(k+1)c_{k+2} + c_k = 0$$

and

$$c_{k+2} = -\frac{1}{(k+2)(k+1)}c_k, \quad k = 0,1,2,\ldots.$$

Iterating we find

$$c_2 = -\frac{1}{2}c_0$$

$$c_3 = -\frac{1}{3 \cdot 2}c_1$$

$$c_4 = -\frac{1}{4 \cdot 3}c_2 = \frac{1}{4 \cdot 3 \cdot 2}c_0$$

$$c_5 = -\frac{1}{5 \cdot 4}c_3 = \frac{1}{5 \cdot 4 \cdot 3 \cdot 2}c_1$$

$$c_6 = -\frac{1}{6 \cdot 5}c_4 = -\frac{1}{6!}c_0$$

$$c_7 = -\frac{1}{7 \cdot 6}c_5 = -\frac{1}{7!}c_1$$

and so on. Therefore

$$y = c_0 + c_1 x - \frac{1}{2}c_0 x^2 - \frac{1}{3!}c_1 x^3 + \frac{1}{4!}c_0 x^4 + \frac{1}{5!}c_1 x^5 - \cdots$$

$$= c_0 \left[ 1 - \frac{1}{2}x^2 + \frac{1}{4!}x^4 - \cdots \right] + c_1 \left[ 1 - \frac{1}{3!}x^3 + \frac{1}{5!}x^5 - \cdots \right]$$

$$= c_0 \sum_{n=0}^{\infty} \frac{(-1)^n x^{2n}}{(2n)!} + c_1 \sum_{n=0}^{\infty} \frac{(-1)^n x^{2n+1}}{(2n+1)!} = c_0 \cos x + c_1 \sin x.$$

**30.** The auxiliary equation is $2m^2 + m = m(2m + 1) = 0$, so $y = c_1 + c_2 e^{-x/2}$. Substituting $y = \sum_{n=0}^{\infty} c_n x^n$ into the differential equation leads to

$$2y'' + y' = 2 \underbrace{\sum_{n=2}^{\infty} n(n-1)c_n x^{n-2}}_{k=n-2} + \underbrace{\sum_{n=1}^{\infty} nc_n x^{n-1}}_{k=n-1}$$

$$= 2 \sum_{k=0}^{\infty} (k+2)(k+1)c_{k+2} x^k + \sum_{k=0}^{\infty} (k+1)c_{k+1} x^k$$

$$= \sum_{k=0}^{\infty} [2(k+2)(k+1)c_{k+2} + (k+1)c_{k+1}]x^k = 0.$$

Thus

$$2(k+2)(k+1)c_{k+2} + (k+1)c_{k+1} = 0$$

and

$$c_{k+2} = -\frac{1}{2(k+2)} c_{k+1}, \quad k = 0, 1, 2, \ldots.$$

Iterating we find

$$c_2 = -\frac{1}{2}\frac{1}{2}c_1$$

$$c_3 = -\frac{1}{2}\frac{1}{3}c_2 = \frac{1}{2^2}\frac{1}{3\cdot 2}c_1$$

$$c_4 = -\frac{1}{2}\frac{1}{4}c_3 = \frac{1}{2^3}\frac{1}{4!}c_1$$

and so on. Therefore

$$y = c_0 + c_1 x - \frac{1}{2}\frac{1}{2}c_1 x^2 + \frac{1}{2^2 3!}c_1 x^3 - \frac{1}{2^3 4!}c_1 x^4 + \cdots$$

$$\boxed{\begin{aligned} c_0 &= C_0 - 2c_1 \\ c_1 &= -\tfrac{1}{2}C_1 \end{aligned}}$$

$$= C_0 + \left[C_1 - \frac{1}{2}C_1 x + \frac{1}{2}\frac{1}{2}\frac{1}{2}C_1 x^2 - \frac{1}{2^2 2\cdot 3!}\frac{1}{2}C_1 x^3 + \cdots\right]$$

$$= C_0 + C_1\left[1 - \frac{x}{2} + \frac{1}{2}\left(\frac{x}{2}\right)^2 - \frac{1}{3!}\left(\frac{x}{3}\right)^3 + \cdots\right]$$

$$= C_0 + C_1 \sum_{n=0}^{\infty}\frac{(-1)^n}{n!}\left(\frac{x}{2}\right)^n = C_0 + C_1 \sum_{n=0}^{\infty}\frac{1}{n!}\left(-\frac{x}{n}\right)^n = C_0 + C_1 e^{-x/2}.$$

## —— Exercises 6.3 ——

**3.** Substituting $y = \sum_{n=0}^{\infty} c_n x^n$ into the differential equation we have

$$y'' - 2xy' + y = \underbrace{\sum_{n=2}^{\infty} n(n-1)c_n x^{n-2}}_{k=n-2} - 2\underbrace{\sum_{n=1}^{\infty} nc_n x^n}_{k=n} + \underbrace{\sum_{n=0}^{\infty} c_n x^n}_{k=n}$$

$$= \sum_{k=0}^{\infty}(k+2)(k+1)c_{k+2}x^k - 2\sum_{k=1}^{\infty} kc_k x^k + \sum_{k=0}^{\infty} c_k x^k$$

$$= 2c_2 + c_0 + \sum_{k=1}^{\infty}[(k+2)(k+1)c_{k+2} - (2k-1)c_k]x^k = 0.$$

Thus

$$2c_2 + c_0 = 0$$

$$(k+2)(k+1)c_{k+2} - (2k-1)c_k = 0$$

**65**

and

$$c_2 = -\frac{1}{2}c_0$$

$$c_{k+2} = \frac{2k-1}{(k+2)(k+1)}c_k, \quad k = 1,2,3,\ldots.$$

Choosing $c_0 = 1$ and $c_1 = 0$ we find

$$c_2 = -\frac{1}{2}$$

$$c_3 = c_5 = c_7 = \cdots = 0$$

$$c_4 = -\frac{1}{8}$$

$$c_6 = -\frac{7}{336}$$

and so on. For $c_0 = 0$ and $c_1 = 1$ we obtain

$$c_2 = c_4 = c_6 = \cdots = 0$$

$$c_3 = \frac{1}{6}$$

$$c_5 = \frac{1}{24}$$

$$c_7 = \frac{1}{112}$$

and so on. Thus, two solutions are

$$y_1 = 1 - \frac{1}{2}x^2 - \frac{1}{8}x^4 - \frac{7}{336}x^6 - \cdots \quad \text{and} \quad y_2 = x + \frac{1}{6}x^3 + \frac{1}{24}x^5 + \frac{1}{112}x^7 + \cdots .$$

6. Substituting $y = \sum_{n=0}^{\infty} c_n x^n$ into the differential equation we have

$$y'' + 2xy' + 2y = \underbrace{\sum_{n=2}^{\infty} n(n-1)c_n x^{n-2}}_{k=n-2} + 2\underbrace{\sum_{n=1}^{\infty} nc_n x^n}_{k=n} + 2\underbrace{\sum_{n=0}^{\infty} c_n x^n}_{k=n}$$

$$= \sum_{k=0}^{\infty} (k+2)(k+1)c_{k+2}x^k + 2\sum_{k=1}^{\infty} kc_k x^k + 2\sum_{k=0}^{\infty} c_k x^k$$

$$= 2c_2 + 2c_0 + \sum_{k=1}^{\infty} [(k+2)(k+1)c_{k+2} + 2(k+1)c_k]x^k = 0.$$

Thus

$$2c_2 + 2c_0 = 0$$

$$(k+2)(k+1)c_{k+2} + 2(k+1)c_k = 0$$

and

$$c_2 = -c_0$$

$$c_{k+2} = -\frac{2}{k+2}\, c_k, \quad k = 1, 2, 3, \dots .$$

Choosing $c_0 = 1$ and $c_1 = 0$ we find

$$c_2 = -1$$

$$c_3 = c_5 = c_7 = \cdots = 0$$

$$c_4 = \frac{1}{2}$$

$$c_6 = -\frac{1}{6}$$

and so on. For $c_0 = 0$ and $c_1 = 1$ we obtain

$$c_2 = c_4 = c_6 = \cdots = 0$$

$$c_3 = -\frac{2}{3}$$

$$c_5 = \frac{4}{15}$$

$$c_7 = -\frac{8}{105}$$

and so on. Thus, two solutions are

$$y_1 = 1 - x^2 + \frac{1}{2}x^4 - \frac{1}{6}x^6 + \cdots \quad \text{and} \quad y_2 = x - \frac{2}{3}x^3 + \frac{4}{15}x^5 - \frac{8}{105}x^7 + \cdots .$$

**9.** Substituting $y = \sum_{n=0}^{\infty} c_n x^n$ into the differential equation we have

$$\left(x^2 - 1\right)y'' + 4xy' + 2y = \underbrace{\sum_{n=2}^{\infty} n(n-1)c_n x^n}_{k=n} - \underbrace{\sum_{n=2}^{\infty} n(n-1)c_n x^{n-2}}_{k=n-2} + 4\underbrace{\sum_{n=1}^{\infty} nc_n x^n}_{k=n} + 2\underbrace{\sum_{n=0}^{\infty} c_n x^n}_{k=n}$$

$$= \sum_{k=2}^{\infty} k(k-1)c_k x^k - \sum_{k=0}^{\infty} (k+2)(k+1)c_{k+2} x^k + 4\sum_{k=1}^{\infty} kc_k x^k + 2\sum_{k=0}^{\infty} c_k x^k$$

$$= -2c_2 + 2c_0 + (-6c_3 + 6c_1)x + \sum_{k=2}^{\infty} \left[\left(k^2 - k + 4k + 2\right)c_k - (k+2)(k+1)c_{k+2}\right] x^k = 0.$$

Thus

$$-2c_2 + 2c_0 = 0$$

**67**

$$-6c_3 + 6c_1 = 0$$

$$\left(k^2 + 3k + 2\right) c_k - (k+2)(k+1)c_{k+2} = 0$$

and

$$c_2 = c_0$$

$$c_3 = c_1$$

$$c_{k+2} = c_k, \quad k = 2, 3, 4, \ldots.$$

Choosing $c_0 = 1$ and $c_1 = 0$ we find

$$c_2 = 1$$

$$c_3 = c_5 = c_7 = \cdots = 0$$

$$c_4 = c_6 = c_8 = \cdots = 1.$$

For $c_0 = 0$ and $c_1 = 1$ we obtain

$$c_2 = c_4 = c_6 = \cdots = 0$$

$$c_3 = c_5 = c_7 = \cdots = 1.$$

Thus, two solutions are

$$y_1 = 1 + x^2 + x^4 + \cdots \quad \text{and} \quad y_2 = x + x^3 + x^5 + \cdots.$$

**12.** Substituting $y = \sum_{n=0}^{\infty} c_n x^n$ into the differential equation we have

$$\left(x^2 - 1\right) y'' + xy' - y = \underbrace{\sum_{n=2}^{\infty} n(n-1)c_n x^n}_{k=n} - \underbrace{\sum_{n=2}^{\infty} n(n-1)c_n x^{n-2}}_{k=n-2} + \underbrace{\sum_{n=1}^{\infty} nc_n x^n}_{k=n} - \underbrace{\sum_{n=0}^{\infty} c_n x^n}_{k=n}$$

$$= \sum_{k=2}^{\infty} k(k-1)c_k x^k - \sum_{k=0}^{\infty} (k+2)(k+1)c_{k+2} x^k + \sum_{k=1}^{\infty} kc_k x^k - \sum_{k=0}^{\infty} c_k x^k$$

$$= (-c_2 - c_0) - 6c_3 x + \sum_{k=2}^{\infty} \left[-(k+2)(k+1)c_{k+2} + \left(k^2 - 1\right) c_k\right] x^k = 0.$$

Thus

$$-2c_2 - c_0 = 0$$

$$-6c_3 = 0$$

$$-(k+2)(k+1)c_{k+2} + (k-1)(k+1)c_k = 0$$

and

$$c_2 = -\frac{1}{2}c_0$$

$$c_3 = 0$$

$$c_{k+2} = \frac{k-1}{k+2}c_k, \quad k = 2, 3, 4, \ldots .$$

Choosing $c_0 = 1$ and $c_1 = 0$ we find

$$c_2 = -\frac{1}{2}$$

$$c_3 = c_5 = c_7 = \cdots = 0$$

$$c_4 = -\frac{1}{8}$$

and so on. For $c_0 = 0$ and $c_1 = 1$ we obtain

$$c_2 = c_4 = c_6 = \cdots = 0$$

$$c_3 = c_5 = c_7 = \cdots = 0.$$

Thus, two solutions are

$$y_1 = 1 - \frac{1}{2}x^2 - \frac{1}{8}x^4 - \cdots \quad \text{and} \quad y_2 = x.$$

**15.** Substituting $y = \sum_{n=0}^{\infty} c_n x^n$ into the differential equation we have

$$(x-1)y'' - xy' + y = \underbrace{\sum_{n=2}^{\infty} n(n-1)c_n x^{n-1}}_{k=n-1} - \underbrace{\sum_{n=2}^{\infty} n(n-1)c_n x^{n-2}}_{k=n-2} - \underbrace{\sum_{n=1}^{\infty} nc_n x^n}_{k=n} + \underbrace{\sum_{n=0}^{\infty} c_n x^n}_{k=n}$$

$$= \sum_{k=1}^{\infty} (k+1)kc_{k+1}x^k - \sum_{k=0}^{\infty} (k+2)(k+1)c_{k+2}x^k - \sum_{k=1}^{\infty} kc_k x^k + \sum_{k=0}^{\infty} c_k x^k$$

$$= -2c_2 + c_0 + \sum_{k=1}^{\infty} [-(k+2)(k+1)c_{k+2} + (k+1)kc_{k+1} - (k-1)c_k]x^k = 0.$$

Thus

$$-2c_2 + c_0 = 0$$

$$-(k+2)(k+1)c_{k+2} + (k-1)kc_{k+1} - (k-1)c_k = 0$$

and

$$c_2 = \frac{1}{2}c_0$$

$$c_{k+2} = \frac{kc_{k+1}}{k+2} - \frac{(k-1)c_k}{(k+2)(k+1)}, \quad k = 1, 2, 3, \ldots .$$

**69**

## Exercises 6.3

Choosing $c_0 = 1$ and $c_1 = 0$ we find

$$c_2 = \frac{1}{2}, \qquad c_3 = \frac{1}{6}, \qquad c_4 = 0$$

and so on. For $c_0 = 0$ and $c_1 = 1$ we obtain $c_2 = c_3 = c_4 = \cdots = 0$. Thus,

$$y = C_1 \left( 1 + \frac{1}{2}x^2 + \frac{1}{6}x^3 + \cdots \right) + C_2 x$$

and

$$y' = C_1 \left( x + \frac{1}{2}x^2 + \cdots \right) + C_2.$$

The initial conditions imply $C_1 = -2$ and $C_2 = 6$, so

$$y = -2 \left( 1 + \frac{1}{2}x^2 + \frac{1}{6}x^3 + \cdots \right) + 6x = 8x - 2e^x.$$

**18.** Substituting $y = \sum_{n=0}^{\infty} c_n x^n$ into the differential equation we have

$$(x^2 + 1)y'' + 2xy' = \underbrace{\sum_{n=2}^{\infty} n(n-1)c_n x^n}_{k=n} + \underbrace{\sum_{n=2}^{\infty} n(n-1)c_n x^{n-2}}_{k=n-2} + \underbrace{\sum_{n=1}^{\infty} 2nc_n x^n}_{k=n}$$

$$= \sum_{k=2}^{\infty} k(k-1)c_k x^k + \sum_{k=0}^{\infty} (k+2)(k+1)c_{k+2} x^k + \sum_{k=1}^{\infty} 2kc_k x^k$$

$$= 2c_2 + (6c_3 + 2c_1)x + \sum_{k=2}^{\infty} [k(k+1)c_k + (k+2)(k+1)c_{k+2}]x^k = 0.$$

Thus

$$2c_2 = 0,$$

$$6c_3 + 2c_1 = 0,$$

$$k(k+1)c_k + (k+2)(k+1)c_{k+2} = 0$$

and

$$c_2 = 0$$

$$c_3 = -\frac{1}{3}c_1$$

$$c_{k+2} = -\frac{k}{k+2}c_k, \qquad k = 2, 3, 4, \ldots.$$

Choosing $c_0 = 1$ and $c_1 = 0$ we find $c_3 = c_4 = c_5 = \cdots = 0$. For $c_0 = 0$ and $c_1 = 1$ we obtain

$$c_3 - \frac{1}{3}$$

$$c_4 = c_6 = c_8 = \cdots = 0$$

$$c_5 = -\frac{1}{5}$$

$$c_7 = \frac{1}{7}$$

and so on. Thus

$$y = c_0 + c_1 \left( x - \frac{1}{3}x^3 + \frac{1}{5}x^5 - \frac{1}{7}x^7 + \cdots \right)$$

and

$$y' = c_1 \left( 1 - x^2 + x^4 - x^6 + \cdots \right).$$

The initial conditions imply $c_0 = 0$ and $c_1 = 1$, so

$$y = x - \frac{1}{3}x^3 + \frac{1}{5}x^5 - \frac{1}{7}x^7 + \cdots.$$

**21.** Substituting $y = \sum_{n=0}^{\infty} c_n x^n$ into the differential equation we have

$$y'' + e^{-x}y = \sum_{n=2}^{\infty} n(n-1)c_n x^{n-2}$$

$$+ \left( 1 - x + \frac{1}{2}x^2 - \frac{1}{6}x^3 + \frac{1}{24}x^4 - \cdots \right) \left( c_0 + c_1 x + c_2 x^2 + c_3 x^3 + \cdots \right)$$

$$= \left[ 2c_2 + 6c_3 x + 12c_4 x^2 + 20c_5 x^3 + \cdots \right] + \left[ c_0 + (c_1 - c_0)x + \left( c_2 - c_1 + \frac{1}{2}c_0 \right) x^2 + \cdots \right]$$

$$= (2c_2 + c_0) + (6c_3 + c_1 - c_0)x + (12c_4 + c_2 - c_1 + \frac{1}{2}c_0)x^2 + \cdots = 0.$$

Then

$$2c_2 + c_0 = 0$$

$$6c_3 + c_1 - c_0 = 0$$

$$12c_4 + c_2 - c_1 + \frac{1}{2}c_0 = 0$$

and

$$c_2 = -\frac{1}{2}c_0$$

$$c_3 = -\frac{1}{6}c_1 + \frac{1}{6}c_0$$

$$c_4 = -\frac{1}{12}c_2 + \frac{1}{12}c_1 - \frac{1}{24}c_0.$$

Choosing $c_0 = 1$ and $c_1 = 0$ we find

$$c_2 = -\frac{1}{2}, \qquad c_3 = \frac{1}{6}, \qquad c_4 = 0$$

and so on. For $c_0 = 0$ and $c_1 = 1$ we obtain

$$c_2 = 0, \qquad c_3 = -\frac{1}{6}, \qquad c_4 = \frac{1}{12}.$$

Thus, two solutions are

$$y_1 = 1 - \frac{1}{2}x^2 + \frac{1}{6}x^3 + \cdots \quad \text{and} \quad y_2 = x - \frac{1}{6}x^3 + \frac{1}{12}x^4 + \cdots.$$

**24.** Substituting $y = \sum_{n=0}^{\infty} c_n x^n$ into the differential equation leads to

$$y'' - 4xy' - 4y = \underbrace{\sum_{n=2}^{\infty} n(n-1)c_n x^{n-2}}_{k=n-2} - \underbrace{\sum_{n=1}^{\infty} 4nc_n x^n}_{k=n} - \underbrace{\sum_{n=0}^{\infty} 4c_n x^n}_{k=n}$$

$$= \sum_{k=0}^{\infty} (k+2)(k+1)c_{k+2}x^k - \sum_{k=1}^{\infty} 4kc_k x^k - \sum_{k=0}^{\infty} 4c_k x^k$$

$$= 2c_2 - 4c_0 + \sum_{k=1}^{\infty} [(k+2)(k+1)c_{k+2} - 4(k+1)c_k]x^k$$

$$= e^x = 1 + \sum_{k=1}^{\infty} \frac{1}{k!}x^k.$$

Thus

$$2c_2 - 4c_0 = 1$$

$$(k+2)(k+1)c_{k+2} - 4(k+1)c_k = \frac{1}{k!}$$

and

$$c_2 = \frac{1}{2} + 2c_0$$

$$c_{k+2} = \frac{1}{(k+2)!} + \frac{4}{k+2}c_k, \qquad k = 1, 2, 3, \ldots.$$

Let $c_0$ and $c_1$ be arbitrary and iterate to find

$$c_2 = \frac{1}{2} + 2c_0$$

$$c_3 = \frac{1}{3!} + \frac{4}{3}c_1 = \frac{1}{3!} + \frac{4}{3}c_1$$

$$c_4 = \frac{1}{4!} + \frac{4}{4}c_2 = \frac{1}{4!} + \frac{1}{2} + 2c_0 = \frac{13}{4!} + 2c_0$$

$$c_5 = \frac{1}{5!} + \frac{4}{5}c_3 = \frac{1}{5!} + \frac{4}{5 \cdot 3!} + \frac{16}{15}c_1 = \frac{17}{5!} + \frac{16}{15}c_1$$

$$c_6 = \frac{1}{6!} + \frac{4}{6}c_4 = \frac{1}{6!} + \frac{4 \cdot 13}{6 \cdot 4!} + \frac{8}{6}c_0 = \frac{261}{6!} + \frac{4}{3}c_0$$

$$c_7 = \frac{1}{7!} + \frac{4}{7}c_5 = \frac{1}{7!} + \frac{4 \cdot 17}{7 \cdot 5!} + \frac{64}{105}c_1 = \frac{409}{7!} + \frac{64}{105}c_1$$

and so on. The solution is

$$y = c_0 + c_1 x + \left(\frac{1}{2} + 2c_0\right)x^2 + \left(\frac{1}{3!} + \frac{4}{3}c_1\right)x^3 - \left(\frac{13}{4!} + 2c_0\right)x^4 + \left(\frac{17}{5!} + \frac{16}{15}c_1\right)x^5$$

$$+ \left(\frac{261}{6!} + \frac{4}{3}c_0\right)x^6 + \left(\frac{409}{7!} + \frac{64}{105}c_1\right)x^7 + \cdots$$

$$= c_0\left[1 + 2x^2 + 2x^4 + \frac{4}{3}x^6 + \cdots\right] + c_1\left[x + \frac{4}{3}x^3 + \frac{16}{15}x^5 + \frac{64}{105}x^7 + \cdots\right]$$

$$+ \frac{1}{2}x^2 + \frac{1}{3!}x^3 + \frac{13}{4!}x^4 + \frac{17}{5!}x^5 + \frac{261}{6!}x^6 + \frac{409}{7!}x^7 + \cdots .$$

## Exercises 6.4

**3.** Irregular singular point: $x = 3$. Regular singular point: $x = -3$.

**6.** Irregular singular point: $x = 5$. Regular singular point: $x = 0$.

**9.** Irregular singular point: $x = 0$. Regular singular points: $x = 2, \pm 5$.

**12.** Substituting $y = \sum_{n=0}^{\infty} c_n x^{n+r}$ into the differential equation and collecting terms, we obtain

$$2xy'' + 5y' + xy = \left(2r^2 + 3r\right)c_0 x^{r-1} + \left(2r^2 + 7r + 5\right)c_1 x^r$$

$$+ \sum_{k=2}^{\infty} [2(k + r)(k + r - 1)c_k + 5(k + r)c_k + c_{k-2}]x^{k+r-1}$$

$$= 0,$$

**73**

which implies

$$2r^2 + 3r = r(2r + 3) = 0,$$

$$\left(2r^2 + 7r + 5\right)c_1 = 0,$$

and

$$(k+r)(2k + 2r + 3)c_k + c_{k-2} = 0.$$

The indicial roots are $r = -3/2$ and $r = 0$, so $c_1 = 0$. For $r = -3/2$ the recurrence relation is

$$c_k = -\frac{c_{k-2}}{(2k-3)k}, \quad k = 2, 3, 4, \ldots,$$

and

$$c_2 = -\frac{1}{2}c_0, \qquad c_3 = 0, \qquad c_4 = \frac{1}{40}c_0.$$

For $r = 0$ the recurrence relation is

$$c_k = -\frac{c_{k-2}}{k(2k+3)}, \quad k = 2, 3, 4, \ldots,$$

and

$$c_2 = -\frac{1}{14}c_0, \qquad c_3 = 0, \qquad c_4 = \frac{1}{616}c_0.$$

The general solution on $(0, \infty)$ is

$$y = C_1 x^{-3/2}\left(1 - \frac{1}{2}x^2 + \frac{1}{40}x^4 + \cdots\right) + C_2\left(1 - \frac{1}{14}x^2 + \frac{1}{616}x^4 + \cdots\right).$$

**15.** Substituting $y = \sum_{n=0}^{\infty} c_n x^{n+r}$ into the differential equation and collecting terms, we obtain

$$3xy'' + (2-x)y' - y = \left(3r^2 - r\right)c_0 x^{r-1}$$

$$+ \sum_{k=1}^{\infty}[3(k+r-1)(k+r)c_k + 2(k+r)c_k - (k+r)c_{k-1}]x^{k+r-1}$$

$$= 0,$$

which implies

$$3r^2 - r = r(3r - 1) = 0$$

and

$$(k+r)(3k + 3r - 1)c_k - (k+r)c_{k-1} = 0.$$

The indicial roots are $r = 0$ and $r = 1/3$. For $r = 0$ the recurrence relation is

$$c_k = \frac{c_{k-1}}{(3k-1)}, \quad k = 1, 2, 3, \ldots,$$

and

$$c_1 = \frac{1}{2}c_0, \qquad c_2 = \frac{1}{10}c_0, \qquad c_3 = \frac{1}{80}c_0.$$

For $r = 1/3$ the recurrence relation is

$$c_k = \frac{c_{k-1}}{3k}, \quad k = 1, 2, 3, \ldots,$$

and

$$c_1 = \frac{1}{3}c_0, \quad c_2 = \frac{1}{18}c_0, \quad c_3 = \frac{1}{162}c_0.$$

The general solution on $(0, \infty)$ is

$$y = C_1\left(1 + \frac{1}{2}x + \frac{1}{10}x^2 + \frac{1}{80}x^3 + \cdots\right) + C_2 x^{1/3}\left(1 + \frac{1}{3}x + \frac{1}{18}x^2 + \frac{1}{162}x^3 + \cdots\right).$$

**18.** Substituting $y = \sum_{n=0}^{\infty} c_n x^{n+r}$ into the differential equation and collecting terms, we obtain

$$2xy'' + xy' + \left(x^2 - \frac{4}{9}\right)y = \left(r^2 - \frac{4}{9}\right)c_0 x^r + \left(r^2 + 2r + \frac{5}{9}\right)c_1 x^{r+1}$$

$$+ \sum_{k=2}^{\infty}[(k+r)(k+r-1)c_k + (k+r)c_k - \frac{4}{9}c_k + c_{k-2}]x^{k+r}$$

$$= 0,$$

which implies

$$r^2 - \frac{4}{9} = \left(r + \frac{2}{3}\right)\left(r - \frac{2}{3}\right) = 0,$$

$$\left(r^2 + 2r + \frac{5}{9}\right)c_1 = 0,$$

and

$$\left[(k+r)^2 - \frac{4}{9}\right]c_k + c_{k-2} = 0.$$

The indicial roots are $r = -2/3$ and $r = 2/3$, so $c_1 = 0$. For $r = -2/3$ the recurrence relation is

$$c_k = -\frac{9c_{k-2}}{3k(3k-4)}, \quad k = 2, 3, 4, \ldots,$$

and

$$c_2 = -\frac{3}{4}c_0, \quad c_3 = 0, \quad c_4 = \frac{9}{128}c_0.$$

For $r = 2/3$ the recurrence relation is

$$c_k = -\frac{9c_{k-2}}{3k(3k+4)}, \quad k = 2, 3, 4, \ldots,$$

and

$$c_2 = -\frac{3}{20}c_0, \quad c_3 = 0, \quad c_4 = \frac{9}{1,280}c_0.$$

The general solution on $(0, \infty)$ is

$$y = C_1 x^{-2/3}\left(1 - \frac{3}{4}x^2 + \frac{9}{128}x^4 + \cdots\right) + C_2 x^{2/3}\left(1 - \frac{3}{20}x^2 + \frac{9}{1,280}x^4 + \cdots\right).$$

**21.** Substituting $y = \sum_{n=0}^{\infty} c_n x^{n+r}$ into the differential equation and collecting terms, we obtain

$$2x^2 y'' - x(x-1)y' - y = \left(2r^2 - r - 1\right) c_0 x^r$$

$$+ \sum_{k=1}^{\infty} [2(k+r)(k+r-1)c_k + (k+r)c_k - c_k - (k+r-1)c_{k-1}]x^{k+r}$$

$$= 0,$$

which implies
$$2r^2 - r - 1 = (2r+1)(r-1) = 0$$

and
$$[(k+r)(2k+2r-1) - 1]c_k - (k+r-1)2c_{k-1} = 0.$$

The indicial roots are $r = -1/2$ and $r = 1$. For $r = -1/2$ the recurrence relation is

$$c_k = \frac{c_{k-1}}{2k}, \quad k = 1, 2, 3, \ldots,$$

and
$$c_1 = \frac{1}{2}c_0, \qquad c_2 = \frac{1}{8}c_0, \qquad c_3 = \frac{1}{48}c_0.$$

For $r = 1$ the recurrence relation is

$$c_k = \frac{c_{k-1}}{2k+3}, \quad k = 1, 2, 3, \ldots,$$

and
$$c_1 = \frac{1}{5}c_0, \qquad c_2 = \frac{1}{35}c_0, \qquad c_3 = \frac{1}{315}c_0.$$

The general solution on $(0, \infty)$ is

$$y = C_1 x^{-1/2}\left(1 + \frac{1}{2}x + \frac{1}{8}x^2 + \frac{1}{48}x^3 + \cdots\right) + C_2 x\left(1 + \frac{1}{5}x + \frac{1}{35}x^2 + \frac{1}{315}x^3 + \cdots\right).$$

**24.** Substituting $y = \sum_{n=0}^{\infty} c_n x^{n+r}$ into the differential equation and collecting terms, we obtain

$$x^2 y'' + xy' + \left(x^2 - \frac{1}{4}\right)y = \left(r^2 - \frac{1}{4}\right)c_0 x^r + \left(r^2 + 2r + \frac{3}{4}\right)c_1 x^{r+1}$$

$$+ \sum_{k=2}^{\infty} [(k+r)(k+r-1)c_k + (k+r)c_k - \frac{1}{4}c_k + c_{k-2}]x^{k+r}$$

$$= 0,$$

which implies

$$r^2 - \frac{1}{4} = \left(r - \frac{1}{2}\right)\left(r + \frac{1}{2}\right) = 0,$$

$$\left(r^2 + 2r + \frac{3}{4}\right)c_1 = 0,$$

and

$$\left[(k+r)^2 - \frac{1}{4}\right]c_k + c_{k-2} = 0.$$

The indicial roots are $r_1 = 1/2$ and $r_2 = -1/2$, so $c_1 = 0$. For $r_1 = 1/2$ the recurrence relation is

$$c_k = -\frac{c_{k-2}}{k(k+1)}, \quad k = 2, 3, 4, \ldots,$$

and

$$c_2 = -\frac{1}{3!}c_0$$

$$c_3 = c_5 = c_7 = \cdots = 0$$

$$c_4 = \frac{1}{5!}c_0$$

$$c_{2n} = \frac{(-1)^n}{(2n+1)!}c_0.$$

For $r_2 = -1/2$ the recurrence relation is

$$c_k = -\frac{c_{k-2}}{k(k-1)}, \quad k = 2, 3, 4, \ldots,$$

and

$$c_2 = -\frac{1}{2!}c_0$$

$$c_3 = c_5 = c_7 = \cdots = 0$$

$$c_4 = \frac{1}{4!}c_0$$

$$c_{2n} = \frac{(-1)^n}{(2n)!}c_0.$$

The general solution on $(0, \infty)$ is

$$y = C_1 x^{1/2} \sum_{n=0}^{\infty} \frac{(-1)^n}{(2n+1)!} x^{2n} + C_2 x^{-1/2} \sum_{n=0}^{\infty} \frac{(-1)^n}{(2n)!} x^{2n}$$

$$= C_1 x^{-1/2} \sum_{n=0}^{\infty} \frac{(-1)^n}{(2n+1)!} x^{2n+1} + C_2 x^{-1/2} \sum_{n=0}^{\infty} \frac{(-1)^n}{(2n)!} x^{2n}$$

$$= x^{-1/2}[C_1 \sin x + C_2 \cos x].$$

**27.** Substituting $y = \sum_{n=0}^{\infty} c_n x^{n+r}$ into the differential equation and collecting terms, we obtain

$$xy'' + (1 - x)y' - y = r^2 c_0 x^{r-1} + \sum_{k=0}^{\infty}[(k+r)(k+r-1)c_k + (k+r)c_k - (k+r)c_{k-1}]x^{k+r-1} = 0,$$

which implies $r^2 = 0$ and

$$(k+r)^2 c_k - (k+r)c_{k-1} = 0.$$

The indicial roots are $r_1 = r_2 = 0$ and the recurrence relation is

$$c_k = \frac{c_{k-1}}{k}, \quad k = 1, 2, 3, \ldots.$$

One solution is

$$y_1 = c_0 \left(1 + x + \frac{1}{2}x^2 + \frac{1}{3!}x^3 + \cdots\right) = c_0 e^x.$$

A second solution is

$$y_2 = y_1 \int \frac{e^{-\int(1/x-1)dx}}{e^{2x}} dx = e^x \int \frac{e^x/x}{e^{2x}} dx = e^x \int \frac{1}{x}e^{-x}dx$$

$$= e^x \int \frac{1}{x}\left(1 - x + \frac{1}{2}x^2 - \frac{1}{3!}x^3 + \cdots\right) dx = e^x \int \left(\frac{1}{x} - 1 + \frac{1}{2}x - \frac{1}{3!}x^2 + \cdots\right) dx$$

$$= e^x \left[\ln x - x + \frac{1}{2 \cdot 2}x^2 - \frac{1}{3 \cdot 3!}x^3 + \cdots\right] = e^x \ln x - e^x \sum_{n=1}^{\infty} \frac{(-1)^{n+1}}{n \cdot n!}x^n.$$

The general solution on $(0, \infty)$ is

$$y = C_1 e^x + C_2 e^x \left(\ln x - \sum_{n=1}^{\infty} \frac{(-1)^{n+1}}{n \cdot n!}x^n\right).$$

**30.** Substituting $y = \sum_{n=0}^{\infty} c_n x^{n+r}$ into the differential equation and collecting terms, we obtain

$$xy'' - xy' + y = \left(r^2 - r\right) c_0 x^{r-1} + \sum_{k=0}^{\infty} [(k+r+1)(k+r)c_{k+1} - (k+r)c_k + c_k]x^{k+r} = 0$$

which implies

$$r^2 - r = r(r-1) = 0$$

and

$$(k+r+1)(k+r)c_{k+1} - (k+r-1)c_k = 0.$$

The indicial roots are $r_1 = 1$ and $r_2 = 0$. For $r_1 = 1$ the recurrence relation is

$$c_{k+1} = \frac{kc_k}{(k+2)(k+1)}, \quad k = 0, 1, 2, \ldots,$$

and one solution is $y_1 = c_0 x$. A second solution is

$$y_2 = x \int \frac{e^{-\int -dx}}{x^2} dx = x \int \frac{e^x}{x^2} dx = x \int \frac{1}{x^2}\left(1 + x + \frac{1}{2}x^2 + \frac{1}{3!}x^3 + \cdots\right) dx$$

$$= x \int \left(\frac{1}{x^2} + \frac{1}{x} + \frac{1}{2} + \frac{1}{3!}x + \frac{1}{4!}x^2 + \cdots\right) dx = x \left[-\frac{1}{x} + \ln x + \frac{1}{2}x + \frac{1}{12}x^2 + \frac{1}{72}x^3 + \cdots\right]$$

$$= x \ln x - 1 + \frac{1}{2}x^2 + \frac{1}{12}x^3 + \frac{1}{72}x^4 + \cdots.$$

The general solution on $(0, \infty)$ is

$$y = C_1 x + C_2 y_2(x).$$

**33.** Substituting $y = \sum_{n=0}^{\infty} c_n x^{n+r}$ into the differential equation and collecting terms, we obtain

$$xy'' + (x-1)y' - 2y = r^2 c_0 x^{r-1} + \sum_{k=1}^{\infty} [(k+r)(k+r-1)c_k$$

$$ - (k+r)c_k + (k+r-3)c_{k-1}]x^{k+r-1}$$

$$ = 0$$

which implies $r^2 = 0$ and

$$(k+r)(k+r-2)c_k + (k+r-3)c_{k-1} = 0.$$

The indicial roots are $r_1 = r_2 = 0$ and the recurrence relation is

$$k(k-2)c_k + (k-3)c_{k-1} = 0, \quad k = 1, 2, 3, \dots.$$

Then

$$-c_1 - 2c_0 = 0 \quad \Rightarrow \quad c_1 = -2c_0$$

$$0c_2 - c_1 = 0 \quad \Rightarrow \quad c_1 = 0 \text{ and } c_2 \text{ is arbitrary}$$

$$3c_3 + 0c_2 = 0 \quad \Rightarrow \quad c_3 = 0$$

and

$$c_k = -\frac{(k-3)c_{k-1}}{k(k-2)}, \quad k = 4, 5, 6, \dots.$$

Since $c_1 = 0$ and $c_1 = -2c_0$, we have $c_0 = 0$. Taking $c_2 = 0$ we obtain $c_3 = c_4 = c_5 = \cdots = 0$. Thus, $y_1 = c_2 x^2$. A second solution is

$$y_2 = x^2 \int \frac{e^{-\int(1-1/x)\,dx}}{x^4}\,dx = x^2 \int \frac{xe^{-x}}{x^4}\,dx = x^2 \int \frac{1}{x^3}\left(1 - x + \frac{1}{2}x^2 - \frac{1}{3!}x^3 + \frac{1}{4!}x^4 - \cdots\right)dx$$

$$ = x^2 \int \left(\frac{1}{x^3} - \frac{1}{x^2} + \frac{1}{2x} - \frac{1}{3!} + \frac{1}{4!}x - \cdots\right)dx = x^2\left[-\frac{1}{2x^2} + \frac{1}{x} + \frac{1}{2}\ln x - \frac{1}{6}x + \frac{1}{48}x^2 - \cdots\right]$$

$$ = \frac{1}{2}x^2 \ln x - \frac{1}{2} + x - \frac{1}{6}x^3 + \frac{1}{48}x^4 - \cdots.$$

**36.** Substituting $y = \sum_{n=0}^{\infty} c_n x^{n+r}$ into the differential equation and collecting terms, we obtain

$$x^2 y'' - y' + y = rc_0 x^{r-1} + \sum_{k=0}^{\infty} \left([(k+r)(k+r-1) + 1]c_k - (k+r+1)c_{k+1}\right)x^{k+r} = 0.$$

Thus $r = 0$ and the recurrence relation is

$$c_{k+1} = \frac{k(k-1) + 1}{k+1}c_k, \quad k = 0, 1, 2, \dots.$$

**79**

Then
$$c_1 = 0, \qquad c_2 = \frac{1}{2}c_0, \qquad c_3 = \frac{1}{2}c_0, \qquad c_4 = \frac{7}{8}c_0,$$

and so on. Therefore, one solution is

$$y(x) = c_0 \left[ 1 + x + \frac{1}{2}x^2 + \frac{1}{2}x^3 + \frac{7}{8}x^4 + \cdots \right].$$

**39.** Identifying $p_0 = 5/3$ and $q_0 = -1/3$, the indicial equation is

$$r(r-1) + \frac{5}{3}r - \frac{1}{3} = r^2 + \frac{2}{3}r - \frac{1}{3} = (r+1)\left(r - \frac{1}{3}\right) = 0.$$

The indicial roots are $-1$ and $1/3$.

## Exercises 6.5

**3.** Since $\nu^2 = 25/4$ the general solution is $y = c_1 J_{5/2}(x) + c_2 J_{-5/2}(x)$.

**6.** Since $\nu^2 = 4$ the general solution is $y = c_1 J_2(x) + c_2 Y_2(x)$.

**9.** If $y = x^{-1/2}v(x)$ then

$$y' = x^{-1/2}v'(x) - \frac{1}{2}x^{-3/2}v(x),$$

$$y'' = x^{-1/2}v''(x) - x^{-3/2}v'(x) + \frac{3}{4}x^{-5/2}v(x),$$

and

$$x^2 y'' + 2xy' + \lambda^2 x^2 y = x^{3/2}v'' + x^{1/2}v' + \left(\lambda^2 x^{3/2} - \frac{1}{4}x^{-1/2}\right)v.$$

Multiplying by $x^{1/2}$ we obtain

$$x^2 v'' + xv' + \left(\lambda^2 x^2 - \frac{1}{4}\right)v = 0,$$

whose solution is $v = c_1 J_{1/2}(\lambda x) + c_2 J_{-1/2}(\lambda x)$. Then $y = c_1 x^{-1/2} J_{1/2}(\lambda x) + c_2 x^{-1/2} J_{-1/2}(\lambda x)$.

**12.** From $y = \sqrt{x}\, J_\nu(\lambda x)$ we find

$$y' = \lambda\sqrt{x}\, J_\nu'(\lambda x) + \frac{1}{2}x^{-1/2} J_\nu(\lambda x)$$

and

$$y'' = \lambda^2 \sqrt{x}\, J_\nu''(\lambda x) + \lambda x^{-1/2} J_\nu'(\lambda x) - \frac{1}{4}x^{-3/2} J_\nu(\lambda x).$$

Substituting into the differential equation, we have

$$x^2 y'' + \left(\lambda^2 x^2 - \nu^2 + \frac{1}{4}\right)y = \sqrt{x}\left[\lambda^2 x^2 J_\nu''(\lambda x) + \lambda x J_\nu'(\lambda x) + \left(\lambda^2 x^2 - \nu^2\right) J_\nu(\lambda x)\right]$$

$$= \sqrt{x} \cdot 0 \qquad \text{(since } J_n \text{ is a solution of Bessel's equation)}$$

$$= 0.$$

Therefore, $\sqrt{x}\,J_\nu(\lambda x)$ is a solution of the original equation.

**15.** From Problem 10 with $n = -1$ we find $y = x^{-1}J_{-1}(x)$. From Problem 11 with $n = 1$ we find
$y = x^{-1}J_1(x) = -x^{-1}J_{-1}(x)$.

**18.** From Problem 10 with $n = 3$ we find $y = x^3 J_3(x)$. From Problem 11 with $n = -3$ we find
$y = x^3 J_{-3}(x) = -x^3 J_3(x)$.

**21.** The recurrence relation follows from

$$xJ_{\nu+1}(x) + xJ_{\nu-1}(x) = \sum_{n=0}^{\infty} \frac{(-1)^{n-1}2n}{n!\Gamma(1+\nu+n)}\left(\frac{x}{2}\right)^{2n+\nu} + \sum_{n=0}^{\infty}\frac{(-1)^n 2(\nu+n)}{n!\Gamma(1+\nu+n)}\left(\frac{x}{2}\right)^{2n+\nu}$$

$$= \sum_{n=0}^{\infty}\frac{(-1)^n 2\nu}{n!\Gamma(1+\nu+n)}\left(\frac{x}{2}\right)^{2n+\nu} = 2\nu J_\nu(x).$$

**24.** By Problem 19 we obtain $J_0'(x) = J_{-1}(x)$ and by Problem 22

$$2J_0'(x) = J_{-1}(x) - J_1(x) = J_0'(x) - J_1(x)$$

so that $J_0'(x) = -J_1(x)$.

**27.** Since

$$\Gamma\left(1 - \frac{1}{2} + n\right) = \frac{(2n-1)!}{(n-1)!2^{2n-1}}$$

we obtain

$$J_{-1/2}(x) = \sum_{n=0}^{\infty}\frac{(-1)^n 2^{1/2}x^{-1/2}}{2n(2n-1)!\sqrt{\pi}}x^{2n} = \sqrt{\frac{2}{\pi x}}\cos x.$$

**30.** By Problem 21 we obtain $3J_{3/2}(x) = xJ_{5/2}(x) + xJ_{1/2}(x)$ so that

$$J_{5/2}(x) = \sqrt{\frac{2}{\pi x}}\left(\frac{3\sin x}{x^2} - \frac{3\cos x}{x} - \sin x\right).$$

**33.** By Problem 21 we obtain $-5J_{-5/2}(x) = xJ_{-3/2}(x) + xJ_{-7/2}(x)$ so that

$$J_{-7/2}(x) = \sqrt{\frac{2}{\pi x}}\left(\frac{-15\cos x}{x^3} - \frac{15\sin x}{x^2} + \frac{6\cos x}{x} + \sin x\right).$$

## Exercises 6.5

**36.** If $y_1 = J_0(x)$ then using equation (35) on Page 299 in the text gives

$$y_2 = J_0(x) \int \frac{e^{-\int dx/x}}{(J_0(x))^2} \, dx$$

$$= J_0(x) \int \frac{dx}{x\left(1 - \frac{x^2}{4} + \frac{x^4}{64} - \frac{x^6}{2304} + \cdots\right)^2} \, dx$$

$$= J_0(x) \int \left(\frac{1}{x} + \frac{x}{2} + \frac{5x^3}{32} + \frac{23x^5}{576} + \cdots\right) dx$$

$$= J_0(x) \left(\ln x + \frac{x^2}{4} + \frac{5x^4}{128} + \frac{23x^6}{3456} + \cdots\right)$$

$$= J_0(x) \ln x + \left(1 - \frac{x^2}{4} + \frac{x^4}{64} - \frac{x^6}{2304} + \cdots\right)\left(\frac{x^2}{4} + \frac{5x^4}{128} + \frac{23x^6}{3456} + \cdots\right)$$

$$= J_0(x) \ln x + \frac{x^2}{4} - \frac{3x^4}{128} + \frac{11x^6}{13824} - \cdots.$$

**39. (a)** Using the formulas on Page 315 in the text we obtain

$$P_6(x) = \frac{1}{16}\left(231x^6 - 315x^4 + 105x^2 - 5\right)$$

and

$$P_7(x) = \frac{1}{16}\left(429x^7 - 693x^5 + 315x^3 - 35x\right).$$

**(b)** $P_6(x)$ satisfies $\left(1 - x^2\right)y'' - 2xy' + 42y = 0$ and $P_7(x)$ satisfies $\left(1 - x^2\right)y'' - 2xy' + 56y = 0$.

**42.** The polynomials are shown in (18) on Page 316 in the text.

**45.** The recurrence relation can be wrtten

$$P_{k+1}(x) = \frac{2k+1}{k+1} x P_k(x) - \frac{k}{k+1} P_{k-1}(x), \qquad k = 2, \, 3, \, 4, \, \ldots \, .$$

$k = 1$: $P_2(x) = \frac{3}{2}x^2 - \frac{1}{2}$

$k = 2$: $P_3(x) = \frac{5}{3}x\left(\frac{3}{2}x^2 - \frac{1}{2}\right) - \frac{2}{3}x = \frac{5}{2}x^3 - \frac{3}{2}x$

$k = 3$: $P_4(x) = \frac{7}{4}x\left(\frac{5}{2}x^3 - \frac{3}{2}x\right) - \frac{3}{4}\left(\frac{3}{2}x^2 - \frac{1}{2}\right) = \frac{35}{8}x^4 - \frac{30}{8}x^2 + \frac{3}{8}$

$k = 4$: $P_5(x) = \frac{9}{5}x\left(\frac{35}{8}x^4 - \frac{30}{8}x^2 + \frac{3}{8}\right) - \frac{4}{5}\left(\frac{5}{2}x^3 - \frac{3}{2}x\right) = \frac{63}{8}x^5 - \frac{35}{4}x^3 + \frac{15}{8}x$

$k = 5$: $P_6(x) = \frac{11}{6}x\left(\frac{63}{8}x^5 - \frac{35}{4}x^3 + \frac{15}{8}x\right) - \frac{5}{6}\left(\frac{35}{8}x^4 - \frac{30}{8}x^2 + \frac{3}{8}\right) = \frac{231}{16}x^6 - \frac{315}{16}x^4 + \frac{105}{16}x - \frac{5}{16}$

**48.** All integrals of the form $\int_{-1}^{1} P_n(x)P_m(x)\,dx$ are 0 for $n \neq m$.

# ━━━━━ Chapter 6 Review Exercises ━━━━━

**3.** The auxiliary equation is $m^2 - 5m + 6 = (m-2)(m-3) = 0$ and a particular solution is $y_p = x^4 - x^2 \ln x$ so that
$$y = c_1 x^2 + c_2 x^3 + x^4 - x^2 \ln x.$$

**6.** Since
$$P(x) = 0 \quad \text{and} \quad Q(x) = \frac{2}{(x^2 - 4)(x^2 + 4)}$$
the singular points are $x = 2$, $x = -2$, $x = 2i$, and $x = -2i$. All others are ordinary points.

**9.** Since
$$P(x) = -\frac{1}{x^2(x^2 - 9)} \quad \text{and} \quad Q(x) = \frac{1}{x(x^2 - 9)^2}$$
the regular singular points are $x = 3$ and $x = -3$. The irregular singular point is $x = 0$.

**12.** Since $P(x) = -2x/\left(x^2 - 4\right)$ and $Q(x) = 9/\left(x^2 - 4\right)$ an interval of convergence is $-2 < x < 2$.

**15.** Substituting $y = \sum_{n=0}^{\infty} c_n x^n$ into the differential equation we obtain
$$(x-1)y'' + 3y = (-2c_2 + 3c_0) + \sum_{k=3}^{\infty}(k-1)(k-2)c_{k-1} - k(k-1)c_k + 3c_{k-2}]x^{k-2} = 0$$
which implies $c_2 = 3c_0/2$ and
$$c_k = \frac{(k-1)(k-2)c_{k-1} + 3c_{k-2}}{k(k-1)}, \quad k = 3, 4, 5, \ldots.$$

Choosing $c_0 = 1$ and $c_1 = 0$ we find
$$c_2 = \frac{3}{2}, \quad c_3 = \frac{1}{2}, \quad c_4 = \frac{5}{8}$$
and so on. For $c_0 = 0$ and $c_1 = 1$ we obtain
$$c_2 = 0, \quad c_3 = \frac{1}{2}, \quad c_4 = \frac{1}{4}$$
and so on. Thus, two solutions are
$$y_1 = C_1\left(1 + \frac{3}{2}x^2 + \frac{1}{2}x^3 + \frac{5}{8}x^4 + \cdots\right)$$
and
$$y_2 = C_2\left(x + \frac{1}{2}x^3 + \frac{1}{4}x^4 + \cdots\right).$$

**83**

## Chapter 6 Review Exercises

**18.** Substituting $y = \sum_{n=0}^{\infty} c_n x^{n+r}$ into the differential equation we obtain

$$2xy'' + y' + y = \left(2r^2 - r\right) c_0 x^{r-1} + \sum_{k=1}^{\infty} [2(k+r)(k+r-1)c_k + (k+r)c_k + c_{k-1}]x^{k+r-1} = 0$$

which implies

$$2r^2 - r = r(2r - 1) = 0$$

and

$$(k+r)(2k + 2r - 1)c_k + c_{k-1} = 0.$$

The indicial roots are $r = 0$ and $r = 1/2$. For $r = 0$ the recurrence relation is

$$c_k = -\frac{c_{k-1}}{k(2k-1)}, \quad k = 1, 2, 3, \ldots,$$

so

$$c_1 = -c_0, \qquad c_2 = \frac{1}{6}c_0, \qquad c_3 = -\frac{1}{90}c_0.$$

For $r = 1/2$ the recurrence relation is

$$c_k = -\frac{c_{k-1}}{k(2k+1)}, \quad k = 1, 2, 3, \ldots,$$

so

$$c_1 = -\frac{1}{3}c_0, \qquad c_2 = \frac{1}{30}c_0, \qquad c_3 = -\frac{1}{630}c_0.$$

Two linearly independent solutions are

$$y_1 = C_1 x \left(1 - x + \frac{1}{6}x^2 - \frac{1}{90}x^3 + \cdots\right)$$

and

$$y_2 = C_2 x^{1/2} \left(1 - \frac{1}{3}x + \frac{1}{30}x^2 - \frac{1}{630}x^3 + \cdots\right).$$

**21.** Substituting $y = \sum_{n=0}^{\infty} c_n x^{n+r}$ into the differential equation we obtain

$$xy'' - (2x - 1)y' + (x - 1)y = r^2 c_0 x^{r-1} + \left[\left(r^2 + 2r + 1\right)c_1 - (2r+1)c_0\right]x^r$$

$$+ \sum_{k=2}^{\infty} [(k+r)(k+r-1)c_k + (k+r)c_k - 2(k+r-1)c_{k-1} - c_{k-1} + c_{k-2}]x^{k+r-1}$$

$$= 0$$

which implies

$$r^2 = 0,$$

$$(r+1)^2 c_1 - (2r+1)c_0 = 0,$$

and

$$(k+r)^2 c_k - (2k + 2r - 1)c_{k-1} + c_{k-2} = 0.$$

The indicial roots are $r_1 = r_2 = 0$, so $c_1 = c_0$ and

$$c_k = \frac{(2k-1)c_{k-1} - c_{k-2}}{k^2}, \quad k = 2, 3, 4, \ldots.$$

Thus

$$c_2 = \frac{1}{2}c_0, \qquad c_3 = \frac{1}{3!}c_0, \qquad c_4 = \frac{1}{4!}c_0$$

and one solution is

$$y_1 = c_0 \left( 1 + x + \frac{1}{2}x^2 + \frac{1}{3!}x^3 + \frac{1}{4!}x^4 + \cdots \right) = c_0 e^x.$$

A second solution is

$$y_2 = e^x \int \frac{e^{\int (2 - 1/x)dx}}{e^{2x}} \, dx = e^x \int \frac{e^{2x} \, dx}{xe^{2x}} = e^x \int \frac{1}{x} \, dx = e^x \ln x.$$

# 7 Laplace Transform

**3.** $\mathscr{L}\{f(t)\} = \int_0^1 te^{-st}dt + \int_1^\infty e^{-st}dt = \left(-\frac{1}{s}te^{-st} - \frac{1}{s^2}e^{-st}\right)\Big|_0^1 - \frac{1}{s}e^{-st}\Big|_1^\infty$

$= \left(-\frac{1}{s}e^{-s} - \frac{1}{s^2}e^{-s}\right) - \left(0 - \frac{1}{s^2}\right) - \frac{1}{s}(0 - e^{-s}) = \frac{1}{s^2}(1 - e^{-s}), \quad s > 0$

**6.** $\mathscr{L}\{f(t)\} = \int_{\pi/2}^\infty (\cos t)e^{-st}dt = \left(-\frac{s}{s^2+1}e^{-st}\cos t + \frac{1}{s^2+1}e^{-st}\sin t\right)\Big|_{\pi/2}^\infty$

$= 0 - \left(0 + \frac{1}{s^2+1}e^{-\pi s/2}\right) = -\frac{1}{s^2+1}e^{-\pi s/2}, \quad s > 0$

**9.** $f(t) = \begin{cases} 1 - t, & 0 < t < 1 \\ 0, & t > 0 \end{cases}$

$\mathscr{L}\{f(t)\} = \int_0^1 (1 - t)e^{-st}dt = \left(-\frac{1}{s}(1 - t)e^{-st} + \frac{1}{s^2}e^{-st}\right)\Big|_0^1 = \frac{1}{s^2}e^{-s} + \frac{1}{s} - \frac{1}{s^2}, \quad s > 0$

**12.** $\mathscr{L}\{f(t)\} = \int_0^\infty e^{-2t-5}e^{-st}dt = e^{-5}\int_0^\infty e^{-(s+2)t}dt = -\frac{e^{-5}}{s+2}e^{-(s+2)t}\Big|_0^\infty = \frac{e^{-5}}{s+2}, \quad s > -2$

**15.** $\mathscr{L}\{f(t)\} = \int_0^\infty e^{-t}(\sin t)e^{-st}dt = \int_0^\infty (\sin t)e^{-(s+1)t}dt$

$= \left(\frac{-(s+1)}{(s+1)^2+1}e^{-(s+1)t}\sin t - \frac{1}{(s+1)^2+1}e^{-(s+1)t}\cos t\right)\Big|_0^\infty$

$= \frac{1}{(s+1)^2+1} = \frac{1}{s^2+2s+2}, \quad s > -1$

**18.** $\mathscr{L}\{f(t)\} = \int_0^\infty t(\sin t)e^{-st}dt$

$= \left[\left(-\frac{t}{s^2+1} - \frac{2s}{(s^2+1)^2}\right)(\cos t)e^{-st} - \left(\frac{st}{s^2+1} + \frac{s^2-1}{(s^2+1)^2}\right)(\sin t)e^{-st}\right]_0^\infty$

$= \frac{2s}{(s^2+1)^2}, \quad s > 0$

**21.** $\mathscr{L}\{4t - 10\} = \frac{4}{s^2} - \frac{10}{s}$

**24.** $\mathscr{L}\{-4t^2 + 16t + 9\} = -4\frac{2}{s^3} + \frac{16}{s^2} + \frac{9}{s}$

**27.** $\mathscr{L}\{1 + e^{4t}\} = \frac{1}{s} + \frac{1}{s - 4}$

**30.** $\mathscr{L}\{e^{2t} - 2 + e^{-2t}\} = \dfrac{1}{s-2} - \dfrac{2}{s} + \dfrac{1}{s+2}$

**33.** $\mathscr{L}\{\sinh kt\} = \dfrac{k}{s^2 - k^2}$

**36.** $\mathscr{L}\{e^{-t}\cosh t\} = \mathscr{L}\left\{e^{-t}\dfrac{e^t + e^{-t}}{2}\right\} = \mathscr{L}\left\{\dfrac{1}{2} + \dfrac{1}{2}e^{-2t}\right\} = \dfrac{1}{2s} + \dfrac{1}{2(s+2)}$

**39.** $\mathscr{L}\{\cos t \cos 2t\} = \mathscr{L}\left\{\dfrac{1}{2}\cos 3t + \dfrac{1}{2}\cos t\right\} = \dfrac{1}{2}\dfrac{s}{s^2+9} + \dfrac{1}{2}\dfrac{s}{s^2+1}$

**42.** $\mathscr{L}\{\sin^3 t\} = \mathscr{L}\left\{\sin t\left(\dfrac{1}{2} - \dfrac{1}{2}\cos 2t\right)\right\} = \mathscr{L}\left\{\dfrac{1}{2}\sin t - \dfrac{1}{2}\left(\dfrac{1}{2}\sin 3t - \dfrac{1}{2}\sin t\right)\right\} = \dfrac{3}{4}\dfrac{1}{s^2+1} - \dfrac{1}{4}\dfrac{3}{s^2+9}$

**45.** $\mathscr{L}\{t^{1/2}\} = \dfrac{\Gamma(3/2)}{s^{3/2}} = \dfrac{\sqrt{\pi}}{2s^{3/2}}$

**48.** Since $f$ and $g$ are of exponential order there exist numbers $c$, $d$, $M$, and $N$ such that $|f(t)| \le Me^{ct}$ and $|g(t)| \le Ne^{dt}$ for $t > T$. Then

$$|(fg)(t)| = |f(t)||g(t)| \le Me^{ct}Ne^{dt} = MNe^{(c+d)t}$$

for $t > T$, and $fg$ is of exponential order.

## Exercises 7.2

**3.** $\mathscr{L}^{-1}\left\{\dfrac{1}{s^2} - \dfrac{48}{s^5}\right\} = \mathscr{L}^{-1}\left\{\dfrac{1}{s^2} - \dfrac{48}{24}\cdot\dfrac{4!}{s^5}\right\} = t - 2t^4$

**6.** $\mathscr{L}^{-1}\left\{\dfrac{(s+2)^3}{s^3}\right\} = \mathscr{L}^{-1}\left\{\dfrac{1}{s} + 4\cdot\dfrac{1}{s^2} + 2\cdot\dfrac{2}{s^3}\right\} = 1 + 4t + 2t^2$

**9.** $\mathscr{L}^{-1}\left\{\dfrac{1}{4s+1}\right\} = \mathscr{L}^{-1}\left\{\dfrac{1}{4}\cdot\dfrac{1}{s+1/4}\right\} = \dfrac{1}{4}e^{-t/4}$

**12.** $\mathscr{L}^{-1}\left\{\dfrac{10s}{s^2+16}\right\} = 10\cos 4t$

**15.** $\mathscr{L}^{-1}\left\{\dfrac{1}{s^2-16}\right\} = \mathscr{L}^{-1}\left\{\dfrac{1/8}{s-4} - \dfrac{1/8}{s+4}\right\} = \dfrac{1}{8}e^{4t} - \dfrac{1}{8}e^{-4t} = \dfrac{1}{4}\sinh 4t$

**18.** $\mathscr{L}^{-1}\left\{\dfrac{s+1}{s^2+2}\right\} = \mathscr{L}^{-1}\left\{\dfrac{s}{s^2+2} + \dfrac{1}{\sqrt{2}}\cdot\dfrac{\sqrt{2}}{s^2+2}\right\} = \cos\sqrt{2}\,t + \dfrac{1}{\sqrt{2}}\sin\sqrt{2}\,t$

**21.** $\mathscr{L}^{-1}\left\{\dfrac{s}{s^2+2s-3}\right\} = \mathscr{L}^{-1}\left\{\dfrac{1}{4}\cdot\dfrac{1}{s-1} + \dfrac{3}{4}\cdot\dfrac{1}{s+3}\right\} = \dfrac{1}{4}e^t + \dfrac{3}{4}e^{-3t}$

**24.** $\mathscr{L}^{-1}\left\{\dfrac{s-3}{(s-\sqrt{3})(s+\sqrt{3})}\right\} = \mathscr{L}^{-1}\left\{\dfrac{s}{s^2-3} - \sqrt{3}\cdot\dfrac{\sqrt{3}}{s^2-3}\right\} = \cosh\sqrt{3}\,t - \sqrt{3}\sinh\sqrt{3}\,t$

**27.** $\mathscr{L}^{-1}\left\{\dfrac{2s+4}{(s-2)(s^2+4s+3)}\right\} = \mathscr{L}^{-1}\left\{\dfrac{8}{15}\cdot\dfrac{1}{s-2} - \dfrac{1}{3}\cdot\dfrac{1}{s+1} - \dfrac{1}{5}\cdot\dfrac{1}{s+3}\right\} = \dfrac{8}{15}e^{2t} - \dfrac{1}{3}e^{-t} - \dfrac{1}{5}e^{-3t}$

## Exercises 7.2

**30.** $\mathcal{L}^{-1}\left\{\dfrac{s-1}{s^2(s^2+1)}\right\} = \mathcal{L}^{-1}\left\{\dfrac{1}{s} - \dfrac{1}{s^2} - \dfrac{s}{s^2+1} + \dfrac{1}{s^2+1}\right\} = 1 - t - \cos t + \sin t$

**33.** $\mathcal{L}^{-1}\left\{\dfrac{1}{(s^2+1)(s^2+4)}\right\} = \mathcal{L}^{-1}\left\{\dfrac{1}{3}\cdot\dfrac{1}{s^2+1} - \dfrac{1}{6}\cdot\dfrac{2}{s^2+4}\right\} = \dfrac{1}{3}\sin t - \dfrac{1}{6}\sin 2t$

**36.** $\mathcal{L}\{f(t)\} = \displaystyle\int_0^\infty e^{(3-s)t}\,dt = \dfrac{1}{s-3}$ for $s > 3$

## Exercises 7.3

**3.** $\mathcal{L}\left\{t^3 e^{-2t}\right\} = \dfrac{3!}{(s+2)^4}$

**6.** $\mathcal{L}\left\{e^{-2t}\cos 4t\right\} = \dfrac{s+2}{(s+2)^2+16}$

**9.** $\mathcal{L}\left\{t\left(e^t + e^{2t}\right)^2\right\} = \mathcal{L}\left\{te^{2t} + 2te^{3t} + te^{4t}\right\} = \dfrac{1}{(s-2)^2} + \dfrac{2}{(s-3)^2} + \dfrac{1}{(s-4)^2}$

**12.** $\mathcal{L}\left\{e^t\cos^2 3t\right\} = \mathcal{L}\left\{\dfrac{1}{2}e^t + \dfrac{1}{2}e^t\cos 6t\right\} = \dfrac{1}{2}\dfrac{1}{s-1} + \dfrac{1}{2}\dfrac{s-1}{(s-1)^2+36}$

**15.** $\mathcal{L}^{-1}\left\{\dfrac{1}{s^2-6s+10}\right\} = \mathcal{L}^{-1}\left\{\dfrac{1}{(s-3)^2+1^2}\right\} = e^{3t}\sin t$

**18.** $\mathcal{L}^{-1}\left\{\dfrac{2s+5}{s^2+6s+34}\right\} = \mathcal{L}^{-1}\left\{2\dfrac{(s+3)}{(s+3)^2+5^2} - \dfrac{1}{5}\dfrac{5}{(s+3)^2+5^2}\right\} = 2e^{-3t}\cos 5t - \dfrac{1}{5}e^{-3t}\sin 5t$

**21.** $\mathcal{L}^{-1}\left\{\dfrac{2s-1}{s^2(s+1)^3}\right\} = \mathcal{L}^{-1}\left\{\dfrac{5}{s} - \dfrac{1}{s^2} - \dfrac{5}{s+1} - \dfrac{4}{(s+1)^2} - \dfrac{3}{2}\dfrac{2}{(s+1)^3}\right\} = 5-t-5e^{-t}-4te^{-t}-\dfrac{3}{2}t^2 e^{-t}$

**24.** $\mathcal{L}\{e^{2-t}\,\mathcal{U}(t-2)\} = \mathcal{L}\left\{e^{-(t-2)}\,\mathcal{U}(t-2)\right\} = \dfrac{e^{-2s}}{s+1}$

**27.** $\mathcal{L}\{\cos 2t\,\mathcal{U}(t-\pi)\} = \mathcal{L}\{\cos 2(t-\pi)\,\mathcal{U}(t-\pi)\} = \dfrac{se^{-\pi s}}{s^2+4}$

**30.** $\mathcal{L}\left\{te^{t-5}\,\mathcal{U}(t-5)\right\} = \mathcal{L}\left\{(t-5)e^{t-5}\,\mathcal{U}(t-5) + 5e^{t-5}\,\mathcal{U}(t-5)\right\} = \dfrac{e^{-5s}}{(s-1)^2} + \dfrac{5e^{-5s}}{s-1}$

**33.** $\mathcal{L}^{-1}\left\{\dfrac{e^{-\pi s}}{s^2+1}\right\} = \sin(t-\pi)\,\mathcal{U}(t-\pi)$

**36.** $\mathcal{L}^{-1}\left\{\dfrac{e^{-2s}}{s^2(s-1)}\right\} = \mathcal{L}^{-1}\left\{-\dfrac{e^{-2s}}{s} - \dfrac{e^{-2s}}{s^2} + \dfrac{e^{-2s}}{s-1}\right\} = -\,\mathcal{U}(t-2) - (t-2)\,\mathcal{U}(t-2) + e^{t-2}\,\mathcal{U}(t-2)$

**39.** $\mathcal{L}\{t^2\sinh t\} = \dfrac{d^2}{ds^2}\left(\dfrac{1}{s^2-1}\right) = \dfrac{6s^2+2}{(s^2-1)^3}$

**42.** $\mathcal{L}\left\{te^{-3t}\cos 3t\right\} = -\dfrac{d}{ds}\left(\dfrac{s+3}{(s+3)^2+9}\right) = \dfrac{(s+3)^2-9}{[(s+3)^2+9]^2}$

**45. (c)**

**48. (b)**

**51.** $\mathcal{L}\{2 - 4\,\mathcal{U}(t-3)\} = \dfrac{2}{s} - \dfrac{4}{s}e^{-3s}$

**54.** $\mathcal{L}\left\{\sin t\,\mathcal{U}\left(t - \dfrac{3\pi}{2}\right)\right\} = \mathcal{L}\left\{-\cos\left(t - \dfrac{3\pi}{2}\right)\mathcal{U}\left(t - \dfrac{3\pi}{2}\right)\right\} = -\dfrac{se^{-3\pi s/2}}{s^2 + 1}$

**57.** $\mathcal{L}\{f(t)\} = \mathcal{L}\{\mathcal{U}(t-a) - \mathcal{U}(t-b)\} = \dfrac{e^{-as}}{s} - \dfrac{e^{-bs}}{s}$

**60.** $\mathcal{L}^{-1}\left\{\dfrac{2}{s} - \dfrac{3e^{-s}}{s^2} + \dfrac{5e^{-2s}}{s^2}\right\} = 2 - 3(t-1)\,\mathcal{U}(t-1) + 5(t-2)\,\mathcal{U}(t-2)$

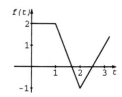

$$= \begin{cases} 2, & 0 \le t < 1 \\ -3t + 5, & 1 \le t < 2 \\ 2t - 5, & t \ge 2 \end{cases}$$

**63.** $f(t) = -\dfrac{1}{t}\mathcal{L}^{-1}\left\{\dfrac{d}{ds}\left(\dfrac{\pi}{2} - \tan^{-1}\dfrac{s}{2}\right)\right\} = -\dfrac{1}{t}\mathcal{L}^{-1}\left\{-\dfrac{2}{s^2 + 2^2}\right\} = \dfrac{\sin 2t}{t}$

## Exercises 7.4

**3.** $\mathcal{L}\{y'' + 3y'\} = \mathcal{L}\{y''\} + 3\mathcal{L}\{y'\} = s^2 Y(s) - sy(0) - y'(0) + 3[sY(s) - y(0)] = (s^2 + 3s)Y(s) - s - 2$

**6.** We solve $\mathcal{L}\{y'' + y\} = \mathcal{L}\{1\} = 1/s$.

$$s^2 Y(s) - sy(0) - y'(0) + Y(s) = \dfrac{1}{s}$$

$$(s^2 + 1)Y(s) - 2s - 3 = \dfrac{1}{s}$$

$$Y(s) = \dfrac{1}{s(s^2 + 1)} + \dfrac{2s + 3}{s^2 + 1}$$

**9.** $\mathcal{L}\left\{\displaystyle\int_0^t e^{-\tau}\cos\tau\,d\tau\right\} = \dfrac{1}{s}\mathcal{L}\{e^{-t}\cos t\} = \dfrac{1}{s}\dfrac{s+1}{(s+1)^2 + 1} = \dfrac{s+1}{s(s^2 + 2s + 2)}$

**12.** $\mathcal{L}\left\{\displaystyle\int_0^t \sin\tau\cos(t-\tau)\,d\tau\right\} = \mathcal{L}\{\sin t\}\mathcal{L}\{\cos t\} = \dfrac{s}{(s^2 + 1)^2}$

**15.** $\mathcal{L}\{1 * t^3\} = \dfrac{1}{s}\dfrac{3!}{s^4} = \dfrac{6}{s^5}$

**18.** $\mathcal{L}\{t^2 * te^t\} = \dfrac{2}{s^3(s-1)^2}$

**21.** $\mathcal{L}^{-1}\left\{\dfrac{1}{s+5}F(s)\right\} = e^{-5t} * f(t) = \displaystyle\int_0^t f(\tau)e^{-5(t-\tau)}d\tau$

**24.** $\mathscr{L}^{-1}\left\{\dfrac{1}{s(s^2+1)}\right\} = 1 * \sin t = \displaystyle\int_0^t \sin(t-\tau)\,d\tau = \cos(t-\tau)\Big|_0^t = 1 - \cos t$

**27.** $\mathscr{L}^{-1}\left\{\dfrac{s}{(s^2+4)^2}\right\} = \cos 2t * \dfrac{1}{2}\sin 2t = \dfrac{1}{2}\displaystyle\int_0^t \cos 2\tau \sin 2(t-\tau)\,d\tau$

$$= \frac{1}{2}\int_0^t \cos 2\tau(\sin 2t \cos 2\tau - \cos 2t \sin 2\tau)\,d\tau = \frac{1}{2}\left[\sin 2t \int_0^t \cos^2 2\tau\,d\tau - \cos 2t \int_0^t \frac{1}{2}\sin 4\tau\,d\tau\right]$$

$$= \frac{1}{2}\sin 2t\left[\frac{1}{2}\tau + \frac{1}{8}\sin 4\tau\right]_0^t - \frac{1}{4}\cos 2t\left[-\frac{1}{4}\cos 4\tau\right]_0^t$$

$$= \frac{1}{2}\sin 2t\left(\frac{1}{2}t + \frac{1}{8}\sin 4t\right) + \frac{1}{16}\cos 2t(\cos 4t - 1)$$

$$= \frac{1}{4}t\sin 2t + \frac{1}{16}\sin 2t \sin 4t + \frac{1}{16}\cos 2t \cos 4t - \frac{1}{16}\cos 2t$$

$$= \frac{1}{4}t\sin 2t + \frac{1}{16}\left[\sin 2t(2\sin 2t \cos 2t) + \cos 2t\left(\cos^2 2t - \sin^2 2t\right) - \cos 2t\right]$$

$$= \frac{1}{4}t\sin 2t + \frac{1}{16}\cos 2t\left[2\sin^2 2t + \cos^2 2t - \sin^2 2t - 1\right] = \frac{1}{4}t\sin 2t$$

**30.** $f * (g+h) = \displaystyle\int_0^t f(\tau)[g(t-\tau) + h(t-\tau)]\,d\tau = \int_0^t f(\tau)g(t-\tau)\,d\tau + \int_0^t f(\tau)h(t-\tau)\,d\tau$

$$= \int_0^t f(\tau)[g(t-\tau) + h(t-\tau)]\,d\tau = f*g + f*h$$

**33.** $\mathscr{L}\{f(t)\} = \dfrac{1}{1-e^{-bs}}\displaystyle\int_0^b \dfrac{a}{b}te^{-st}\,dt = \dfrac{a}{s}\left(\dfrac{1}{bs} - \dfrac{1}{e^{bs}-1}\right)$

**36.** $\mathscr{L}\{f(t)\} = \dfrac{1}{1-e^{-2\pi s}}\displaystyle\int_0^\pi e^{-st}\sin t\,dt = \dfrac{1}{s^2+1} \cdot \dfrac{1}{1-e^{-\pi s}}$

———————— **Exercises 7.5** ————————————

**3.** The Laplace transform of the differential equation is

$$s\mathscr{L}\{y\} - y(0) + 4\mathscr{L}\{y\} = \frac{1}{s+4}.$$

Solving for $\mathscr{L}\{y\}$ we obtain $\mathscr{L}\{y\} = \dfrac{1}{(s+4)^2} + \dfrac{2}{s+4}.$

Thus, $\qquad\qquad\qquad\qquad y = te^{-4t} + 2e^{-4t}.$

**6.** The Laplace transform of the differential equation is

$$s^2\mathscr{L}\{y\} - sy(0) - y'(0) - 6\left[s\mathscr{L}\{y\} - y(0)\right] + 13\mathscr{L}\{y\} = 0.$$

Solving for $\mathcal{L}\{y\}$ we obtain

$$\mathcal{L}\{y\} = -\frac{3}{s^2 - 6s + 13} = -\frac{3}{2}\frac{2}{(s-3)^2 + 2^2}.$$

Thus,

$$y = -\frac{3}{2}e^{3t}\sin 2t.$$

**9.** The Laplace transform of the differential equation is

$$s^2\,\mathcal{L}\{y\} - sy(0) - y'(0) - 4\left[s\,\mathcal{L}\{y\} - y(0)\right] + 4\,\mathcal{L}\{y\} = \frac{6}{(s-2)^4}.$$

Solving for $\mathcal{L}\{y\}$ we obtain $\quad \mathcal{L}\{y\} = \frac{1}{20}\frac{5!}{(s-2)^6}$. Thus, $y = \frac{1}{20}t^5 e^{2t}$.

**12.** The Laplace transform of the differential equation is

$$s^2\,\mathcal{L}\{y\} - sy(0) - y'(0) + 16\,\mathcal{L}\{y\} = \frac{1}{s}.$$

Solving for $\mathcal{L}\{y\}$ we obtain

$$\mathcal{L}\{y\} = \frac{s^2 + 2s + 1}{s(s^2 + 16)} = \frac{1}{16}\frac{1}{s} + \frac{15}{16}\frac{s}{s^2 + 4^2} + \frac{1}{2}\frac{4}{s^2 + 4^2}.$$

Thus,

$$y = \frac{1}{16} + \frac{15}{16}\cos 4t + \frac{1}{2}\sin 4t.$$

**15.** The Laplace transform of the differential equation is

$$2\left[s^3\,\mathcal{L}\{y\} - s^2(0) - sy'(0) - y''(0)\right] + 3[s^2\,\mathcal{L}\{y\} - sy(0) - y'(0)] - 3[s\,\mathcal{L}\{y\} - y(0)] - 2\,\mathcal{L}\{y\} = \frac{1}{s+1}.$$

Solving for $\mathcal{L}\{y\}$ we obtain

$$\mathcal{L}\{y\} = \frac{2s + 3}{(s+1)(s-1)(2s+1)(s+2)} = \frac{1}{2}\frac{1}{s+1} + \frac{5}{18}\frac{1}{s-1} - \frac{8}{9}\frac{1}{s+1/2} + \frac{1}{9}\frac{1}{s+2}.$$

Thus,

$$y = \frac{1}{2}e^{-t} + \frac{5}{18}e^t - \frac{8}{9}e^{-t/2} + \frac{1}{9}e^{-2t}.$$

**18.** The Laplace transform of the differential equation is

$$s^4\,\mathcal{L}\{y\} - s^3 y(0) - s^2 y'(0) - sy''(0) - y'''(0) - \mathcal{L}\{y\} = \frac{1}{s^2}.$$

Solving for $\mathcal{L}\{y\}$ we obtain

$$\mathcal{L}\{y\} = \frac{1}{s^2(s^4 - 1)} = -\frac{1}{s^2} + \frac{1}{4}\frac{1}{s-1} - \frac{1}{4}\frac{1}{s+1} + \frac{1}{2}\frac{1}{s^2 + 1}.$$

Thus,

$$y = -t + \frac{1}{4}e^t - \frac{1}{4}e^{-t} + \frac{1}{2}\sin t.$$

**21.** The Laplace transform of the differential equation is

$$s\mathcal{L}\{y\} - y(0) + 2\mathcal{L}\{y\} = \frac{1}{s^2} - e^{-s}\frac{s+1}{s^2}.$$

Solving for $\mathcal{L}\{y\}$ we obtain

$$\mathcal{L}\{y\} = \frac{1}{s^2(s+2)} - e^{-s}\frac{s+1}{s^2(s+1)} = -\frac{1}{4}\frac{1}{s} + \frac{1}{2}\frac{1}{s^2} + \frac{1}{4}\frac{1}{s+2} - e^{-s}\left[\frac{1}{4}\frac{1}{s} + \frac{1}{2}\frac{1}{s^2} - \frac{1}{4}\frac{1}{s+2}\right].$$

Thus,

$$y = -\frac{1}{4} + \frac{1}{2}t + \frac{1}{4}e^{-2t} - \left[\frac{1}{4} + \frac{1}{2}(t-1) - \frac{1}{4}e^{-2(t-1)}\right]\mathcal{U}(t-1).$$

**24.** The Laplace transform of the differential equation is

$$s^2\mathcal{L}\{y\} - sy(0) - y'(0) - 5\left[s\mathcal{L}\{y\} - y(0)\right] + 6\mathcal{L}\{y\} = \frac{e^{-s}}{s}.$$

Solving for $\mathcal{L}\{y\}$ we obtain

$$\mathcal{L}\{y\} = e^{-s}\frac{1}{s(s-2)(s-3)} + \frac{1}{(s-2)(s-3)}$$

$$= e^{-s}\left[\frac{1}{6}\frac{1}{s} - \frac{1}{2}\frac{1}{s-2} + \frac{1}{3}\frac{1}{s-3}\right] - \frac{1}{s-2} + \frac{1}{s-3}.$$

Thus,

$$y = \left[\frac{1}{6} - \frac{1}{2}e^{2(t-1)} + \frac{1}{3}e^{3(t-1)}\right]\mathcal{U}(t-1) + e^{3t} - e^{2t}.$$

**27.** Taking the Laplace transform of both sides of the differential equation and letting $c = y(0)$ we obtain

$$\mathcal{L}\{y''\} + \mathcal{L}\{2y'\} + \mathcal{L}\{y\} = 0$$

$$s^2\mathcal{L}\{y\} - sy(0) - y'(0) + 2s\mathcal{L}\{y\} - 2y(0) + \mathcal{L}\{y\} = 0$$

$$s^2\mathcal{L}\{y\} - cs - 2 + 2s\mathcal{L}\{y\} - 2c + \mathcal{L}\{y\} = 0$$

$$\left(s^2 + 2s + 1\right)\mathcal{L}\{y\} = cs + 2c + 2$$

$$\mathcal{L}\{y\} = \frac{cs}{(s+1)^2} + \frac{2c+2}{(s+1)^2}$$

$$= c\frac{s+1-1}{(s+1)^2} + \frac{2c+2}{(s+1)^2}$$

$$= \frac{c}{s+1} + \frac{c+2}{(s+1)^2}.$$

Therefore,

$$y(t) = c\mathcal{L}^{-1}\left\{\frac{1}{s+1}\right\} + (c+2)\mathcal{L}^{-1}\left\{\frac{1}{(s+1)^2}\right\} = ce^{-t} + (c+2)te^{-t}.$$

To find $c$ we let $y(1) = 2$. Then $2 = ce^{-1} + (c+2)e^{-1} = 2(c+1)e^{-1}$ and $c = e - 1$. Thus,

$$y(t) = (e-1)e^{-t} + (e+1)te^{-t}.$$

**30.** The Laplace transform of the given equation is

$$\mathscr{L}\{f\} = \mathscr{L}\{2t\} - 4\,\mathscr{L}\{\sin t\}\,\mathscr{L}\{f\}.$$

Solving for $\mathscr{L}\{f\}$ we obtain

$$\mathscr{L}\{f\} = \frac{2s^2 + 2}{s^2(s^2 + 5)} = \frac{2}{5}\frac{1}{s^2} + \frac{8}{5\sqrt{5}}\frac{\sqrt{5}}{s^2 + 5}.$$

Thus,

$$f(t) = \frac{2}{5}t + \frac{8}{5\sqrt{5}}\sin\sqrt{5}\,t.$$

**33.** The Laplace transform of the given equation is

$$\mathscr{L}\{f\} + \mathscr{L}\{1\}\,\mathscr{L}\{f\} = \mathscr{L}\{1\}.$$

Solving for $\mathscr{L}\{f\}$ we obtain $\mathscr{L}\{f\} = \dfrac{1}{s+1}$. Thus, $f(t) = e^{-t}$.

**36.** The Laplace transform of the given equation is

$$\mathscr{L}\{t\} - 2\,\mathscr{L}\{f\} = \mathscr{L}\{e^t - e^{-t}\}\,\mathscr{L}\{f\}.$$

Solving for $\mathscr{L}\{f\}$ we obtain

$$\mathscr{L}\{f\} = \frac{s^2 - 1}{2s^4} = \frac{1}{2}\frac{1}{s^2} - \frac{1}{12}\frac{3!}{s^4}.$$

Thus,

$$f(t) = \frac{1}{2}t - \frac{1}{12}t^3.$$

**39.** From equation (3) in the text the differential equation is

$$0.005\frac{di}{dt} + i + 50\int_0^t i(\tau)\,d\tau = 100[1 - \mathscr{U}(t-1)], \quad i(0) = 0.$$

The Laplace transform of this equation is

$$0.005[s\,\mathscr{L}\{i\} - i(0)] + \mathscr{L}\{i\} + 50\frac{1}{s}\mathscr{L}\{i\} = 100\left[\frac{1}{s} - \frac{1}{s}e^{-s}\right].$$

Solving for $\mathscr{L}\{i\}$ we obtain

$$\mathscr{L}\{i\} = \frac{20{,}000}{(s+100)^2}(1 - e^{-s}]).$$

Thus,

$$i(t) = 20{,}000te^{-100t} - 20{,}000(t-1)e^{-100(t-1)}\,\mathscr{U}(t-1).$$

**42.** The differential equation is

$$10\frac{dq}{dt} + 10q = 30e^t - 30e^t \,\mathcal{U}(t - 1.5).$$

The Laplace transform of this equation is

$$s\mathscr{L}\{q\} - q_0 + \mathscr{L}\{q\} = \frac{3}{s-1} - \frac{3e^{1.5}}{s-1.5}e^{-1.5s}.$$

Solving for $\mathscr{L}\{q\}$ we obtain

$$\mathscr{L}\{q\} = \left(q_0 - \frac{3}{2}\right)\cdot\frac{1}{s+1} + \frac{3}{2}\cdot\frac{1}{s-1} - 3e^{1.5}\left(\frac{-2/5}{s+1} + \frac{2/5}{s-1.5}\right)e^{-1.55}.$$

Thus,

$$q(t) = \left(q_0 - \frac{3}{2}\right)e^{-t} + \frac{3}{2}e^t + \frac{6}{5}e^{1.5}\left(e^{-(t-1.5)} - e^{1.5(t-1.5)}\right)\mathcal{U}(t-1.5).$$

**45.** The differential equation is

$$\frac{di}{dt} + 10i = \sin t + \cos\left(t - \frac{3\pi}{2}\right)\mathcal{U}\left(t - \frac{3\pi}{2}\right), \quad i(0) = 0.$$

The Laplace transform of this equation is

$$s\mathscr{L}\{i\} + 10\mathscr{L}\{i\} = \frac{1}{s^2+1} + \frac{se^{-3\pi s/2}}{s^2+1}.$$

Solving for $\mathscr{L}\{i\}$ we obtain

$$\mathscr{L}\{i\} = \frac{1}{(s^2+1)(s+10)} + \frac{s}{(s^2+1)(s+10)}e^{-3\pi s/2}$$

$$= \frac{1}{101}\left(\frac{1}{s+10} - \frac{s}{s^2+1} + \frac{10}{s^2+1}\right) + \frac{1}{101}\left(\frac{-10}{s+10} + \frac{10s}{s^2+1} + \frac{1}{s^2+1}\right)e^{-3\pi s/2}.$$

Thus,

$$i(t) = \frac{1}{101}\left(e^{-10t} - \cos t + 10\sin t\right)$$

$$+ \frac{1}{101}\left(-10e^{-10(t-3\pi/2)} + 10\cos\left(t - \frac{3\pi}{2}\right) + \sin\left(t - \frac{3\pi}{2}\right)\right)\mathcal{U}\left(t - \frac{3\pi}{2}\right).$$

**48.** The differential equation is

$$\frac{d^2q}{dt^2} + 20\frac{dq}{dt} + 200q = 150, \quad q(0) = q'(0) = 0.$$

The Laplace transform of this equation is

$$s^2\mathscr{L}\{q\} + 20s\mathscr{L}\{q\} + 200\,\mathscr{L}\{q\} = \frac{150}{s}.$$

Solving for $\mathcal{L}\{q\}$ we obtain

$$\mathcal{L}\{q\} = \frac{150}{s(s^2 + 20s + 200)} = \frac{3}{4}\frac{1}{s} - \frac{3}{4}\frac{s+10}{(s+10)^2 + 10^2} - \frac{3}{4}\frac{10}{(s+10)^2 + 10^2}.$$

Thus,

$$q(t) = \frac{3}{4} - \frac{3}{4}e^{-10t}\cos 10t - \frac{3}{4}e^{-10t}\sin 10t$$

and

$$i(t) = q'(t) = 15e^{-10t}\sin 10t.$$

If $E(t) = 150 - 150\,\mathcal{U}(t - 2)$, then

$$\mathcal{L}\{q\} = \frac{150}{s(s^2 + 20s + 200)}\left(1 - e^{-2s}\right)$$

$$q(t) = \frac{3}{4} - \frac{3}{4}e^{-10t}\cos 10t - \frac{3}{4}e^{-10t}\sin 10t - \left[\frac{3}{4} - \frac{3}{4}e^{-10(t-2)}\cos 10(t-2)\right.$$

$$\left. - \frac{3}{4}e^{-10(t-2)}\sin 10(t-2)\right]\mathcal{U}(t-2).$$

**51.** The differential equation is

$$\frac{d^2q}{dt^2} + \frac{1}{LC}q = \frac{E_0}{L}e^{-kt}, \quad q(0) = q'(0) = 0.$$

The Laplace transform of this equation is

$$s^2\mathcal{L}\{q\} + \frac{1}{LC}\mathcal{L}\{q\} = \frac{E_0}{L}\frac{1}{s+k}.$$

Solving for $\mathcal{L}\{q\}$ we obtain

$$\mathcal{L}\{q\} = \frac{E_0}{L}\frac{1}{(s+k)(s^2 + 1/LC)} = \frac{E_0}{L}\left(\frac{1/(k^2 + 1/LC)}{s+k} - \frac{s/(k^2 + 1/LC)}{s^2 + 1/LC} + \frac{k/(k^2 + 1/LC)}{s^2 + 1/LC}\right).$$

Thus,

$$q(t) = \frac{E_0}{L(k^2 + 1/LC)}\left[e^{-kt} - \cos\left(t/\sqrt{LC}\right) + k\sqrt{LC}\sin\left(t/\sqrt{LC}\right)\right].$$

**54.** Recall from Chapter 5 that $mx'' = -kx + f(t)$. Now $m = W/g = 16/32 = 1/2$ slug, and $k = 4.5$, so the differential equation is

$$\frac{1}{2}x'' + 4.5x = 4\sin 3t + 2\cos 3t \quad \text{or} \quad x'' + 9x = 8\sin 3t + 4\cos 3t.$$

The initial conditions are $x(0) = x'(0) = 0$. The Laplace transform of the differential equation is

$$s^2\mathcal{L}\{x\} + 9\mathcal{L}\{x\} = \frac{24}{s^2 + 9} + \frac{4s}{s^2 + 9}.$$

## Exercises 7.5

Solving for $\mathscr{L}\{x\}$ we obtain

$$\mathscr{L}\{x\} = \frac{4s + 24}{(s^2 + 9)^2} = \frac{2}{3}\frac{2(3)s}{(s^2 + 9)^2} + \frac{12}{27}\frac{2(3)^3}{(s^2 + 9)^2}.$$

Thus,

$$x(t) = \frac{2}{3}t\sin 3t + \frac{4}{9}(\sin 3t - 3t\cos 3t) = \frac{2}{3}t\sin 3t + \frac{4}{9}\sin 3t - \frac{4}{3}t\cos 3t.$$

**57.** The differential equation is

$$EI\frac{d^4y}{dx^4} = w_0[1 - \mathscr{U}(x - L/2)].$$

Taking the Laplace transform of both sides and using $y(0) = y'(0) = 0$ we obtain

$$s^4\mathscr{L}\{y\} - sy''(0) - y'''(0) = \frac{w_0}{EI}\frac{1}{s}\left(1 - e^{-Ls/2}\right).$$

Letting $y''(0) = c_1$ and $y'''(0) = c_2$ we have

$$\mathscr{L}\{y\} = \frac{c_1}{s^3} + \frac{c_2}{s^4} + \frac{w_0}{EI}\frac{1}{s^5}\left(1 - e^{-Ls/2}\right)$$

so that

$$y(x) = \frac{1}{2}c_1 x^2 + \frac{1}{6}c_2 x^3 + \frac{1}{24}\frac{w_0}{EI}\left[x^4 - \left(x - \frac{L}{2}\right)^4\mathscr{U}\left(x - \frac{L}{2}\right)\right].$$

To find $c_1$ and $c_2$ we compute

$$y''(x) = c_1 + c_2 x + \frac{1}{2}\frac{w_0}{EI}\left[x^2 - \left(x - \frac{L}{2}\right)^2\mathscr{U}\left(x - \frac{L}{2}\right)\right]$$

and

$$y'''(x) = c_2 + \frac{w_0}{EI}\left[x - \left(x - \frac{L}{2}\right)\mathscr{U}\left(x - \frac{L}{2}\right)\right].$$

Then $y''(L) = y'''(L) = 0$ yields the system

$$c_1 + c_2 L + \frac{1}{2}\frac{w_0}{EI}\left[L^2 - \left(\frac{L}{2}\right)^2\right] = c_1 + c_2 L + \frac{3}{8}\frac{w_0 L^2}{EI} = 0$$

$$c_2 + \frac{w_0}{EI}\left(\frac{L}{2}\right) = c_2 + \frac{1}{2}\frac{w_0 L}{EI} = 0.$$

Solving for $c_1$ and $c_2$ we obtain $c_1 = \frac{1}{8}w_0 L^2/EI$ and $c_2 = -\frac{1}{2}w_0 L/EI$. Thus,

$$y(x) = \frac{w_0}{EI}\left(\frac{1}{16}L^2 x^2 - \frac{1}{12}Lx^3 + \frac{1}{24}x^4 - \frac{1}{24}\left(x - \frac{L}{2}\right)^4\mathscr{U}\left(x - \frac{L}{2}\right)\right).$$

**60.** The Laplace transform of the differential equation is

$$-\frac{d}{ds}\left[s^2\mathscr{L}\{y\} - y'(0)\right] - 2\frac{d}{ds}[s\mathscr{L}\{y\}] + 2\mathscr{L}\{y\} = 0.$$

Then

$$-s^2\left(\frac{d}{ds}\mathscr{L}\{y\}\right) - 2s\,\mathscr{L}\{y\} - 2s\left(\frac{d}{ds}\mathscr{L}\{y\}\right) - 2\mathscr{L}\{y\} + 2\mathscr{L}\{y\} = 0$$

and

$$\frac{d}{ds}\mathscr{L}\{y\} + \frac{2}{s+2}\mathscr{L}\{y\} = 0.$$

This is a separable differential equation so

$$\frac{d\mathscr{L}\{y\}}{\mathscr{L}\{y\}} = -\frac{2\,ds}{s+2} \implies \ln\mathscr{L}\{y\} = -2\ln(s+2) + c \implies \mathscr{L}\{y\} = c_1 e^{-2\ln(s+2)} = c_1(s+2)^{-1}$$

and $y(t) = c_1 t e^{-2t}$.

## Exercises 7.6

**3.** The Laplace transform of the differential equation is

$$\mathscr{L}\{y\} = \frac{1}{s^2+1}\left(1 + e^{-2\pi s}\right)$$

so that

$$y = \sin t + \sin t\,\mathscr{U}(t - 2\pi).$$

**6.** The Laplace transform of the differential equation is

$$\mathscr{L}\{y\} = \frac{s}{s^2+1} + \frac{1}{s^2+1}(e^{-2\pi s} + e^{-4\pi s})$$

so that

$$y = \cos t + \sin t[\mathscr{U}(t - 2\pi) + \mathscr{U}(t - 4\pi)].$$

**9.** The Laplace transform of the differential equation is

$$\mathscr{L}\{y\} = \frac{1}{(s+2)^2+1}e^{-2\pi s}$$

so that

$$y = e^{-2(t-2\pi)}\sin t\,\mathscr{U}(t - 2\pi).$$

**12.** The Laplace transform of the differential equation is

$$\mathscr{L}\{y\} = \frac{1}{(s-1)^2(s-6)} + \frac{e^{-2s} + e^{-4s}}{(s-1)(s-6)}$$

$$= -\frac{1}{25}\frac{1}{s-1} - \frac{1}{5}\frac{1}{(s-1)^2} + \frac{1}{25}\frac{1}{s-6} + \left[-\frac{1}{5}\frac{1}{s-1} + \frac{1}{5}\frac{1}{s-6}\right]\left(e^{-2s} + e^{-4s}\right)$$

so that

$$y = -\frac{1}{25}e^t - \frac{1}{5}te^t + \frac{1}{25}e^{6t} + \left[-\frac{1}{5}e^{t-2} + \frac{1}{5}e^{6(t-2)}\right]\mathcal{U}(t-2)$$

$$+ \left[-\frac{1}{5}e^{t-4} + \frac{1}{5}e^{6(t-4)}\right]\mathcal{U}(t-4).$$

**15.** Letting $f(t) = e^{-st}$, we have by (7)

$$\mathcal{L}\{\delta(t - t_0)\} = \int_0^\infty e^{-st}\delta(t - t_0)\, dt = f(t_0) = e^{-st_0}.$$

**18.** The Laplace transform of the differential equation is

$$\mathcal{L}\{y\} = \frac{1}{w}\frac{s}{s^2 + w^2}$$

so that

$$y = \frac{1}{w}\sin wt.$$

Note that $y'(0) = 1$.

## ——————— Chapter 7 Review Exercises ———————

**3.** False; consider $f(t) = t^{-1/2}$.

**6.** False; consider $f(t) = 1$ and $g(t) = 1$.

**9.** $\mathcal{L}\{\sin 2t\} = \dfrac{2}{s^2 + 4}$

**12.** $\mathcal{L}\{\sin 2t\, \mathcal{U}(t - \pi)\} = \mathcal{L}\{\sin 2(t - \pi)\mathcal{U}(t - \pi)\} = \dfrac{2}{s^2 + 4}e^{-\pi s}$

**15.** $\mathcal{L}^{-1}\left\{\dfrac{1}{(s - 5)^3}\right\} = \mathcal{L}^{-1}\left\{\dfrac{1}{2}\dfrac{2}{(s - 5)^3}\right\} = \dfrac{1}{2}t^2 e^{5t}$

**18.** $\mathcal{L}^{-1}\left\{\dfrac{1}{s^2}e^{-5s}\right\} = (t - 5)\mathcal{U}(t - 5)$

**21.** $\mathcal{L}\{e^{-5t}\}$ exists for $s > -5$.

**24.** $1 * 1 = \displaystyle\int_0^t d\tau = t$

**27.** **(a)** $f(t) = 2 - 2\mathcal{U}(t - 2) + [(t - 2) + 2]\mathcal{U}(t - 2) = 2 + (t - 2)\mathcal{U}(t - 2)$

**(b)** $\mathcal{L}\{f(t)\} = \dfrac{2}{s} + \dfrac{1}{s^2}e^{-2s}$

**(c)** $\mathcal{L}\{e^t f(t)\} = \dfrac{2}{s - 1} + \dfrac{1}{(s - 1)^2}e^{-2(s-1)}$

**30.** Taking the Laplace transform of the differential equation we obtain

$$\mathcal{L}\{y\} = \frac{1}{(s-1)^2(s^2-8s+20)}$$

$$= \frac{6}{169}\frac{1}{s-1} + \frac{1}{13}\frac{1}{(s-1)^2} - \frac{6}{169}\frac{s-4}{(s-4)^2+2^2} + \frac{5}{338}\frac{2}{(s-4)^2+2^2}$$

so that

$$y = \frac{6}{169}e^t + \frac{1}{13}te^t - \frac{6}{169}e^{4t}\cos 2t + \frac{5}{338}e^{4t}\sin 2t.$$

**33.** Taking the Laplace transform of the differential equation we obtain

$$\mathcal{L}\{y\} = \frac{s^3+2}{s^3(s-5)} - \frac{2+2s+s^2}{s^3(s-5)}e^{-s}$$

$$= -\frac{2}{125}\frac{1}{s} - \frac{2}{25}\frac{1}{s^2} - \frac{1}{5}\frac{2}{s^3} + \frac{127}{125}\frac{1}{s-5} - \left[-\frac{37}{125}\frac{1}{s} - \frac{12}{25}\frac{1}{s^2} - \frac{1}{5}\frac{2}{s^3} + \frac{37}{125}\frac{1}{s-5}\right]e^{-s}$$

so that

$$y = -\frac{2}{125} - \frac{2}{25}t - \frac{1}{5}t^2 + \frac{127}{125}e^{5t} - \left[-\frac{37}{125} - \frac{12}{25}(t-1) - \frac{1}{5}(t-1)^2 + \frac{37}{125}e^{5(t-1)}\right]\mathcal{U}(t-1).$$

**36.** Taking the Laplace transform of the integral equation we obtain

$$(\mathcal{L}\{f\})^2 = 6 \cdot \frac{6}{s^4} \quad \text{or} \quad \mathcal{L}\{f\} = \pm 6 \cdot \frac{1}{s^2}$$

so that $f(t) = \pm 6t$.

**39.** Taking the Laplace transform of the given differential equation we obtain

$$\mathcal{L}\{y\} = \frac{2w_0}{EIL}\left(\frac{L}{48} \cdot \frac{4!}{s^5} - \frac{1}{120} \cdot \frac{5!}{s^6} + \frac{1}{120} \cdot \frac{5!}{s^6}e^{-sL/2}\right) + \frac{c_1}{2} \cdot \frac{2!}{s^3} + \frac{c_2}{6} \cdot \frac{3!}{s^4}$$

so that

$$y = \frac{2w_0}{EIL}\left[\frac{L}{48}x^4 - \frac{1}{120}x^5 + \frac{1}{120}\left(x-\frac{L}{2}\right)^5\mathcal{U}\left(x-\frac{L}{2}\right) + \frac{c_1}{2}x^2 + \frac{c_2}{6}x^3\right]$$

where $y''(0) = c_1$ and $y'''(0) = c_2$. Using $y''(L) = 0$ and $y'''(L) = 0$ we find

$$c_1 = w_0L^2/24EI, \qquad c_2 = -w_0L/4EI.$$

Hence

$$y = \frac{w_0}{12EIL}\left[-\frac{1}{5}x^5 + \frac{L}{2}x^4 - \frac{L^2}{2}x^3 + \frac{L^3}{4}x^2 + \frac{1}{5}\left(x-\frac{L}{2}\right)^5\mathcal{U}\left(x-\frac{L}{2}\right)\right].$$

# 8 Systems of Linear Differential Equations

————— **Exercises 8.1** —————————————————

**3.** From $Dx = -y + t$ and $Dy = x - t$ we obtain $y = t - Dx$, $Dy = 1 - D^2x$, and $(D^2 + 1)x = 1 + t$. Then

$$x = c_1 \cos t + c_2 \sin t + 1 + t$$

and

$$y = c_1 \sin t - c_2 \cos t + t - 1.$$

**6.** From $(D + 1)x + (D - 1)y = 2$ and $3x + (D + 2)y = -1$ we obtain $x = -\frac{1}{3} - \frac{1}{3}(D + 2)y$, $Dx = -\frac{1}{3}(D^2 + 2D)y$, and $(D^2 + 5)y = -7$. Then

$$y = c_1 \cos \sqrt{5}\, t + c_2 \sin \sqrt{5}\, t - \frac{7}{5}$$

and

$$x = \left(-\frac{2}{3}c_1 - \frac{\sqrt{5}}{3}c_2\right) \cos \sqrt{5}\, t + \left(\frac{\sqrt{5}}{3}c_1 - \frac{2}{3}c_2\right) \sin \sqrt{5}\, t + \frac{3}{5}.$$

**9.** From $Dx + D^2y = e^{3t}$ and $(D + 1)x + (D - 1)y = 4e^{3t}$ we obtain $D(D^2 + 1)x = 34e^{3t}$ and $D(D^2 + 1)y = -8e^{3t}$. Then

$$y = c_1 + c_2 \sin t + c_3 \cos t - \frac{4}{15}e^{3t}$$

and

$$x = c_4 + c_5 \sin t + c_6 \cos t + \frac{17}{15}e^{3t}.$$

Substituting into $(D + 1)x + (D - 1)y = 4e^{3t}$ gives

$$(c_4 - c_1) + (c_5 - c_6 - c_3 - c_2) \sin t + (c_6 + c_5 + c_2 - c_3) \cos t = 0$$

so that $c_4 = c_1$, $c_5 = c_3$, $c_6 = -c_2$, and

$$x = c_1 - c_2 \cos t + c_3 \sin t + \frac{17}{15}e^{3t}.$$

**12.** From $(2D^2 - D - 1)x - (2D + 1)y = 1$ and $(D - 1)x + Dy = -1$ we obtain $(2D + 1)(D - 1)(D + 1)x = -1$ and $(2D + 1)(D + 1)y = -2$. Then

$$x = c_1 e^{-t/2} + c_2 e^{-t} + c_3 e^{t} + 1$$

and

$$y = c_4 e^{-t/2} + c_5 e^{-t} - 2.$$

Substituting into $(D - 1)x + Dy = -1$ gives

$$\left(-\frac{3}{2}c_1 - \frac{1}{2}c_4\right) e^{-t/2} + (-2c_2 - c_5)e^{-t} = 0$$

so that $c_4 = -3c_1$, $c_5 = -2c_2$, and

$$y = -3c_1 e^{-t/2} - 2c_2 e^{-t} - 2.$$

**15.** From $(D - 1)x + (D^2 + 1)y = 1$ and $(D^2 - 1)x + (D + 1)y = 2$ we obtain $D^2(D - 1)(D + 1)x = 1$ and $D^2(D - 1)(D + 1)y = 1$. Then

$$x = c_1 + c_2 t + c_3 e^t + c_4 e^{-t} - \frac{1}{2}t^2$$

and

$$y = c_5 + c_6 t + c_7 e^t + c_8 e^{-t} - \frac{1}{2}t^2.$$

Substituting into $(D - 1)x + (D^2 + 1)y = 1$ gives

$$(c_2 - c_1 - 1 + c_5) + (c_6 - c_2 - 1)t + (2c_8 - 2c_4)e^{-t} + (2c_7)e^t = 1$$

so that $c_6 = c_2 + 1$, $c_8 = c_4$, $c_7 = 0$, $c_5 = c_1 - c_2 + 2$, and

$$y = (c_1 - c_2 + 2) + (c_2 + 1)t + c_4 e^{-t} - \frac{1}{2}t^2.$$

**18.** From $Dx + z = e^t$, $(D - 1)x + Dy + Dz = 0$, and $x + 2y + Dz = e^t$ we obtain $z = -Dx + e^t$, $Dz = -D^2 x + e^t$, and the system $(-D^2 + D - 1)x + Dy = -e^t$ and $(-D^2 + 1)x + 2y = 0$. Then $y = \frac{1}{2}(D^2 - 1)x$, $Dy = \frac{1}{2}D(D^2 - 1)x$, and $(D - 2)(D^2 + 1)x = -2e^t$ so that

$$x = c_1 e^{2t} + c_2 \cos t + c_3 \sin t + e^t,$$

$$y = \frac{3}{2}c_1 e^{2t} - c_2 \cos t - c_3 \sin t,$$

and

$$z = -2c_1 e^{2t} - c_3 \cos t + c_2 \sin t.$$

**21.** From $2Dx + (D - 1)y = t$ and $Dx + Dy = t^2$ we obtain $(D + 1)y = 2t^2 - t$. Then

$$y = c_1 e^{-t} + 2t^2 - 5t + 5$$

and $Dx = c_1 e^{-t} + t^2 - 4t + 5$ so that

$$x = -c_1 e^{-t} + c_2 + \frac{1}{3}t^3 - 2t^2 + 5t.$$

**24.** From $Dx - y = -1$ and $3x + (D-2)y = 0$ we obtain $x = -\frac{1}{3}(D-2)y$ so that $Dx = -\frac{1}{3}(D^2 - 2D)y$. Then $-\frac{1}{3}(D^2 - 2D)y = y - 1$ and $(D^2 - 2D + 3)y = 3$. Thus

$$y = e^t \left( c_1 \cos \sqrt{2}\, t + c_2 \sin \sqrt{2}\, t \right) + 1$$

and

$$x = \frac{1}{3} e^t \left[ \left( c_1 - \sqrt{2}\, c_2 \right) \cos \sqrt{2}\, t + \left( \sqrt{2}\, c_1 + c_2 \right) \sin \sqrt{2}\, t \right] + \frac{2}{3}.$$

Using $x(0) = y(0) = 0$ we obtain

$$c_1 + 1 = 0$$

$$\frac{1}{3}\left( c_1 - \sqrt{2}\, c_2 \right) + \frac{2}{3} = 0.$$

Thus $c_1 = -1$ and $c_2 = \sqrt{2}/2$. The solution of the initial value problem is

$$x = e^t \left( -\frac{2}{3} \cos \sqrt{2}\, t - \frac{\sqrt{2}}{6} \sin \sqrt{2}\, t \right) + \frac{2}{3}$$

$$y = e^t \left( -\cos \sqrt{2}\, t + \frac{\sqrt{2}}{2} \sin \sqrt{2}\, t \right) + 1.$$

## Exercises 8.2

**3.** Taking the Laplace transform of the system gives

$$s \mathscr{L}\{x\} + 1 = \mathscr{L}\{x\} - 2\mathscr{L}\{y\}$$

$$s \mathscr{L}\{y\} - 2 = 5 \mathscr{L}\{x\} - \mathscr{L}\{y\}$$

so that

$$\mathscr{L}\{x\} = \frac{-s-5}{s^2+9} = -\frac{s}{s^2+9} - \frac{5}{3}\frac{3}{s^2+9}$$

and

$$x = -\cos 3t - \frac{5}{3} \sin 3t.$$

Then

$$y = \frac{1}{2}x - \frac{1}{2}x' = 2\cos 3t - \frac{7}{3} \sin 3t.$$

**6.** Taking the Laplace transform of the system gives

$$(s+1) \mathscr{L}\{x\} - (s-1)\mathscr{L}\{y\} = -1$$

$$s \mathscr{L}\{x\} + (s+2) \mathscr{L}\{y\} = 1$$

so that

$$\mathcal{L}\{y\} = \frac{s + 1/2}{s^2 + s + 1} = \frac{s + 1/2}{(s + 1/2)^2 + (\sqrt{3}/2)^2}$$

and

$$\mathcal{L}\{x\} = \frac{-3/2}{s^2 + s + 1} = \frac{-3/2}{(s + 1/2)^2 + (\sqrt{3}/2)^2}.$$

Then

$$y = e^{-t/2} \cos \frac{\sqrt{3}}{2}t \quad \text{and} \quad x = e^{-t/2} \sin \frac{\sqrt{3}}{2}t.$$

9. Adding the equations and then subtracting them gives

$$\frac{d^2 x}{dt^2} = \frac{1}{2}t^2 + 2t$$

$$\frac{d^2 y}{dt^2} = \frac{1}{2}t^2 - 2t.$$

Taking the Laplace transform of the system gives

$$\mathcal{L}\{x\} = 8\frac{1}{s} + \frac{1}{24}\frac{4!}{s^5} + \frac{1}{3}\frac{3!}{s^4}$$

and

$$\mathcal{L}\{y\} = \frac{1}{24}\frac{4!}{s^5} - \frac{1}{3}\frac{3!}{s^4}$$

so that

$$x = 8 + \frac{1}{24}t^4 + \frac{1}{3}t^3 \quad \text{and} \quad y = \frac{1}{24}t^4 - \frac{1}{3}t^3.$$

12. Taking the Laplace transform of the system gives

$$(s - 4)\,\mathcal{L}\{x\} + 2\mathcal{L}\{y\} = \frac{2e^{-s}}{s}$$

$$-3\,\mathcal{L}\{x\} + (s + 1)\,\mathcal{L}\{y\} = \frac{1}{2} + \frac{e^{-s}}{s}$$

so that

$$\mathcal{L}\{x\} = \frac{-1/2}{(s - 1)(s - 2)} + e^{-s}\frac{1}{(s - 1)(s - 2)}$$

$$= \left[\frac{1}{2}\frac{1}{s - 1} - \frac{1}{2}\frac{1}{s - 2}\right] + e^{-s}\left[-\frac{1}{s - 1} + \frac{1}{s - 2}\right]$$

and

$$\mathcal{L}\{y\} = \frac{e^{-s}}{s} + \frac{s/4 - 1}{(s - 1)(s - 2)} + e^{-s}\frac{-s/2 + 2}{(s - 1)(s - 2)}$$

$$= \frac{3}{4}\frac{1}{s - 1} - \frac{1}{2}\frac{1}{s - 2} + e^{-s}\left[\frac{1}{s} - \frac{3}{2}\frac{1}{s - 1} + \frac{1}{s - 2}\right].$$

**103**

Then

$$x = \frac{1}{2}e^t - \frac{1}{2}e^{2t} + \left[-e^{t-1} + e^{2(t-1)}\right]\mathcal{U}(t-1)$$

and

$$y = \frac{3}{4}e^t - \frac{1}{2}e^{2t} + \left[1 - \frac{3}{2}e^{t-1} + e^{2(t-1)}\right]\mathcal{U}(t-1).$$

15. (a) By Kirchoff's first law we have $i_1 = i_2 + i_3$. By Kirchoff's second law, on each loop we have
$E(t) = Ri_1 + L_1 i_2'$ and $E(t) = Ri_1 + L_2 i_3'$ or $L_1 i_2' + Ri_2 + Ri_3 = E(t)$ and $L_2 i_3' + Ri_2 + Ri_3 = E(t)$.

(b) Taking the Laplace transform of the system

$$0.01i_2' + 5i_2 + 5i_3 = 100$$

$$0.0125i_3' + 5i_2 + 5i_3 = 100$$

gives

$$(s + 500)\mathcal{L}\{i_2\} + 500\mathcal{L}\{i_3\} = \frac{10,000}{s}$$

$$400\mathcal{L}\{i_2\} + (s + 400)\mathcal{L}\{i_3\} = \frac{8,000}{s}$$

so that

$$\mathcal{L}\{i_3\} = \frac{8,000}{s^2 + 900s} = \frac{80}{9}\frac{1}{s} - \frac{80}{9}\frac{1}{s + 900}.$$

Then

$$i_3 = \frac{80}{9} - \frac{80}{9}e^{-900t} \quad \text{and} \quad i_2 = 20 - 0.0025i_3' - i_3 = \frac{100}{9} - \frac{100}{9}e^{-900t}.$$

(c) $i_1 = i_2 + i_3 = 20 - 20e^{-900t}$.

18. By Kirchoff's first law we have $i_1 = i_2 + i_3$. By Kirchoff's second law, on each loop we have
$E(t) = Li_1' + Ri_2$ and $E(t) = Li_1' + \frac{1}{C}q$ so that $q = CRi_2$. Then $i_3 = q' = CRi_2'$ so that system is

$$Li' + Ri_2 = E(t)$$

$$CRi_2' + i_2 - i_1 = 0.$$

21. (a) Using Kirchoff's first law we write $i_1 = i_2 + i_3$. Since $i_2 = dq/dt$ we have $i_1 - i_3 = dq/dt$. Using
Kirchoff's second law and summing the voltage drops across the shorter loop gives

$$E(t) = iR_1 + \frac{1}{C}q, \tag{1}$$

so that

$$i_1 = \frac{1}{R_1}E(t) - \frac{1}{R_1 C}q.$$

Then

$$\frac{dq}{dt} = i_1 - i_3 = \frac{1}{R_1}E(t) - \frac{1}{R_1C}q - i_3$$

and

$$R_1\frac{dq}{dt} + \frac{1}{C}q + R_1i_3 = E(t).$$

Summing the voltage drops across the longer loop gives

$$E(t) = i_1R_1 + L\frac{di_3}{dt} + R_2i_3.$$

Combining this with (1) we obtain

$$i_1R_1 + L\frac{di_3}{dt} + R_2i_3 = i_1R_1 + \frac{1}{C}q$$

or

$$L\frac{di_3}{dt} + R_2i_3 - \frac{1}{C}q = 0.$$

(b) Using $L = R_1 = R_2 = C = 1$, $E(t) = 50e^{-t}\,\mathcal{U}(t-1) = 50e^{-1}e^{-(t-1)}\,\mathcal{U}(t-1)$, $q(0) = i_3(0) = 0$, and taking the Laplace transform of the system we obtain

$$(s+1)\,\mathcal{L}\{q\} + \mathcal{L}\{i_3\} = \frac{50e^{-1}}{s+1}e^{-s}$$

$$(s+1)\,\mathcal{L}\{i_3\} - \mathcal{L}\{q\} = 0,$$

so that

$$\mathcal{L}\{q\} = \frac{50e^{-1}e^{-s}}{(s+1)^2 + 1}$$

and

$$q(t) = 50e^{-1}e^{-(t-1)}\sin(t-1)\mathcal{U}(t-1) = 50e^{-t}\sin(t-1)\mathcal{U}(t-1).$$

**105**

—————— **Exercises 8.3** ——————

**3.** Let $x_1 = y$, $x_2 = y'$, $x_3 = y''$, and $y''' = 3y'' - 6y' + 10y + t^2 + 1$ so that

$$x_1' = x_2$$
$$x_2' = x_3$$
$$x_3' = 3x_3 - 6x_2 + 10x_1 + t^2 + 1.$$

**6.** Let $x_1 = y$, $x_2 = y'$, $x_3 = y''$, $x_4 = y'''$, and $y^{(4)} = -\frac{1}{2}y''' + 4y + 10$ so that

$$x_1' = x_2$$
$$x_2' = x_3$$
$$x_3' = x_4$$
$$x_4' = -\frac{1}{2}x_4 + 4x_1 + 10.$$

**9.** From

$$x' + 4x - y' = 7t$$
$$x' + y' - 2y = 3t$$

we obtain

$$2x' + 4x - 2y = 10t$$
$$2y' - 4x - 2y = -4t$$

so that

$$x' = -2x + y + 5t$$
$$y' = 2x + y - 2t.$$

**12.** From $x'' - 2y'' = \sin t$ and $x'' + y'' = \cos t$ we obtain

$$3x'' = 2\cos t + \sin t$$
$$3y'' = \cos t - \sin t.$$

Let $x_1 = x$, $x_2 = x'$, $x_3 = y$, and $x_4 = y'$. Then

$$x_1' = x_2$$

$$x_2' = \frac{2}{3}\cos t + \frac{1}{3}\sin t$$

$$x_3' = x_4$$

$$x_4' = \frac{1}{3}\cos t - \frac{1}{3}\sin t.$$

**15.** Let $z_1 = x$, $z_2 = x'$, $z_3 = x''$, $z_4 = y$, and $z_5 = y'$ so that

$$z_1' = z_2$$

$$z_2' = z_3$$

$$z_3' = 4z_1 - 3z_3 + 4z_5$$

$$z_4' = z_5$$

$$z_5' = 10t^2 - 4z_2 + 3z_5.$$

**18.** The system is

$$x_1' = 2 \cdot 3 + \frac{1}{50}x_2 - \frac{1}{50}x_1 \cdot 4$$

$$x_2' = \frac{1}{50}x_1 \cdot 4 - \frac{1}{50}x_2 - \frac{1}{50}x_2 \cdot 3.$$

**21.** Since $Dx + Dy = -x - y$ and $Dx + Dy = -\frac{1}{2}y$ we obtain $y = -2x$ and $Dx = -x$. Then $x = c_1 e^{-t}$ and $y = -2c_1 e^{-t}$.

## —————— Exercises 8.4 ——————

**3. (a)** $\mathbf{AB} = \begin{pmatrix} -2-9 & 12-6 \\ 5+12 & -30+8 \end{pmatrix} = \begin{pmatrix} -11 & 6 \\ 17 & -22 \end{pmatrix}$

**(b)** $\mathbf{BA} = \begin{pmatrix} -2-30 & 3+24 \\ 6-10 & -9+8 \end{pmatrix} = \begin{pmatrix} -32 & 27 \\ -4 & -1 \end{pmatrix}$

**(c)** $\mathbf{A}^2 = \begin{pmatrix} 4+15 & -6-12 \\ -10-20 & 15+16 \end{pmatrix} = \begin{pmatrix} 19 & -18 \\ -30 & 31 \end{pmatrix}$

**(d)** $\mathbf{B}^2 = \begin{pmatrix} 1+18 & -6+12 \\ -3+6 & 18+4 \end{pmatrix} = \begin{pmatrix} 19 & 6 \\ 3 & 22 \end{pmatrix}$

## Exercises 8.4

**6. (a)** $\mathbf{AB} = \begin{pmatrix} 5 & -6 & 7 \end{pmatrix} \begin{pmatrix} 3 \\ 4 \\ -1 \end{pmatrix} = (-16)$

**(b)** $\mathbf{BA} = \begin{pmatrix} 3 \\ 4 \\ -1 \end{pmatrix} \begin{pmatrix} 5 & -6 & 7 \end{pmatrix} = \begin{pmatrix} 15 & -18 & 21 \\ 20 & -24 & 28 \\ -5 & 6 & -7 \end{pmatrix}$

**(c)** $(\mathbf{BA})\mathbf{C} = \begin{pmatrix} 15 & -18 & 21 \\ 20 & -24 & 28 \\ -5 & 6 & -7 \end{pmatrix} \begin{pmatrix} 1 & 2 & 4 \\ 0 & 1 & -1 \\ 3 & 2 & 1 \end{pmatrix} = \begin{pmatrix} 78 & 54 & 99 \\ 104 & 72 & 132 \\ -26 & -18 & -33 \end{pmatrix}$

**(d)** Since $\mathbf{AB}$ is $1 \times 1$ and $\mathbf{C}$ is $3 \times 3$ the product $(\mathbf{AB})\mathbf{C}$ is not defined.

**9. (a)** $(\mathbf{AB})^T = \begin{pmatrix} 7 & 10 \\ 38 & 75 \end{pmatrix}^T = \begin{pmatrix} 7 & 38 \\ 10 & 75 \end{pmatrix}$

**(b)** $\mathbf{B}^T\mathbf{A}^T = \begin{pmatrix} 5 & -2 \\ 10 & -5 \end{pmatrix} \begin{pmatrix} 3 & 8 \\ 4 & 1 \end{pmatrix} = \begin{pmatrix} 7 & 38 \\ 10 & 75 \end{pmatrix}$

**12.** $\begin{pmatrix} 6t \\ 3t^2 \\ -3t \end{pmatrix} + \begin{pmatrix} -t+1 \\ -t^2+t \\ 3t-3 \end{pmatrix} - \begin{pmatrix} 6t \\ 8 \\ -10t \end{pmatrix} = \begin{pmatrix} -t+1 \\ 2t^2+t-8 \\ 10t-3 \end{pmatrix}$

**15.** Since $\det \mathbf{A} = 0$, $\mathbf{A}$ is singular.

**18.** Since $\det \mathbf{A} = -6$, $\mathbf{A}$ is nonsingular.

$$\mathbf{A}^{-1} = -\frac{1}{6} \begin{pmatrix} 2 & -10 \\ -2 & 7 \end{pmatrix}$$

**21.** Since $\det \mathbf{A} = -9$, $\mathbf{A}$ is nonsingular. The cofactors are

$$\begin{array}{lll} A_{11} = -2 & A_{12} = -13 & A_{13} = 8 \\ A_{21} = -2 & A_{22} = 5 & A_{23} = -1 \\ A_{31} = -1 & A_{32} = 7 & A_{33} = -5. \end{array}$$

Then

$$\mathbf{A}^{-1} = -\frac{1}{9} \begin{pmatrix} -2 & -13 & 8 \\ -2 & 5 & -1 \\ -1 & 7 & -5 \end{pmatrix}^T = -\frac{1}{9} \begin{pmatrix} -2 & -2 & -1 \\ -13 & 5 & 7 \\ 8 & -1 & -5 \end{pmatrix}.$$

**24.** Since $\det \mathbf{A}(t) = 2e^{2t} \neq 0$, $\mathbf{A}$ is nonsingular.

$$\mathbf{A}^{-1} = \frac{1}{2}e^{-2t} \begin{pmatrix} e^t \sin t & 2e^t \cos t \\ -e^t \cos t & 2e^t \sin t \end{pmatrix}$$

**27.** $\mathbf{X} = \begin{pmatrix} 2e^{2t} + 8e^{-3t} \\ -2e^{2t} + 4e^{-3t} \end{pmatrix}$ so that $\dfrac{d\mathbf{X}}{dt} = \begin{pmatrix} 4e^{2t} - 24e^{-3t} \\ -4e^{2t} - 12e^{-3t} \end{pmatrix}$.

**30. (a)** $\dfrac{d\mathbf{A}}{dt} = \begin{pmatrix} -2t/(t^2+1)^2 & 3 \\ 2t & 1 \end{pmatrix}$

**(b)** $\dfrac{d\mathbf{B}}{dt} = \begin{pmatrix} 6 & 0 \\ -1/t^2 & 4 \end{pmatrix}$

**(c)** $\displaystyle\int_0^1 \mathbf{A}(t)\,dt = \begin{pmatrix} \tan^{-1} t & \frac{3}{2}t^2 \\ \frac{1}{3}t^3 & \frac{1}{2}t^2 \end{pmatrix} \Big|_{t=0}^{t=1} = \begin{pmatrix} \frac{\pi}{4} & \frac{3}{2} \\ \frac{1}{3} & \frac{1}{2} \end{pmatrix}$

**(d)** $\displaystyle\int_1^2 \mathbf{B}(t)\,dt = \begin{pmatrix} 3t^2 & 2t \\ \ln t & 2t^2 \end{pmatrix} \Big|_{t=1}^{t=2} = \begin{pmatrix} 9 & 2 \\ \ln 2 & 6 \end{pmatrix}$

**(e)** $\mathbf{A}(t)\mathbf{B}(t) = \begin{pmatrix} 6t/(t^2+1)+3 & 2/(t^2+1)+12t^2 \\ 6t^3+1 & 2t^2+4t^2 \end{pmatrix}$

**(f)** $\dfrac{d}{dt}\mathbf{A}(t)\mathbf{B}(t) = \begin{pmatrix} (6-6t^2)/(t^2+1)^2 & -4t/(t^2+1)^2+24t \\ 18t^2 & 12t \end{pmatrix}$

**(g)** $\displaystyle\int_1^t \mathbf{A}(s)\mathbf{B}(s)\,ds = \begin{pmatrix} 6s/(s^2+1)+3 & 2/(s^2+1)+12s^2 \\ 6s^3+1 & 6s^2 \end{pmatrix} \Big|_{s=1}^{s=t}$

$= \begin{pmatrix} 3t+3\ln(t^2+1)-3-3\ln 2 & 4t^3+2\tan^{-1}t-4-\pi/2 \\ (3/2)t^4+t-(5/2) & 2t^3-2 \end{pmatrix}$

**33.** $\begin{pmatrix} 1 & -1 & -5 & | & 7 \\ 5 & 4 & -16 & | & -10 \\ 0 & 1 & 1 & | & -5 \end{pmatrix} \Longrightarrow \begin{pmatrix} 1 & -1 & -5 & | & 7 \\ 0 & 1 & 1 & | & -5 \\ 0 & 9 & 9 & | & -45 \end{pmatrix} \Longrightarrow \begin{pmatrix} 1 & 0 & -4 & | & 2 \\ 0 & 1 & 1 & | & -5 \\ 0 & 0 & 0 & | & 0 \end{pmatrix}$

Letting $z = t$ we find $y = -5 - t$, and $x = 2 + 4t$.

**36.** $\begin{pmatrix} 1 & 0 & 2 & | & 8 \\ 1 & 2 & -2 & | & 4 \\ 2 & 5 & -6 & | & 6 \end{pmatrix} \Longrightarrow \begin{pmatrix} 1 & 0 & 2 & | & 8 \\ 0 & 2 & -4 & | & -4 \\ 0 & 5 & -10 & | & -10 \end{pmatrix} \Longrightarrow \begin{pmatrix} 1 & 0 & 2 & | & 8 \\ 0 & 1 & -2 & | & -2 \\ 0 & 0 & 0 & | & 0 \end{pmatrix}$

Letting $z = t$ we find $y = -2 + 2t$, and $x = 8 - 2t$.

**39.** $\begin{pmatrix} 1 & 2 & 4 & | & 2 \\ 2 & 4 & 3 & | & 1 \\ 1 & 2 & -1 & | & 7 \end{pmatrix} \Longrightarrow \begin{pmatrix} 1 & 2 & 4 & | & 2 \\ 0 & 0 & -5 & | & -3 \\ 0 & 0 & -5 & | & 5 \end{pmatrix} \Longrightarrow \begin{pmatrix} 1 & 2 & 0 & | & -2/5 \\ 0 & 0 & 1 & | & 3/5 \\ 0 & 0 & 0 & | & 8 \end{pmatrix}$

## Exercises 8.4

There is no solution.

**42.** We solve

$$\det(\mathbf{A} - \lambda\mathbf{I}) = \begin{vmatrix} 2 - \lambda & 1 \\ 2 & 1 - \lambda \end{vmatrix} = \lambda(\lambda - 3) = 0.$$

For $\lambda_1 = 0$ we have

$$\begin{pmatrix} 2 & 1 & | & 0 \\ 2 & 1 & | & 0 \end{pmatrix} \Longrightarrow \begin{pmatrix} 1 & 1/2 & | & 0 \\ 0 & 0 & | & 0 \end{pmatrix}$$

so that $k_1 = -\frac{1}{2}k_2$. If $k_2 = 2$ then

$$\mathbf{K}_1 = \begin{pmatrix} -1 \\ 2 \end{pmatrix}.$$

For $\lambda_2 = 3$ we have

$$\begin{pmatrix} -1 & 1 & | & 0 \\ 2 & -2 & | & 0 \end{pmatrix} \Longrightarrow \begin{pmatrix} 1 & -1 & | & 0 \\ 0 & 0 & | & 0 \end{pmatrix}$$

so that $k_1 = k_2$. If $k_2 = 1$ then

$$\mathbf{K}_2 = \begin{pmatrix} 1 \\ 1 \end{pmatrix}.$$

**45.** We solve

$$\det(\mathbf{A} - \lambda\mathbf{I}) = \begin{vmatrix} 5 - \lambda & -1 & 0 \\ 0 & -5 - \lambda & 9 \\ 5 & -1 & -\lambda \end{vmatrix} = \begin{vmatrix} 4 - \lambda & -1 & 0 \\ 4 - \lambda & -5 - \lambda & 9 \\ 4 - \lambda & -1 & -\lambda \end{vmatrix} = \lambda(4 - \lambda)(\lambda + 4) = 0.$$

If $\lambda_1 = 0$ then

$$\begin{pmatrix} 5 & -1 & 0 & | & 0 \\ 0 & -5 & 9 & | & 0 \\ 5 & -1 & 0 & | & 0 \end{pmatrix} \Longrightarrow \begin{pmatrix} 1 & 0 & -9/25 & | & 0 \\ 0 & 1 & -9/5 & | & 0 \\ 0 & 0 & 0 & | & 0 \end{pmatrix}$$

so that $k_1 = \frac{9}{25}k_3$ and $k_2 = \frac{9}{5}k_3$. If $k_3 = 25$ then

$$\mathbf{K}_1 = \begin{pmatrix} 9 \\ 45 \\ 25 \end{pmatrix}.$$

If $\lambda_2 = 4$ then

$$\begin{pmatrix} 1 & -1 & 0 & | & 0 \\ 0 & -9 & 9 & | & 0 \\ 5 & -1 & -4 & | & 0 \end{pmatrix} \Longrightarrow \begin{pmatrix} 1 & 0 & -1 & | & 0 \\ 0 & 1 & -1 & | & 0 \\ 0 & 0 & 0 & | & 0 \end{pmatrix}$$

**110**

so that $k_1 = k_3$ and $k_2 = k_3$. If $k_3 = 1$ then

$$\mathbf{K}_2 = \begin{pmatrix} 1 \\ 1 \\ 1 \end{pmatrix}.$$

If $\lambda_3 = -4$ then

$$\begin{pmatrix} 9 & -1 & 0 & | & 0 \\ 0 & -1 & 9 & | & 0 \\ 5 & -1 & 4 & | & 0 \end{pmatrix} \implies \begin{pmatrix} 1 & 0 & -1 & | & 0 \\ 0 & 1 & -9 & | & 0 \\ 0 & 0 & 0 & | & 0 \end{pmatrix}$$

so that $k_1 = k_3$ and $k_2 = 9k_3$. If $k_3 = 1$ then

$$\mathbf{K}_3 = \begin{pmatrix} 1 \\ 9 \\ 1 \end{pmatrix}.$$

**48.** We solve

$$\det(\mathbf{A} - \lambda\mathbf{I}) = \begin{vmatrix} 1-\lambda & 6 & 0 \\ 0 & 2-\lambda & 1 \\ 0 & 1 & 2-\lambda \end{vmatrix} = \begin{vmatrix} 1-\lambda & 6 & 0 \\ 0 & 3-\lambda & 3-\lambda \\ 0 & 1 & 2-\lambda \end{vmatrix} = (3-\lambda)(1-\lambda)^2 = 0.$$

For $\lambda = 3$ we have

$$\begin{pmatrix} -2 & 6 & 0 & | & 0 \\ 0 & 0 & 0 & | & 0 \\ 0 & 1 & -1 & | & 0 \end{pmatrix} \implies \begin{pmatrix} 1 & 0 & -3 & | & 0 \\ 0 & 1 & -1 & | & 0 \\ 0 & 0 & 0 & | & 0 \end{pmatrix}$$

so that $k_1 = 3k_3$ and $k_2 = k_3$. If $k_3 = 1$ then

$$\mathbf{K}_1 = \begin{pmatrix} 3 \\ 1 \\ 1 \end{pmatrix}.$$

For $\lambda_2 = \lambda_3 = 1$ we have

$$\begin{pmatrix} 0 & 6 & 0 & | & 0 \\ 0 & 1 & 1 & | & 0 \\ 0 & 1 & 1 & | & 0 \end{pmatrix} \implies \begin{pmatrix} 0 & 1 & 0 & | & 0 \\ 0 & 0 & 1 & | & 0 \\ 0 & 0 & 0 & | & 0 \end{pmatrix}$$

so that $k_2 = 0$ and $k_3 = 0$. If $k_1 = 1$ then

$$\mathbf{K}_2 = \begin{pmatrix} 1 \\ 0 \\ 0 \end{pmatrix}.$$

**51.** Let
$$A = \begin{pmatrix} a_{11} & a_{12} \\ a_{21} & a_{22} \end{pmatrix}.$$

Then
$$\frac{d}{dt}[\mathbf{A}(t)\mathbf{X}(t)] = \frac{d}{dt}\begin{pmatrix} a_1 & a_2 \\ a_3 & a_4 \end{pmatrix}\begin{pmatrix} x_1 \\ x_2 \end{pmatrix} = \frac{d}{dt}\begin{pmatrix} a_1x_1 + a_2x_2 \\ a_3x_1 + a_4x_2 \end{pmatrix} = \begin{pmatrix} a_1x_1' + a_1'x_1 + a_2x_2' + a_2'x_2 \\ a_3x_1' + a_3'x_1 + a_4x_2' + a_4'x_2 \end{pmatrix}$$

$$= \begin{pmatrix} a_1 & a_2 \\ a_3 & a_4 \end{pmatrix}\begin{pmatrix} x_1' \\ x_2' \end{pmatrix} + \begin{pmatrix} a_1' & a_2' \\ a_3' & a_4' \end{pmatrix}\begin{pmatrix} x_1 \\ x_2 \end{pmatrix} = \mathbf{A}(t)\mathbf{X}'(t) + \mathbf{A}'(t)\mathbf{X}(t).$$

**54.** Since
$$(\mathbf{AB})(\mathbf{B}^{-1}\mathbf{A}^{-1}) = \mathbf{A}(\mathbf{BB}^{-1})\mathbf{A}^{-1} = \mathbf{AIA}^{-1} = \mathbf{AA}^{-1} = \mathbf{I}$$

and
$$(\mathbf{B}^{-1}\mathbf{A}^{-1})(\mathbf{AB}) = \mathbf{B}^{-1}(\mathbf{A}^{-1}\mathbf{A})\mathbf{B} = \mathbf{B}^{-1}\mathbf{IB} = \mathbf{B}^{-1}\mathbf{B} = \mathbf{I}$$

we have
$$(\mathbf{AB})^{-1} = \mathbf{B}^{-1}\mathbf{A}^{-1}.$$

## Exercises 8.5

**3.** Let $\mathbf{X} = \begin{pmatrix} x \\ y \\ z \end{pmatrix}$. Then

$$\mathbf{X}' = \begin{pmatrix} -3 & 4 & -9 \\ 6 & -1 & 0 \\ 10 & 4 & 3 \end{pmatrix}\mathbf{X}.$$

**6.** Let $\mathbf{X} = \begin{pmatrix} x \\ y \end{pmatrix}$. Then

$$\mathbf{X}' = \begin{pmatrix} -3 & 4 \\ 5 & 9 \end{pmatrix}\mathbf{X} + \begin{pmatrix} e^{-t}\sin 2t \\ 4e^{-t}\cos 2t \end{pmatrix}.$$

**9.** $\dfrac{dx}{dt} = x - y + 2z + e^{-t} - 3t; \quad \dfrac{dy}{dt} = 3x - 4y + z + 2e^{-t} + t; \quad \dfrac{dz}{dt} = -2x + 5y + 6z + 2e^{-t} - t$

**12.** Since
$$\mathbf{X}' = \begin{pmatrix} 5\cos t - 5\sin t \\ 2\cos t - 4\sin t \end{pmatrix}e^t \quad \text{and} \quad \begin{pmatrix} -2 & 5 \\ -2 & 4 \end{pmatrix}\mathbf{X} = \begin{pmatrix} 5\cos t - 5\sin t \\ 2\cos t - 4\sin t \end{pmatrix}e^t$$

we see that
$$\mathbf{X}' = \begin{pmatrix} -2 & 5 \\ -2 & 4 \end{pmatrix}\mathbf{X}.$$

**15.** Since

$$\mathbf{X}' = \begin{pmatrix} 0 \\ 0 \\ 0 \end{pmatrix} \quad \text{and} \quad \begin{pmatrix} 1 & 2 & 1 \\ 6 & -1 & 0 \\ -1 & -2 & -1 \end{pmatrix} \mathbf{X} = \begin{pmatrix} 0 \\ 0 \\ 0 \end{pmatrix}$$

we see that

$$\mathbf{X}' = \begin{pmatrix} 1 & 2 & 1 \\ 6 & -1 & 0 \\ -1 & -2 & -1 \end{pmatrix} \mathbf{X}.$$

**18.** Yes, since $W(\mathbf{X}_1, \mathbf{X}_2) = 8e^{2t} \neq 0$ and $\mathbf{X}_1$ and $\mathbf{X}_2$ are linearly independent on $-\infty < t < \infty$.

**21.** Since

$$\mathbf{X}_p' = \begin{pmatrix} 2 \\ -1 \end{pmatrix} \quad \text{and} \quad \begin{pmatrix} 1 & 4 \\ 3 & 2 \end{pmatrix} \mathbf{X}_p + \begin{pmatrix} 2 \\ -4 \end{pmatrix} t + \begin{pmatrix} -7 \\ -18 \end{pmatrix} = \begin{pmatrix} 2 \\ -1 \end{pmatrix}$$

we see that

$$\mathbf{X}_p' = \begin{pmatrix} 1 & 4 \\ 3 & 2 \end{pmatrix} \mathbf{X}_p + \begin{pmatrix} 2 \\ -4 \end{pmatrix} t + \begin{pmatrix} -7 \\ -18 \end{pmatrix}.$$

**24.** Since

$$\mathbf{X}_p' = \begin{pmatrix} 3\cos 3t \\ 0 \\ -3\sin 3t \end{pmatrix} \quad \text{and} \quad \begin{pmatrix} 1 & 2 & 3 \\ -4 & 2 & 0 \\ -6 & 1 & 0 \end{pmatrix} \mathbf{X}_p + \begin{pmatrix} -1 \\ 4 \\ 3 \end{pmatrix} \sin 3t = \begin{pmatrix} 3\cos 3t \\ 0 \\ -3\sin 3t \end{pmatrix}$$

we see that

$$\mathbf{X}_p' = \begin{pmatrix} 1 & 2 & 3 \\ -4 & 2 & 0 \\ -6 & 1 & 0 \end{pmatrix} \mathbf{X}_p + \begin{pmatrix} -1 \\ 4 \\ 3 \end{pmatrix} \sin 3t.$$

**27.** $\Phi(t) = \begin{pmatrix} e^{2t} & e^{7t} \\ -2e^{2t} & 3e^{7t} \end{pmatrix}$ and $\Phi^{-1}(t) = \dfrac{1}{5e^{9t}} \begin{pmatrix} 3e^{7t} & -e^{7t} \\ 2e^{2t} & e^{2t} \end{pmatrix}.$

**30.** $\Phi(t) = \begin{pmatrix} 2\cos t & -2\sin t \\ 3\cos t + \sin t & -3\sin t + \cos t \end{pmatrix}$ and $\Phi^{-1}(t) = \dfrac{1}{2} \begin{pmatrix} -3\sin t + \cos t & 2\sin t \\ -3\cos t - \sin t & 2\cos t \end{pmatrix}.$

**33.** We have

$$\mathbf{X}(t) = c_1 \begin{pmatrix} -1 \\ 3 \end{pmatrix} e^t + c_2 \begin{pmatrix} -1 \\ 3 \end{pmatrix} te^t + c_2 \begin{pmatrix} 0 \\ -1 \end{pmatrix} e^t$$

so that

$$\mathbf{X}(0) = c_1 \begin{pmatrix} -1 \\ 3 \end{pmatrix} + c_2 \begin{pmatrix} -1 \\ 3 \end{pmatrix} + c_2 \begin{pmatrix} 0 \\ -1 \end{pmatrix} = \begin{pmatrix} 1 \\ 0 \end{pmatrix}$$

and

$$\mathbf{X}(0) = c_1 \begin{pmatrix} -1 \\ 3 \end{pmatrix} + c_2 \begin{pmatrix} -1 \\ 3 \end{pmatrix} + c_2 \begin{pmatrix} 0 \\ -1 \end{pmatrix} = \begin{pmatrix} 0 \\ 1 \end{pmatrix}$$

**113**

which give $c_1 = -1$, $c_2 = -3$, and $c_1 = 0$, $c_2 = -1$. Then

$$\Psi(t) = \begin{pmatrix} 3te^t + e^t & te^t \\ -9te^t & -3te^t + e^t \end{pmatrix}.$$

**36.** Since the column vectors of $\Psi(t)$ solve $X' = AX$ we know that $X = \Psi(t)X_0$ solves $X' = AX$, and $X(t_0) = \Psi(t_0)X_0 = IX_0 = X_0$.

## —————— Exercises 8.6 ——————

**3.** The system is

$$X' = \begin{pmatrix} -4 & 2 \\ -5/2 & 2 \end{pmatrix} X$$

and $\det(A - \lambda I) = (\lambda - 1)(\lambda + 3) = 0$. For $\lambda_1 = 1$ we obtain

$$\begin{pmatrix} -5 & 2 & | & 0 \\ -5/2 & 1 & | & 0 \end{pmatrix} \implies \begin{pmatrix} -5 & 2 & | & 0 \\ 0 & 0 & | & 0 \end{pmatrix} \quad \text{so that} \quad K_1 = \begin{pmatrix} 2 \\ 5 \end{pmatrix}.$$

For $\lambda_2 = -3$ we obtain

$$\begin{pmatrix} -1 & 2 & | & 0 \\ -5/2 & 5 & | & 0 \end{pmatrix} \implies \begin{pmatrix} -1 & 2 & | & 0 \\ 0 & 0 & | & 0 \end{pmatrix} \quad \text{so that} \quad K_2 = \begin{pmatrix} 2 \\ 1 \end{pmatrix}.$$

Then

$$X = c_1 \begin{pmatrix} 2 \\ 5 \end{pmatrix} e^t + c_2 \begin{pmatrix} 2 \\ 1 \end{pmatrix} e^{-3t}.$$

**6.** The system is

$$X' = \begin{pmatrix} -6 & 2 \\ -3 & 1 \end{pmatrix} X$$

and $\det(A - \lambda I) = \lambda(\lambda + 5) = 0$. For $\lambda_1 = 0$ we obtain

$$\begin{pmatrix} -6 & 2 & | & 0 \\ -3 & 1 & | & 0 \end{pmatrix} \implies \begin{pmatrix} 1 & -1/3 & | & 0 \\ 0 & 0 & | & 0 \end{pmatrix} \quad \text{so that} \quad K_1 = \begin{pmatrix} 1 \\ 3 \end{pmatrix}.$$

For $\lambda_2 = -5$ we obtain

$$\begin{pmatrix} -1 & 2 & | & 0 \\ -3 & 6 & | & 0 \end{pmatrix} \implies \begin{pmatrix} 1 & -2 & | & 0 \\ 0 & 0 & | & 0 \end{pmatrix} \quad \text{so that} \quad K_2 = \begin{pmatrix} 2 \\ 1 \end{pmatrix}.$$

Then

$$X = c_1 \begin{pmatrix} 1 \\ 3 \end{pmatrix} + c_2 \begin{pmatrix} 2 \\ 1 \end{pmatrix} e^{-5t}.$$

**9.** We have $\det(\mathbf{A} - \lambda\mathbf{I}) = -(\lambda+1)(\lambda-3)(\lambda+2) = 0$. For $\lambda_1 = -1$, $\lambda_2 = 3$, and $\lambda_3 = -2$ we obtain

$$\mathbf{K}_1 = \begin{pmatrix} -1 \\ 0 \\ 1 \end{pmatrix}, \quad \mathbf{K}_2 = \begin{pmatrix} 1 \\ 4 \\ 3 \end{pmatrix}, \quad \text{and} \quad \mathbf{K}_3 = \begin{pmatrix} 1 \\ -1 \\ 3 \end{pmatrix},$$

so that

$$\mathbf{X} = c_1 \begin{pmatrix} -1 \\ 0 \\ 1 \end{pmatrix} e^{-t} + c_2 \begin{pmatrix} 1 \\ 4 \\ 3 \end{pmatrix} e^{3t} + c_3 \begin{pmatrix} 1 \\ -1 \\ 3 \end{pmatrix} e^{-2t}.$$

**12.** We have $\det(\mathbf{A} - \lambda\mathbf{I}) = (\lambda-3)(\lambda+5)(6-\lambda) = 0$. For $\lambda_1 = 3$, $\lambda_2 = -5$, and $\lambda_3 = 6$ we obtain

$$\mathbf{K}_1 = \begin{pmatrix} 1 \\ 1 \\ 0 \end{pmatrix}, \quad \mathbf{K}_2 = \begin{pmatrix} 1 \\ -1 \\ 0 \end{pmatrix}, \quad \text{and} \quad \mathbf{K}_3 = \begin{pmatrix} 2 \\ -2 \\ 11 \end{pmatrix},$$

so that

$$\mathbf{X} = c_1 \begin{pmatrix} 1 \\ 1 \\ 0 \end{pmatrix} e^{3t} + c_2 \begin{pmatrix} 1 \\ -1 \\ 0 \end{pmatrix} e^{-5t} + c_3 \begin{pmatrix} 2 \\ -2 \\ 11 \end{pmatrix} e^{6t}.$$

In Problems 15-27 the form of the answer will vary according to the choice of eigenvector. For example, in Problem 15, if $\mathbf{K}_1$ is chosen to be $\begin{pmatrix} 1 \\ -2i \end{pmatrix}$ the solution has the form

$$\mathbf{X} = c_1 \begin{pmatrix} \cos t \\ 2\cos t + \sin t \end{pmatrix} e^{4t} + c_2 \begin{pmatrix} \sin t \\ 2\sin t - \cos t \end{pmatrix} e^{4t}.$$

**15.** We have $\det(\mathbf{A} - \lambda\mathbf{I}) = \lambda^2 - 8\lambda + 17 = 0$. For $\lambda_1 = 4+i$ we obtain

$$\mathbf{K}_1 = \begin{pmatrix} 2+i \\ 5 \end{pmatrix}$$

so that

$$\mathbf{X}_1 = \begin{pmatrix} 2+i \\ 5 \end{pmatrix} e^{(4+i)t} = \begin{pmatrix} 2\cos t - \sin t \\ 5\cos t \end{pmatrix} e^{4t} + i \begin{pmatrix} \cos t + 2\sin t \\ 5\sin t \end{pmatrix} e^{4t}.$$

Then

$$\mathbf{X} = c_1 \begin{pmatrix} 2\cos t - \sin t \\ 5\cos t \end{pmatrix} e^{4t} + c_2 \begin{pmatrix} 2\sin t + \cos t \\ 5\sin t \end{pmatrix} e^{4t}.$$

**18.** We have $\det(\mathbf{A} - \lambda\mathbf{I}) = \lambda^2 - 10\lambda + 34 = 0$. For $\lambda_1 = 5+3i$ we obtain

$$\mathbf{K}_1 = \begin{pmatrix} 1-3i \\ 2 \end{pmatrix}$$

**115**

so that

$$\mathbf{X_1} = \begin{pmatrix} 1 - 3i \\ 2 \end{pmatrix} e^{(5+3i)t} = \begin{pmatrix} \cos 3t + 3 \sin 3t \\ 2 \cos 3t \end{pmatrix} e^{5t} + i \begin{pmatrix} \sin 3t - 3 \cos 3t \\ 2 \cos 3t \end{pmatrix} e^{5t}.$$

Then

$$\mathbf{X} = c_1 \begin{pmatrix} \cos 3t + 3 \sin 3t \\ 2 \cos 3t \end{pmatrix} e^{5t} + c_2 \begin{pmatrix} \sin 3t - 3 \cos 3t \\ 2 \cos 3t \end{pmatrix} e^{5t}.$$

**21.** We have $\det(\mathbf{A} - \lambda\mathbf{I}) = -\lambda \left(\lambda^2 + 1\right) = 0$. For $\lambda_1 = 0$ we obtain

$$\mathbf{K_1} = \begin{pmatrix} 1 \\ 0 \\ 0 \end{pmatrix}.$$

For $\lambda_2 = i$ we obtain

$$\mathbf{K_2} = \begin{pmatrix} -i \\ i \\ 1 \end{pmatrix}$$

so that

$$\mathbf{X_2} = \begin{pmatrix} -i \\ i \\ 1 \end{pmatrix} e^{it} = \begin{pmatrix} \sin t \\ -\sin t \\ \cos t \end{pmatrix} + i \begin{pmatrix} -\cos t \\ \cos t \\ \sin t \end{pmatrix}.$$

Then

$$\mathbf{X} = c_1 \begin{pmatrix} 1 \\ 0 \\ 0 \end{pmatrix} + c_2 \begin{pmatrix} \sin t \\ -\sin t \\ \cos t \end{pmatrix} + c_3 \begin{pmatrix} -\cos t \\ \cos t \\ \sin t \end{pmatrix}.$$

**24.** We have $\det(\mathbf{A} - \lambda\mathbf{I}) = -(\lambda - 6)(\lambda^2 - 8\lambda + 20) = 0$. For $\lambda_1 = 6$ we obtain

$$\mathbf{K_1} = \begin{pmatrix} 0 \\ 1 \\ 0 \end{pmatrix}.$$

For $\lambda_2 = 4 + 2i$ we obtain

$$\mathbf{K_2} = \begin{pmatrix} -i \\ 0 \\ 2 \end{pmatrix}$$

so that

$$\mathbf{X_2} = \begin{pmatrix} -i \\ 0 \\ 2 \end{pmatrix} e^{(4+2i)t} = \begin{pmatrix} \sin 2t \\ 0 \\ 2 \cos 2t \end{pmatrix} e^{4t} + i \begin{pmatrix} -\cos 2t \\ 0 \\ 2 \sin 2t \end{pmatrix} e^{4t}.$$

**116**

Then

$$\mathbf{X} = c_1 \begin{pmatrix} 0 \\ 1 \\ 0 \end{pmatrix} e^{6t} + c_2 \begin{pmatrix} \sin 2t \\ 0 \\ 2\cos 2t \end{pmatrix} e^{4t} + c_3 \begin{pmatrix} -\cos 2t \\ 0 \\ 2\sin 2t \end{pmatrix} e^{4t}.$$

**27.** We have $\det(\mathbf{A} - \lambda\mathbf{I}) = (1-\lambda)(\lambda^2 + 25) = 0$. For $\lambda_1 = 1$ we obtain

$$\mathbf{K}_1 = \begin{pmatrix} 25 \\ -7 \\ 6 \end{pmatrix}.$$

For $\lambda_2 = 5i$ we obtain

$$\mathbf{K}_2 = \begin{pmatrix} 1+5i \\ 1 \\ 1 \end{pmatrix}$$

so that

$$\mathbf{X}_2 = \begin{pmatrix} 1+5i \\ 1 \\ 1 \end{pmatrix} e^{5it} = \begin{pmatrix} \cos 5t - 5\sin 5t \\ \cos 5t \\ \cos 5t \end{pmatrix} + i \begin{pmatrix} \sin 5t + 5\cos 5t \\ \sin 5t \\ \sin 5t \end{pmatrix}.$$

Then

$$\mathbf{X} = c_1 \begin{pmatrix} 25 \\ -7 \\ 6 \end{pmatrix} e^t + c_2 \begin{pmatrix} \cos 5t - 5\sin 5t \\ \cos 5t \\ \cos 5t \end{pmatrix} + c_3 \begin{pmatrix} \sin 5t + 5\cos 5t \\ \sin 5t \\ \sin 5t \end{pmatrix}.$$

If

$$\mathbf{X}(0) = \begin{pmatrix} 4 \\ 6 \\ -7 \end{pmatrix}$$

then $c_1 = c_2 = -1$ and $c_3 = 6$.

**30.** We have $\det(\mathbf{A} - \lambda\mathbf{I}) = (\lambda + 1)^2 = 0$. For $\lambda_1 = -1$ we obtain

$$\mathbf{K} = \begin{pmatrix} 1 \\ 1 \end{pmatrix}.$$

A solution of $(\mathbf{A} - \lambda_1\mathbf{I})\mathbf{P} = \mathbf{K}$ is

$$\mathbf{P} = \begin{pmatrix} 0 \\ 1/5 \end{pmatrix}$$

so that

$$\mathbf{X} = c_1 \begin{pmatrix} 1 \\ 1 \end{pmatrix} e^{-t} + c_2 \left[ \begin{pmatrix} 1 \\ 1 \end{pmatrix} te^{-t} + \begin{pmatrix} 0 \\ 1/5 \end{pmatrix} e^{-t} \right].$$

**117**

**33.** We have $\det(\mathbf{A} - \lambda\mathbf{I}) = (1 - \lambda)(\lambda - 2)^2 = 0$. For $\lambda_1 = 1$ we obtain

$$\mathbf{K}_1 = \begin{pmatrix} 1 \\ 1 \\ 1 \end{pmatrix}.$$

For $\lambda_2 = 2$ we obtain

$$\mathbf{K}_2 = \begin{pmatrix} 1 \\ 0 \\ 1 \end{pmatrix} \quad \text{and} \quad \mathbf{K}_3 = \begin{pmatrix} 1 \\ 1 \\ 0 \end{pmatrix}.$$

Then

$$\mathbf{X} = c_1 \begin{pmatrix} 1 \\ 1 \\ 1 \end{pmatrix} e^t + c_2 \begin{pmatrix} 1 \\ 0 \\ 1 \end{pmatrix} e^{2t} + c_3 \begin{pmatrix} 1 \\ 1 \\ 0 \end{pmatrix} e^{2t}.$$

**36.** We have $\det(\mathbf{A} - \lambda\mathbf{I}) = (1 - \lambda)(\lambda - 2)^2 = 0$. For $\lambda_1 = 1$ we obtain

$$\mathbf{K}_1 = \begin{pmatrix} 1 \\ 0 \\ 0 \end{pmatrix}.$$

For $\lambda_2 = 2$ we obtain

$$\mathbf{K} = \begin{pmatrix} 0 \\ -1 \\ 1 \end{pmatrix}.$$

A solution of $(\mathbf{A} - \lambda_2\mathbf{I})\mathbf{P} = \mathbf{K}$ is

$$\mathbf{P} = \begin{pmatrix} 0 \\ -1 \\ 0 \end{pmatrix}$$

so that

$$\mathbf{X} = c_1 \begin{pmatrix} 1 \\ 0 \\ 0 \end{pmatrix} e^t + c_2 \begin{pmatrix} 0 \\ -1 \\ 1 \end{pmatrix} e^{2t} + c_3 \left[ \begin{pmatrix} 0 \\ -1 \\ 1 \end{pmatrix} te^{2t} + \begin{pmatrix} 0 \\ -1 \\ 0 \end{pmatrix} e^{2t} \right].$$

**39.** We have $\det(\mathbf{A} - \lambda\mathbf{I}) = (\lambda - 4)^2 = 0$. For $\lambda_1 = 4$ we obtain

$$\mathbf{K} = \begin{pmatrix} 2 \\ 1 \end{pmatrix}.$$

A solution of $(\mathbf{A} - \lambda_1\mathbf{I})\mathbf{P} = \mathbf{K}$ is

$$\mathbf{P} = \begin{pmatrix} 1 \\ 1 \end{pmatrix}$$

so that

$$\mathbf{X} = c_1 \begin{pmatrix} 2 \\ 1 \end{pmatrix} e^{4t} + c_2 \left[ \begin{pmatrix} 2 \\ 1 \end{pmatrix} te^{4t} + \begin{pmatrix} 1 \\ 1 \end{pmatrix} e^{4t} \right].$$

If

$$\mathbf{X}(0) = \begin{pmatrix} -1 \\ 6 \end{pmatrix}$$

then $c_1 = -7$ and $c_2 = 13$.

**42.** We have $\det(\mathbf{A} - \lambda \mathbf{I}) = (\lambda + 3/25)(\lambda + 1/25) = 0$. For $\lambda_1 = -3/25$ and $\lambda_2 = -1/25$ we obtain

$$\mathbf{K}_1 = \begin{pmatrix} -1 \\ 2 \end{pmatrix} \quad \text{and} \quad \mathbf{K}_2 = \begin{pmatrix} 1 \\ 2 \end{pmatrix}$$

so that

$$\mathbf{\Phi}(t) = \begin{pmatrix} -e^{-3t/25} & e^{-t/25} \\ 2e^{-3t/25} & 2e^{-t/25} \end{pmatrix} \quad \text{and} \quad \mathbf{\Phi}^{-1}(t) = -\frac{1}{4}e^{4t/25} \begin{pmatrix} 2e^{-t/25} & e^{-t/25} \\ -2e^{-3t/25} & -e^{-3t/25} \end{pmatrix}.$$

Then

$$\mathbf{X} = \mathbf{\Phi}(t)\mathbf{\Phi}^{-1}(0)\mathbf{X}(0) = \begin{pmatrix} \frac{25}{2}e^{-3t/25} + \frac{25}{2}e^{-t/25} \\ -25e^{-3t/25} + 25e^{-t/25} \end{pmatrix}.$$

## ———— Exercises 8.7 ————

**3.** Solving

$$\begin{vmatrix} 1 - \lambda & 3 \\ 3 & 1 - \lambda \end{vmatrix} = \lambda^2 - 2\lambda - 8 = (\lambda - 4)(\lambda + 2) = 0$$

we obtain eigenvalues $\lambda_1 = -2$ and $\lambda_2 = 4$. Corresponding eigenvectors are

$$\mathbf{K}_1 = \begin{pmatrix} 1 \\ -1 \end{pmatrix} \quad \text{and} \quad \mathbf{K}_2 = \begin{pmatrix} 1 \\ 1 \end{pmatrix}.$$

Thus

$$\mathbf{X}_c = c_1 \begin{pmatrix} 1 \\ -1 \end{pmatrix} e^{-2t} + c_2 \begin{pmatrix} 1 \\ 1 \end{pmatrix} e^{4t}.$$

Substituting

$$\mathbf{X}_p = \begin{pmatrix} a_3 \\ b_3 \end{pmatrix} t^2 + \begin{pmatrix} a_2 \\ b_2 \end{pmatrix} t + \begin{pmatrix} a_1 \\ b_1 \end{pmatrix}$$

into the system yields

$$a_3 + 3b_3 = 2 \qquad a_2 + 3b_2 = 2a_3 \qquad a_1 + 3b_1 = a_2$$

$$3a_3 + b_3 = 0 \qquad 3a_2 + b_2 + 1 = 2b_3 \qquad 3a_1 + b_1 + 5 = b_2$$

**119**

from which we obtain $a_3 = -1/4$, $b_3 = 3/4$, $a_2 = 1/4$, $b_2 = -1/4$, $a_1 = -2$, and $b_1 = 3/4$. Then

$$\mathbf{X}(t) = c_1 \begin{pmatrix} 1 \\ -1 \end{pmatrix} e^{-2t} + c_2 \begin{pmatrix} 1 \\ 1 \end{pmatrix} e^{4t} + \begin{pmatrix} -1/4 \\ 3/4 \end{pmatrix} t^2 + \begin{pmatrix} 1/4 \\ -1/4 \end{pmatrix} t + \begin{pmatrix} -2 \\ 3/4 \end{pmatrix}.$$

**6.** Solving

$$\begin{vmatrix} -1-\lambda & 5 \\ -1 & 1-\lambda \end{vmatrix} = \lambda^2 + 4 = 0$$

we obtain the eigenvalues $\lambda_1 = 2i$ and $\lambda_2 = -2i$. Corresponding eigenvectors are

$$\mathbf{K}_1 = \begin{pmatrix} 5 \\ 1+2i \end{pmatrix} \quad \text{and} \quad \mathbf{K}_2 = \begin{pmatrix} 5 \\ 1-2i \end{pmatrix}.$$

Thus

$$\mathbf{X}_c = c_1 \begin{pmatrix} 5\cos 2t \\ \cos 2t - 2\sin 2t \end{pmatrix} + c_2 \begin{pmatrix} 5\sin 2t \\ 2\cos 2t + \sin 2t \end{pmatrix}.$$

Substituting

$$\mathbf{X}_p = \begin{pmatrix} a_2 \\ b_2 \end{pmatrix} \cos t + \begin{pmatrix} a_1 \\ b_1 \end{pmatrix} \sin t$$

into the system yields

$$-a_2 + 5b_2 - a_1 = 0$$

$$-a_2 + b_2 - b_1 - 2 = 0$$

$$-a_1 + 5b_1 + a_2 + 1 = 0$$

$$-a_1 + b_1 + b_2 = 0$$

from which we obtain $a_2 = -3$, $b_2 = -2/3$, $a_1 = -1/3$, and $b_1 = 1/3$. Then

$$\mathbf{X}(t) = c_1 \begin{pmatrix} 5\cos 2t \\ \cos 2t - 2\sin 2t \end{pmatrix} + c_2 \begin{pmatrix} 5\sin 2t \\ 2\cos 2t + \sin 2t \end{pmatrix} + \begin{pmatrix} -3 \\ -2/3 \end{pmatrix} \cos t + \begin{pmatrix} -1/3 \\ 1/3 \end{pmatrix} \sin t.$$

**9.** Solving

$$\begin{vmatrix} -1-\lambda & -2 \\ 3 & 4-\lambda \end{vmatrix} = \lambda^2 - 3\lambda + 2 = (\lambda - 1)(\lambda - 2) = 0$$

we obtain the eigenvalues $\lambda_1 = 1$ and $\lambda_2 = 2$. Corresponding eigenvectors are

$$\mathbf{K}_1 = \begin{pmatrix} 1 \\ -1 \end{pmatrix} \quad \text{and} \quad \mathbf{K}_2 = \begin{pmatrix} -4 \\ 6 \end{pmatrix}.$$

Thus

$$\mathbf{X}_c = c_1 \begin{pmatrix} 1 \\ -1 \end{pmatrix} e^t + c_2 \begin{pmatrix} -4 \\ 6 \end{pmatrix} e^{2t}.$$

Substituting

$$\mathbf{X}_p = \begin{pmatrix} a_1 \\ b_1 \end{pmatrix}$$

into the system yields

$$-a_1 - 2b_1 = -3$$

$$3a_1 + 4b_1 = -3$$

from which we obtain $a_1 = -9$ and $b_1 = 6$. Then

$$\mathbf{X}(t) = c_1 \begin{pmatrix} 1 \\ -1 \end{pmatrix} e^t + c_2 \begin{pmatrix} -4 \\ 6 \end{pmatrix} e^{2t} + \begin{pmatrix} -9 \\ 6 \end{pmatrix}.$$

Setting

$$\mathbf{X}(0) = \begin{pmatrix} -4 \\ 5 \end{pmatrix}$$

we obtain

$$c_1 - 4c_2 - 9 = -4$$

$$-c_1 + 6c_2 + 6 = 5.$$

Then $c_1 = 13$ and $c_2 = 2$ so

$$\mathbf{X}(t) = 13 \begin{pmatrix} 1 \\ -1 \end{pmatrix} e^t + 2 \begin{pmatrix} -4 \\ 6 \end{pmatrix} e^{2t} + \begin{pmatrix} -9 \\ 6 \end{pmatrix}.$$

## Exercises 8.8

**3.** From

$$\mathbf{X}' = \begin{pmatrix} 3 & -5 \\ 3/4 & -1 \end{pmatrix} \mathbf{X} + \begin{pmatrix} 1 \\ -1 \end{pmatrix} e^{t/2}$$

we obtain

$$\mathbf{X}_c = c_1 \begin{pmatrix} 10 \\ 3 \end{pmatrix} e^{3t/2} + c_2 \begin{pmatrix} 2 \\ 1 \end{pmatrix} e^{t/2}.$$

Then

$$\mathbf{\Phi} = \begin{pmatrix} 10e^{3t/2} & 2e^{t/2} \\ 3e^{3t/2} & e^{t/2} \end{pmatrix} \quad \text{and} \quad \mathbf{\Phi}^{-1} = \begin{pmatrix} \frac{1}{4}e^{-3t/2} & -\frac{1}{2}e^{-3t/2} \\ -\frac{3}{4}e^{-t/2} & \frac{5}{2}e^{-t/2} \end{pmatrix}$$

so that

$$\mathbf{U} = \int \mathbf{\Phi}^{-1}\mathbf{F}\,dt = \int \begin{pmatrix} \frac{3}{4}e^{-t} \\ -\frac{13}{4} \end{pmatrix} dt = \begin{pmatrix} -\frac{3}{4}e^{-t} \\ -\frac{13}{4}t \end{pmatrix}$$

and

$$\mathbf{X}_p = \mathbf{\Phi}\mathbf{U} = \begin{pmatrix} -13/2 \\ -13/4 \end{pmatrix} te^{t/2} + \begin{pmatrix} -15/2 \\ -9/4 \end{pmatrix} e^{t/2}.$$

**6.** From

$$\mathbf{X}' = \begin{pmatrix} 0 & 2 \\ -1 & 3 \end{pmatrix} \mathbf{X} + \begin{pmatrix} 2 \\ e^{-3t} \end{pmatrix}$$

we obtain

$$\mathbf{X}_c = c_1 \begin{pmatrix} 2 \\ 1 \end{pmatrix} e^t + c_2 \begin{pmatrix} 1 \\ 1 \end{pmatrix} e^{2t}.$$

Then

$$\mathbf{\Phi} = \begin{pmatrix} 2e^t & e^{2t} \\ e^t & e^{2t} \end{pmatrix} \quad \text{and} \quad \mathbf{\Phi}^{-1} = \begin{pmatrix} e^{-t} & -e^{-t} \\ -e^{-2t} & 2e^{-2t} \end{pmatrix}$$

so that

$$\mathbf{U} = \int \mathbf{\Phi}^{-1} \mathbf{F} \, dt = \int \begin{pmatrix} 2e^{-t} - e^{-4t} \\ -2e^{-2t} + 2e^{-5t} \end{pmatrix} dt = \begin{pmatrix} -2e^{-t} + \frac{1}{4}e^{-4t} \\ e^{-2t} - \frac{2}{5}e^{-5t} \end{pmatrix}$$

and

$$\mathbf{X}_p = \mathbf{\Phi}\mathbf{U} = \begin{pmatrix} \frac{1}{10}e^{-3t} - 3 \\ -\frac{3}{20}e^{-3t} - 1 \end{pmatrix}.$$

**9.** From

$$\mathbf{X}' = \begin{pmatrix} 3 & 2 \\ -2 & -1 \end{pmatrix} \mathbf{X} + \begin{pmatrix} 2 \\ 1 \end{pmatrix} e^{-t}$$

we obtain

$$\mathbf{X}_c = c_1 \begin{pmatrix} 1 \\ -1 \end{pmatrix} e^t + c_2 \left[ \begin{pmatrix} 1 \\ -1 \end{pmatrix} te^t + \begin{pmatrix} 0 \\ 1/2 \end{pmatrix} e^t \right].$$

Then

$$\mathbf{\Phi} = \begin{pmatrix} e^t & te^t \\ -e^t & \frac{1}{2}e^t - te^t \end{pmatrix} \quad \text{and} \quad \mathbf{\Phi}^{-1} = \begin{pmatrix} e^{-t} - 2te^{-t} & -2te^{-t} \\ 2e^{-t} & 2e^{-t} \end{pmatrix}$$

so that

$$\mathbf{U} = \int \mathbf{\Phi}^{-1} \mathbf{F} \, dt = \int \begin{pmatrix} 2e^{-2t} - 6te^{-2t} \\ 6e^{-2t} \end{pmatrix} dt = \begin{pmatrix} \frac{1}{2}e^{-2t} + 3te^{-2t} \\ -3e^{-2t} \end{pmatrix}$$

and

$$\mathbf{X}_p = \mathbf{\Phi}\mathbf{U} = \begin{pmatrix} 1/2 \\ -2 \end{pmatrix} e^{-t}.$$

**12.** From

$$\mathbf{X}' = \begin{pmatrix} 1 & -1 \\ 1 & 1 \end{pmatrix} \mathbf{X} + \begin{pmatrix} 3 \\ 3 \end{pmatrix} e^t$$

we obtain

$$\mathbf{X}_c = c_1 \begin{pmatrix} -\sin t \\ \cos t \end{pmatrix} e^t + c_2 \begin{pmatrix} \cos t \\ \sin t \end{pmatrix} e^t.$$

Then

$$\mathbf{\Phi} = \begin{pmatrix} -\sin t & \cos t \\ \cos t & \sin t \end{pmatrix} e^t \quad \text{and} \quad \mathbf{\Phi}^{-1} = \begin{pmatrix} -\sin t & \cos t \\ \cos t & \sin t \end{pmatrix} e^{-t}$$

so that

$$\mathbf{U} = \int \Phi^{-1}\mathbf{F}\, dt = \int \begin{pmatrix} -3\sin t + 3\cos t \\ 3\cos t + 3\sin t \end{pmatrix} dt = \begin{pmatrix} 3\cos t + 3\sin t \\ 3\sin t - 3\cos t \end{pmatrix}$$

and

$$\mathbf{X}_p = \Phi\mathbf{U} = \begin{pmatrix} -3 \\ 3 \end{pmatrix} e^t.$$

**15.** From

$$\mathbf{X}' = \begin{pmatrix} 0 & 1 \\ -1 & 0 \end{pmatrix} \mathbf{X} + \begin{pmatrix} 0 \\ \sec t \tan t \end{pmatrix}$$

we obtain

$$\mathbf{X}_c = c_1 \begin{pmatrix} \cos t \\ -\sin t \end{pmatrix} + c_2 \begin{pmatrix} \sin t \\ \cos t \end{pmatrix}.$$

Then

$$\Phi = \begin{pmatrix} \cos t & \sin t \\ -\sin t & \cos t \end{pmatrix} t \quad \text{and} \quad \Phi^{-1} = \begin{pmatrix} \cos t & -\sin t \\ \sin t & \cos t \end{pmatrix}$$

so that

$$\mathbf{U} = \int \Phi^{-1}\mathbf{F}\, dt = \int \begin{pmatrix} -\tan^2 t \\ \tan t \end{pmatrix} dt = \begin{pmatrix} t - \tan t \\ \ln|\sec t| \end{pmatrix}$$

and

$$\mathbf{X}_p = \Phi\mathbf{U} = \begin{pmatrix} \cos t \\ -\sin t \end{pmatrix} t + \begin{pmatrix} -\sin t \\ \sin t \tan t \end{pmatrix} + \begin{pmatrix} \sin t \\ \cos t \end{pmatrix} \ln|\sec t|.$$

**18.** From

$$\mathbf{X}' = \begin{pmatrix} 1 & -2 \\ 1 & -1 \end{pmatrix} \mathbf{X} + \begin{pmatrix} \tan t \\ 1 \end{pmatrix}$$

we obtain

$$\mathbf{X}_c = c_1 \begin{pmatrix} \cos t - \sin t \\ \cos t \end{pmatrix} + c_2 \begin{pmatrix} \cos t + \sin t \\ \sin t \end{pmatrix}.$$

Then

$$\Phi = \begin{pmatrix} \cos t - \sin t & \cos t + \sin t \\ \cos t & \sin t \end{pmatrix} \quad \text{and} \quad \Phi^{-1} = \begin{pmatrix} -\sin t & \cos t + \sin t \\ \cos t & \sin t - \cos t \end{pmatrix}$$

so that

$$\mathbf{U} = \int \Phi^{-1}\mathbf{F}\, dt = \int \begin{pmatrix} 2\cos t + \sin t - \sec t \\ 2\sin t - \cos t \end{pmatrix} dt = \begin{pmatrix} 2\sin t - \cos t - \ln|\sec t + \tan t| \\ -2\cos t - \sin t \end{pmatrix}$$

and

$$\mathbf{X}_p = \Phi\mathbf{U} = \begin{pmatrix} 3\sin t \cos t - \cos^2 t - 2\sin^2 t + (\sin t - \cos t)\ln|\sec t + \tan t| \\ \sin^2 t - \cos^2 t - \cos t(\ln|\sec t + \tan t|) \end{pmatrix}.$$

**21.** From

$$\mathbf{X}' = \begin{pmatrix} 3 & -1 \\ -1 & 3 \end{pmatrix} \mathbf{X} + \begin{pmatrix} 4e^{2t} \\ 4e^{4t} \end{pmatrix}$$

we obtain

$$\Phi = \begin{pmatrix} -e^{4t} & e^{2t} \\ e^{4t} & e^{2t} \end{pmatrix}, \quad \Phi^{-1} = \begin{pmatrix} -\frac{1}{2}e^{-4t} & \frac{1}{2}e^{4t} \\ \frac{1}{2}e^{-2t} & \frac{1}{2}e^{2t} \end{pmatrix},$$

and

$$X = \Phi\Phi^{-1}(0)X(0) + \Phi \int_0^t \Phi^{-1}F \, ds = \Phi \cdot \begin{pmatrix} 0 \\ 1 \end{pmatrix} + \Phi \cdot \begin{pmatrix} e^{-2t} + 2t - 1 \\ e^{2t} + 2t - 1 \end{pmatrix}$$

$$= \begin{pmatrix} 2 \\ 2 \end{pmatrix} te^{2t} + \begin{pmatrix} -1 \\ 1 \end{pmatrix} e^{2t} + \begin{pmatrix} -2 \\ 2 \end{pmatrix} te^{4t} + \begin{pmatrix} 2 \\ 0 \end{pmatrix} e^{4t}.$$

**24.** From

$$X' = \begin{pmatrix} 3 & -2 \\ 5 & -3 \end{pmatrix} X + \begin{pmatrix} 2 \\ 3 \end{pmatrix}$$

we obtain

$$\Psi = \begin{pmatrix} \sin t - 3\cos t & 2\cos t \\ -5\cos t & \sin t + 3\cos t \end{pmatrix}, \quad \Psi^{-1} = \begin{pmatrix} \sin t + 3\cos t & -2\cos t \\ 5\cos t & \sin t - 3\cos t \end{pmatrix},$$

and

$$X = \Psi X(\pi/2) + \Psi \int_{\pi/2}^t \Psi^{-1}F \, ds = \Psi \cdot \begin{pmatrix} 0 \\ 0 \end{pmatrix} + \Psi \cdot \begin{pmatrix} -2\cos t \\ \sin t - 3\cos t - 1 \end{pmatrix}$$

$$= \begin{pmatrix} 0 \\ 1 \end{pmatrix} - \begin{pmatrix} 2 \\ 3 \end{pmatrix} \cos t - \begin{pmatrix} 0 \\ 1 \end{pmatrix} \sin t.$$

## Exercises 8.9

**3.** Using the result of Problem 1

$$X = \begin{pmatrix} \cosh t & \sinh t \\ \sinh t & \cosh t \end{pmatrix} \begin{pmatrix} c_1 \\ c_2 \end{pmatrix} = c_1 \begin{pmatrix} \cosh t \\ \sinh t \end{pmatrix} + c_2 \begin{pmatrix} \sinh t \\ \cosh t \end{pmatrix}.$$

**6.** To solve

$$X' = \begin{pmatrix} 0 & 1 \\ 1 & 0 \end{pmatrix} X + \begin{pmatrix} \cosh t \\ \sinh t \end{pmatrix}$$

we identify $t_0 = 0$, $F(s) = \begin{pmatrix} \cosh t \\ \sinh t \end{pmatrix}$, and use the results of Problem 1 and equation (3) in the text.

$$\mathbf{X}(t) = e^{t\mathbf{A}}\mathbf{C} + e^{t\mathbf{A}}\int_{t_0}^{t} e^{-s\mathbf{A}}\mathbf{F}(s)\,ds$$

$$= \begin{pmatrix} \cosh t & \sinh t \\ \sinh t & \cosh t \end{pmatrix}\begin{pmatrix} c_1 \\ c_2 \end{pmatrix} + \begin{pmatrix} \cosh t & \sinh t \\ \sinh t & \cosh t \end{pmatrix}\int_0^t \begin{pmatrix} \cosh s & -\sinh s \\ -\sinh s & \cosh s \end{pmatrix}\begin{pmatrix} \cosh s \\ \sinh s \end{pmatrix}ds$$

$$= \begin{pmatrix} c_1\cosh t + c_2\sinh t \\ c_1\sinh t + c_2\cosh t \end{pmatrix} + \begin{pmatrix} \cosh t & \sinh t \\ \sinh t & \cosh t \end{pmatrix}\int_0^t \begin{pmatrix} 1 \\ 0 \end{pmatrix}ds$$

$$= \begin{pmatrix} c_1\cosh t + c_2\sinh t \\ c_1\sinh t + c_2\cosh t \end{pmatrix} + \begin{pmatrix} \cosh t & \sinh t \\ \sinh t & \cosh t \end{pmatrix}\begin{pmatrix} s \\ 0 \end{pmatrix}\Big|_0^t$$

$$= \begin{pmatrix} c_1\cosh t + c_2\sinh t \\ c_1\sinh t + c_2\cosh t \end{pmatrix} + \begin{pmatrix} \cosh t & \sinh t \\ \sinh t & \cosh t \end{pmatrix}\begin{pmatrix} t \\ 0 \end{pmatrix}$$

$$= \begin{pmatrix} c_1\cosh t + c_2\sinh t \\ c_1\sinh t + c_2\cosh t \end{pmatrix} + \begin{pmatrix} t\cosh t \\ t\sinh t \end{pmatrix} = c_1\begin{pmatrix} \cosh t \\ \sinh t \end{pmatrix} + c_2\begin{pmatrix} \sinh t \\ \cosh t \end{pmatrix} + t\begin{pmatrix} \cosh t \\ \sinh t \end{pmatrix}.$$

**9.** Solving

$$\begin{vmatrix} 2-\lambda & 1 \\ -3 & 6-\lambda \end{vmatrix} = \lambda^2 - 8\lambda + 15 = (\lambda - 3)(\lambda - 5) = 0$$

we find eigenvalues $\lambda_1 = 3$ and $\lambda_2 = 5$. Corresponding eigenvectors are

$$\mathbf{K}_1 = \begin{pmatrix} 1 \\ 1 \end{pmatrix} \quad \text{and} \quad \mathbf{K}_2 = \begin{pmatrix} 1 \\ 3 \end{pmatrix}.$$

Then

$$\mathbf{P} = \begin{pmatrix} 1 & 1 \\ 1 & 3 \end{pmatrix}, \quad \mathbf{P}^{-1} = \begin{pmatrix} 3/2 & -1/2 \\ -1/2 & 1/2 \end{pmatrix}, \quad \text{and} \quad \mathbf{D} = \begin{pmatrix} 3 & 0 \\ 0 & 5 \end{pmatrix},$$

so

$$\mathbf{PDP}^{-1} = \begin{pmatrix} 2 & 1 \\ -3 & 6 \end{pmatrix}.$$

**12.** From equation (2) in the text

**125**

$$e^{t\mathbf{D}} = \begin{pmatrix} 1 & 0 & \cdots & 0 \\ 0 & 1 & \cdots & 0 \\ \vdots & \vdots & \ddots & \vdots \\ 0 & 0 & \cdots & 1 \end{pmatrix} + \begin{pmatrix} \lambda_1 & 0 & \cdots & 0 \\ 0 & \lambda_2 & \cdots & 0 \\ \vdots & \vdots & \ddots & \vdots \\ 0 & 0 & \cdots & \lambda_n \end{pmatrix} + \frac{1}{2!}t^2 \begin{pmatrix} \lambda_1^2 & 0 & \cdots & 0 \\ 0 & \lambda_2^2 & \cdots & 0 \\ \vdots & \vdots & \ddots & \vdots \\ 0 & 0 & \cdots & \lambda_n^2 \end{pmatrix}$$

$$+ \frac{1}{3!}t^3 \begin{pmatrix} \lambda_1^3 & 0 & \cdots & 0 \\ 0 & \lambda_2^3 & \cdots & 0 \\ \vdots & \vdots & \ddots & \vdots \\ 0 & 0 & \cdots & \lambda_n^3 \end{pmatrix} + \cdots$$

$$= \begin{pmatrix} 1 + \lambda_1 t + \frac{1}{2!}(\lambda_1 t)^2 + \cdots & 0 & \cdots & 0 \\ 0 & 1 + \lambda_2 t + \frac{1}{2!}(\lambda_2 t)^2 + \cdots & \cdots & 0 \\ \vdots & \vdots & \ddots & \vdots \\ 0 & 0 & \cdots & 1 + \lambda_n t + \frac{1}{2!}(\lambda_n t)^2 + \cdots \end{pmatrix}$$

$$= \begin{pmatrix} e^{\lambda_1 t} & 0 & \cdots & 0 \\ 0 & e^{\lambda_2 t} & \cdots & 0 \\ \vdots & \vdots & \ddots & \vdots \\ 0 & 0 & \cdots & e^{\lambda_n t} \end{pmatrix}$$

# Chapter 8 Review Exercises

**3.** $\mathbf{A}^{-1} = -\frac{1}{2}\begin{pmatrix} 4 & -2 \\ -3 & 1 \end{pmatrix} = \begin{pmatrix} -2 & 1 \\ 3/2 & -1/2 \end{pmatrix}$

**6.** True, by Theorem 8.8.

**9.** True, by the definition of an eigenvector.

**12.** False;

$$\begin{pmatrix} 1 & 1 & 1 & | & 2 \\ 0 & 1 & 0 & | & 3 \\ 0 & 0 & 0 & | & 0 \end{pmatrix} \implies \begin{pmatrix} 1 & 0 & 1 & | & -1 \\ 0 & 1 & 0 & | & 3 \\ 0 & 0 & 0 & | & 0 \end{pmatrix}.$$

**15.** From $(D-2)x - y = -e^t$ and $-3x + (D-4)y = -7e^t$ we obtain $(D-1)(D-5)x = -4e^t$ so that

$$x = c_1 e^t + c_2 e^{5t} + te^t.$$

Then

$$y = (D-2)x + e^t = -c_1 e^t + 3c_2 e^{5t} - te^t + 2e^t.$$

**18.** Taking the Laplace transform of the system gives

$$s^2 \mathcal{L}\{x\} + s^2 \mathcal{L}\{y\} = \frac{1}{s-2}$$

$$2s \mathcal{L}\{x\} + s^2 \mathcal{L}\{y\} = -\frac{1}{s-2}$$

so that

$$\mathcal{L}\{x\} = \frac{2}{s(s-2)^2} = \frac{1}{2}\frac{1}{s} - \frac{1}{2}\frac{1}{s-2} + \frac{1}{(s-2)^2}$$

and

$$\mathcal{L}\{y\} = \frac{-s-2}{s^2(s-2)^2} = -\frac{3}{4}\frac{1}{s} - \frac{1}{2}\frac{1}{s^2} + \frac{3}{4}\frac{1}{s-2} - \frac{1}{(s-2)^2}.$$

Then

$$x = \frac{1}{2} - \frac{1}{2}e^{2t} + te^{2t} \quad \text{and} \quad y = -\frac{3}{4} - \frac{1}{2}t + \frac{3}{4}e^{2t} - te^{2t}.$$

**21.** Let $x_1 = x$, $x_2 = y$, $x_3 = Dx$, and $x_4 = Dy$ so that

$$Dx_1 = x_3$$

$$Dx_2 = x_4$$

$$Dx_3 = x_4 - 2x_3 - 2x_1 - \ln t + 10t - 4$$

$$Dx_4 = -x_3 - x_1 + 5t - 2.$$

**24.** We have $\det(\mathbf{A} - \lambda\mathbf{I}) = (\lambda+6)(\lambda+2) = 0$ so that

$$\mathbf{X} = c_1 \begin{pmatrix} 1 \\ -1 \end{pmatrix} e^{-6t} + c_2 \begin{pmatrix} 1 \\ 1 \end{pmatrix} e^{-2t}.$$

**27.** We have $\det(\mathbf{A} - \lambda\mathbf{I}) = \lambda^2(3-\lambda) = 0$ so that

$$\mathbf{X} = c_1 \begin{pmatrix} -1 \\ 1 \\ 0 \end{pmatrix} + c_2 \begin{pmatrix} -1 \\ 0 \\ 1 \end{pmatrix} + c_3 \begin{pmatrix} 1 \\ 1 \\ 1 \end{pmatrix} e^{3t}.$$

**30.** We have

$$\mathbf{X}_c = c_1 \begin{pmatrix} 2\cos t \\ -\sin t \end{pmatrix} e^t + c_2 \begin{pmatrix} 2\sin t \\ \cos t \end{pmatrix} e^t.$$

Then

$$\Phi = \begin{pmatrix} 2\cos t & 2\sin t \\ -\sin t & \cos t \end{pmatrix} e^t, \quad \Phi^{-1} = \begin{pmatrix} \frac{1}{2}\cos t & -\sin t \\ \frac{1}{2}\sin t & \cos t \end{pmatrix} e^{-t},$$

and

$$\mathbf{U} = \int \Phi^{-1}\mathbf{F}\, dt = \int \begin{pmatrix} \cos t - \sec t \\ \sin t \end{pmatrix} dt = \begin{pmatrix} \sin t - \ln|\sec t + \tan t| \\ -\cos t \end{pmatrix},$$

so that

$$\mathbf{X}_p = \Phi\mathbf{U} = \begin{pmatrix} -2\cos t \ln|\sec t + \tan t| \\ -1 + \sin t \ln|\sec t + \tan t| \end{pmatrix}.$$

# 9 Numerical Methods for Ordinary Differential Equations

─────── **Exercises 9.1** ───────

**3.**

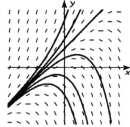

**6.** Setting $2x + y = c$ we obtain $y = -2x + c$; a family of lines with slope $-2$.

**9.** Setting $\sqrt{x^2 + y^2 + 2y + 1} = c$ we obtain $x^2 + (y + 1)^2 = c^2$; a family of circles centered at $(0, -1)$.

**12.** Setting $y + e^x = c$ we obtain $y = c - e^x$; a family of exponential curves.

**15.** Setting $x = c$ we see that the isoclines form a family of vertical lines.

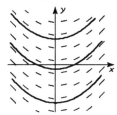

**18.** Setting $1/y = c$ we obtain the isoclines $y = 1/c$.

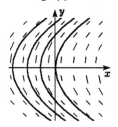

## Exercises 9.1

**21.** Setting $y - \cos\frac{\pi}{2}x = c$ we obtain the isoclines $y = \cos\frac{\pi}{2}x + c$.

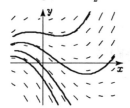

**24.** $y = cx$ is a solution of the differential equation if and only if

$$y' = c = \frac{\alpha x + \beta cx}{\gamma x + \delta cx}$$

if and only if

$$[\delta c^2 + (\gamma - \beta)c - \alpha]x = 0$$

if and only if

$$\delta c^2 + (\gamma - \beta)c - \alpha = 0$$

if and only if

$$c = \frac{\beta - \gamma \pm \sqrt{(\beta - \gamma)^2 + 4\alpha\delta}}{2\delta}$$

if and only if

$$(\beta - \gamma)^2 + 4\alpha\delta \geq 0.$$

**27.** The isoclines of $y' = 2x/y$ are $2x/y = c$ or

$$y = \frac{2}{c}x.$$

Setting $2/c = c$ we obtain $c = \pm\sqrt{2}$. Thus $y = \pm\sqrt{2}\,x$ are solutions of the differential equation .

**30.** The isoclines of $y' = (5x + 10y)/(-4x + 3y)$ are $(5x + 10y)/(-4x + 3y) = c$ or

$$y = \frac{4c + 5}{3c - 10}x.$$

Setting $(4c + 5)/(3c - 10) = c$ we obtain $c = 3c^2 - 14c - 5 = (3c + 1)(c - 5) = 0$. Thus $c = -1/3$ and $c = 5$ and $y = -\frac{1}{3}x$ and $y = 5x$ are solutions of the differential equation .

# Exercises 9.2

All tables in this chapter were constructed in a spreadsheet program which does not support subscripts. Consequently, $x_n$ and $y_n$ will be indicated as $x(n)$ and $y(n)$, respectively.

**3.**

| $h$ = 0.1 | |
|---|---|
| $x(n)$ | $y(n)$ |
| 1.00 | 5.0000 |
| 1.10 | 3.8000 |
| 1.20 | 2.9800 |
| 1.30 | 2.4260 |
| 1.40 | 2.0582 |
| 1.50 | 1.8207 |

| $h$ = 0.05 | |
|---|---|
| $x(n)$ | $y(n)$ |
| 1.00 | 5.0000 |
| 1.05 | 4.4000 |
| 1.10 | 3.8950 |
| 1.15 | 3.4708 |
| 1.20 | 3.1151 |
| 1.25 | 2.8179 |
| 1.30 | 2.5702 |
| 1.35 | 2.3647 |
| 1.40 | 2.1950 |
| 1.45 | 2.0557 |
| 1.50 | 1.9424 |

**6.**

| $h$ = 0.1 | |
|---|---|
| $x(n)$ | $y(n)$ |
| 0.00 | 1.0000 |
| 0.10 | 1.1000 |
| 0.20 | 1.2220 |
| 0.30 | 1.3753 |
| 0.40 | 1.5735 |
| 0.50 | 1.8371 |

| $h$ = 0.05 | |
|---|---|
| $x(n)$ | $y(n)$ |
| 0.00 | 1.0000 |
| 0.05 | 1.0500 |
| 0.10 | 1.1053 |
| 0.15 | 1.1668 |
| 0.20 | 1.2360 |
| 0.25 | 1.3144 |
| 0.30 | 1.4039 |
| 0.35 | 1.5070 |
| 0.40 | 1.6267 |
| 0.45 | 1.7670 |
| 0.50 | 1.9332 |

**9.**

| $h$ = 0.1 | |
|---|---|
| $x(n)$ | $y(n)$ |
| 0.00 | 0.5000 |
| 0.10 | 0.5250 |
| 0.20 | 0.5431 |
| 0.30 | 0.5548 |
| 0.40 | 0.5613 |
| 0.50 | 0.5639 |

| $h$ = 0.05 | |
|---|---|
| $x(n)$ | $y(n)$ |
| 0.00 | 0.5000 |
| 0.05 | 0.5125 |
| 0.10 | 0.5232 |
| 0.15 | 0.5322 |
| 0.20 | 0.5395 |
| 0.25 | 0.5452 |
| 0.30 | 0.5496 |
| 0.35 | 0.5527 |
| 0.40 | 0.5547 |
| 0.45 | 0.5559 |
| 0.50 | 0.5565 |

**12.**

| h = 0.1 | | | h = 0.05 | |
|---|---|---|---|---|
| x(n) | y(n) | | x(n) | y(n) |
| 1.00 | 0.5000 | | 1.00 | 0.5000 |
| 1.10 | 0.5250 | | 1.05 | 0.5125 |
| 1.20 | 0.5499 | | 1.10 | 0.5250 |
| 1.30 | 0.5747 | | 1.15 | 0.5375 |
| 1.40 | 0.5991 | | 1.20 | 0.5499 |
| 1.50 | 0.6231 | | 1.25 | 0.5623 |
| | | | 1.30 | 0.5746 |
| | | | 1.35 | 0.5868 |
| | | | 1.40 | 0.5989 |
| | | | 1.45 | 0.6109 |
| | | | 1.50 | 0.6228 |

**15.**

| h=0.1 | EULER | IMPROVED EULER |
|---|---|---|
| x(n) | y(n) | y(n) |
| 1.00 | 1.0000 | 1.0000 |
| 1.10 | 1.2000 | 1.2469 |
| 1.20 | 1.4938 | 1.6668 |
| 1.30 | 1.9711 | 2.6427 |
| 1.40 | 2.9060 | 8.7988 |

## —————— Exercises 9.3 ——————————————

**3.** We use

$$y'' = 2yy'$$

$$= 2y(1 + y^2)$$

$$= 2y + 2y^3$$

so that

$$y_{n+1} = y_n + (1 + y_n^2)h + (2y_n + 2y_n^3)\frac{1}{2}h^2.$$

| h = 0.1 | | | h = 0.05 | |
|---|---|---|---|---|
| x(n) | y(n) | | x(n) | y(n) |
| 0.00 | 0.0000 | | 0.00 | 0.0000 |
| 0.10 | 0.1000 | | 0.05 | 0.0500 |
| 0.20 | 0.2020 | | 0.10 | 0.1003 |
| 0.30 | 0.3082 | | 0.15 | 0.1510 |
| 0.40 | 0.4211 | | 0.20 | 0.2025 |
| 0.50 | 0.5438 | | 0.25 | 0.2551 |
| | | | 0.30 | 0.3090 |
| | | | 0.35 | 0.3647 |
| | | | 0.40 | 0.4223 |
| | | | 0.45 | 0.4825 |
| | | | 0.50 | 0.5456 |

**6.** We use

$$y'' = 1 + 2yy'$$

$$= 1 + 2y(x + y^2)$$

$$= 1 + 2xy + 2y^3$$

so that

$$y_{n+1} = y_n + (x_n + y_n^2)h + (1 + 2x_ny_n + 2y_n^3)\frac{1}{2}h^2.$$

| h = 0.1 | | h = 0.05 | |
|---|---|---|---|
| x(n) | y(n) | x(n) | y(n) |
| 0.00 | 0.0000 | 0.00 | 0.0000 |
| 0.10 | 0.0050 | 0.05 | 0.0013 |
| 0.20 | 0.0200 | 0.10 | 0.0050 |
| 0.30 | 0.0451 | 0.15 | 0.0113 |
| 0.40 | 0.0804 | 0.20 | 0.0200 |
| 0.50 | 0.1264 | 0.25 | 0.0313 |
| | | 0.30 | 0.0451 |
| | | 0.35 | 0.0615 |
| | | 0.40 | 0.0805 |
| | | 0.45 | 0.1021 |
| | | 0.50 | 0.1265 |

**9.** We use

$$y'' = 2xyy' + y^2 - \frac{xy' - y}{x^2}$$

$$= 2xy\left(xy^2 - \frac{y}{x}\right) + y^2$$

$$- \frac{1}{x}\left(xy^2 - \frac{y}{x}\right) + \frac{y}{x^2}$$

$$= 2x^2y^3 - 2y^2 + \frac{2y}{x^2}$$

so that

$$y_{n+1} = y_n + \left(x_ny_n^2 - \frac{y_n}{x_n}\right)h + \left(2x_n^2y_n^3 - 2y_n^2 + \frac{2y_n}{x_n^2}\right)\frac{1}{2}h^2.$$

| h = 0.1 | | h = 0.05 | |
|---|---|---|---|
| x(n) | y(n) | x(n) | y(n) |
| 1.00 | 1.0000 | 1.00 | 1.0000 |
| 1.10 | 1.0100 | 1.05 | 1.0025 |
| 1.20 | 1.0410 | 1.10 | 1.0101 |
| 1.30 | 1.0969 | 1.15 | 1.0229 |
| 1.40 | 1.1857 | 1.20 | 1.0415 |
| 1.50 | 1.3226 | 1.25 | 1.0663 |
| | | 1.30 | 1.0983 |
| | | 1.35 | 1.1387 |
| | | 1.40 | 1.1891 |
| | | 1.45 | 1.2518 |
| | | 1.50 | 1.3301 |

**12.** Let $f(x,y) = \alpha x + \beta y$ so that $f_x = \alpha, f_y = \beta$, and all higher derivatives are 0. Using the Taylor series expansion for $f(x,y)$ we have

$$f(x_{n+1}, y_{n+1}^*) = f(x_n + h, y_n + hf(x_n, y_n)$$

$$= f(x_n, y_n) + f_x(x_n, y_n)h + f_y(x_n, y_n)hf(x_n, y_n)$$

$$= f(x_n, y_n) + \alpha h + \beta hf(x_n, y_n).$$

Since $f(x_n, y_n) = y_n'$ and $\alpha + \beta y_n' = y_n''$ we have

$$y_{n+1} = y_n + \frac{1}{2}h[f(x_n, y_n) + f(x_{n+1}, y_{n+1}^*)]$$

$$= y_n + \frac{1}{2}h[f(x_n, y_n) + f(x_n, y_n) + \alpha h + \beta hf(x_n, y_n)]$$

$$= y_n + \frac{1}{2}h[2y_n' + h(\alpha + \beta y_n')]$$

$$= y_n + hy_n' + \frac{1}{2}h^2y_n''.$$

—————— **Exercises 9.4** —————————————————

**3.**

| x(n) | y(n) |
|------|--------|
| 0.00 | 0.0000 |
| 0.10 | 0.1003 |
| 0.20 | 0.2027 |
| 0.30 | 0.3093 |
| 0.40 | 0.4228 |
| 0.50 | 0.5463 |

**6.**

| x(n) | y(n) |
|------|--------|
| 0.00 | 0.0000 |
| 0.10 | 0.0050 |
| 0.20 | 0.0200 |
| 0.30 | 0.0451 |
| 0.40 | 0.0805 |
| 0.50 | 0.1266 |

**9.**

| x(n) | y(n) |
|------|--------|
| 1.00 | 1.0000 |
| 1.10 | 1.0101 |
| 1.20 | 1.0417 |
| 1.30 | 1.0989 |
| 1.40 | 1.1905 |
| 1.50 | 1.3333 |

**12.** Separating variables and using partial fractions we have

$$\frac{1}{2\sqrt{32}}\left(\frac{1}{\sqrt{32}-\sqrt{0.025}\,v}+\frac{1}{\sqrt{32}+\sqrt{0.025}\,v}\right)dv = dt$$

and

$$\frac{1}{2\sqrt{32}\sqrt{0.025}}\left(\ln\left|\sqrt{32}+\sqrt{0.025}\,v\right|-\ln\left|\sqrt{32}-\sqrt{0.025}\,v\right|\right) = t + c.$$

Since $v(0) = 0$ we find $c = 0$. Solving for $v$ we obtain

$$v(t) = \frac{16\sqrt{5}\left(e^{\sqrt{3.2}\,t}-1\right)}{e^{\sqrt{3.2}\,t}+1}$$

and $v(5) \approx 35.7678$.

**15.**

| x(n) | y(n) |
|------|----------|
| 1.00 | 1.0000 |
| 1.10 | 1.2511 |
| 1.20 | 1.6934 |
| 1.30 | 2.9425 |
| 1.40 | 903.0282 |

# Exercises 9.5

**3.**

| x(n) | y(n) | |
|------|--------|--------------------|
| 0.00 | 1.0000 | initial condition |
| 0.20 | 0.7328 | Runge-Kutta |
| 0.40 | 0.6461 | Runge-Kutta |
| 0.60 | 0.6585 | Runge-Kutta |
|      | *0.7332* | *predictor* |
| 0.80 | 0.7232 | corrector |

**6.**

| x(n) | y(n) | |
|------|--------|--------------------|
| 0.00 | 1.0000 | initial condition |
| 0.20 | 1.4414 | Runge-Kutta |
| 0.40 | 1.9719 | Runge-Kutta |
| 0.60 | 2.6028 | Runge-Kutta |
|      | *3.3483* | *predictor* |
| 0.80 | 3.3486 | corrector |
|      | *4.2276* | *predictor* |
| 1.00 | 4.2280 | corrector |

| x(n) | y(n) | |
|------|--------|--------------------|
| 0.00 | 1.0000 | initial condition |
| 0.10 | 1.2102 | Runge-Kutta |
| 0.20 | 1.4414 | Runge-Kutta |
| 0.30 | 1.6949 | Runge-Kutta |
|      | *1.9719* | *predictor* |
| 0.40 | 1.9719 | corrector |
|      | *2.2740* | *predictor* |
| 0.50 | 2.2740 | corrector |
|      | *2.6028* | *predictor* |
| 0.60 | 2.6028 | corrector |
|      | *2.9603* | *predictor* |
| 0.70 | 2.9603 | corrector |
|      | *3.3486* | *predictor* |
| 0.80 | 3.3486 | corrector |
|      | *3.7703* | *predictor* |
| 0.90 | 3.7703 | corrector |
|      | *4.2280* | *predictor* |
| 1.00 | 4.2280 | corrector |

**9.**

| x(n) | y(n) | |
|------|--------|--------------------|
| 0.00 | 1.0000 | initial condition |
| 0.10 | 1.0052 | Runge-Kutta |
| 0.20 | 1.0214 | Runge-Kutta |
| 0.30 | 1.0499 | Runge-Kutta |
|      | 1.0918 | predictor |
| 0.40 | 1.0918 | corrector |

**3.** Since the fourth-order Runge-Kutta formula agrees with the Taylor polynomial through $k = 4$, the local truncation error is

$$y^{(5)}(c)\frac{h^5}{5!} \quad \text{where} \quad x_n < c < x_{n+1}.$$

**6. (a)** Using the three-term Taylor method we obtain $y(0.1) \approx y_1 = 1.22$.

**(b)** Using $y''' = 8e^{2x}$ we see that the local truncation error is

$$y'''(c)\frac{h^3}{6} = 8e^{2c}\frac{(0.1)^3}{6} = 0.001333e^{2c}.$$

Since $e^{2x}$ is an increasing function, $e^{2c} \leq e^{2(0.1)} = e^{0.2}$ for $0 \leq c \leq 0.1$. Thus an upper bound for the local truncation error is $0.001333e^{0.2} = 0.001628$.

**(c)** Since $y(0.1) = e^{0.2} = 1.221403$, the actual error is $y(0.1) - y_1 = 0.001403$ which is less than $0.001628$.

**(d)** Using the three-term Taylor method with $h = 0.05$ we obtain $y(0.1) \approx y_2 = 1.221025$.

**(e)** The error in (d) is $1.221403 - 1.221025 = 0.000378$. With global truncation error $O(h^2)$, when the step size is halved we expect the error for $h = 0.05$ to be one-fourth the error for $h = 0.1$. Comparing $0.000378$ with $0.001403$ we see that this is the case.

**9. (a)** Using the improved Euler method we obtain $y(0.1) \approx y_1 = 0.825$.

**(b)** Using $y''' = -10e^{-2x}$ we see that the local truncation error is

$$10e^{-2c}\frac{(0.1)^3}{6} = 0.001667e^{-2c}.$$

Since $e^{-2x}$ is a decreasing function, $e^{-2c} \leq e^0 = 1$ for $0 \leq c \leq 0.1$. Thus an upper bound for the local truncation error is $0.001667(1) = 0.001667$.

**(c)** Since $y(0.1) = 0.823413$, the actual error is $y(0.1) - y_1 = 0.001587$, which is less than $0.001667$.

**(d)** Using the improved Euler method with $h = 0.05$ we obtain $y(0.1) \approx y_2 = 0.823781$.

**(e)** The error in (d) is $|0.823413 - 0.8237181| = 0.000305$. With global truncation error $O(h^2)$, when the step size is halved we expect the error for $h = 0.05$ to be one-fourth the error when $h = 0.1$. Comparing $0.000305$ with $0.001587$ we see that this is the case.

**12. (a)** Using $y'' = 38e^{-3(x-1)}$ we see that the local truncation error is

$$y''(c)\frac{h^2}{2} = 38e^{-3(c-1)}\frac{h^2}{2} = 19h^2e^{-3(c-1)}.$$

**(b)** Since $e^{-3(x-1)}$ is a decreasing function for $1 \leq x \leq 1.5$, $e^{-3(c-1)} \leq e^{-3(1-1)} = 1$ for $1 \leq c \leq 1.5$ and

$$y''(c)\frac{h^2}{2} \leq 19(0.1)^2(1) = 0.19.$$

**(c)** Using the Euler method with $h = 0.1$ we obtain $y(1.5) \approx 1.8207$. With $h = 0.05$ we obtain $y(1.5) \approx 1.9424$.

**(d)** Since $y(1.5) = 2.0532$, the error for $h = 0.1$ is $E_{0.1} = 0.2325$, while the error for $h = 0.05$ is $E_{0.05} = 0.1109$. With global truncation error $O(h)$ we expect $E_{0.1}/E_{0.05} \approx 2$. We actually have $E_{0.1}/E_{0.05} = 2.10$.

**15. (a)** Using $y^{(5)} = -1026e^{-3(x-1)}$ we see that the local truncation error is

$$\left| y^{(5)}(c)\frac{h^5}{120} \right| = 8.55h^5 e^{-3(c-1)}.$$

**(b)** Since $e^{-3(x-1)}$ is a decreasing function for $1 \leq x \leq 1.5$, $e^{-3(c-1)} \leq e^{-3(1-1)} = 1$ for $1 \leq c \leq 1.5$ and

$$y^{(5)}(c)\frac{h^5}{120} \leq 8.55(0.1)^5(1) = 0.0000855.$$

**(c)** Using the fourth-order Runge-Kutta method with $h = 0.1$ we obtain $y(1.5) \approx 2.053338827$. With $h = 0.05$ we obtain $y(1.5) \approx 2.053222989$.

**(d)** Since $y(1.5) = 2.053216232$, the error for $h = 0.1$ is $E_{0.1} = 0.000122595$, while the error for $h = 0.05$ is $E_{0.05} = 0.000006757$. With global truncation error $O(h^4)$ we expect $E_{0.1}/E_{0.05} \approx 16$. We actually have $E_{0.1}/E_{0.05} = 18.14$.

**18. (a)** Using $y''' = \dfrac{2}{(x+1)^3}$ we see that the local truncation error is

$$y'''(c)\frac{h^3}{6} = \frac{1}{(c+1)^3}\frac{h^3}{3}.$$

**(b)** Since $\dfrac{1}{(x+1)^3}$ is a decreasing function for $0 \leq x \leq 0.5$, $\dfrac{1}{(c+1)^3} \leq \dfrac{1}{(0+1)^3} = 1$ for $0 \leq c \leq 0.5$ and

$$y'''(c)\frac{h^3}{6} \leq (1)\frac{(0.1)^3}{3} = 0.000333.$$

**(c)** Using the three-term Taylor method with $h = 0.1$ we obtain $y(0.5) \approx 0.404643$. With $h = 0.05$ we obtain $y(0.5) \approx 0.405270$.

**(d)** Since $y(0.5) = 0.405465$, the error for $h = 0.1$ is $E_{0.1} = 0.000823$, while the error for $h = 0.05$ is $E_{0.05} = 0.000195$. With global truncation error $O(h^2)$ we expect $E_{0.1}/E_{0.05} \approx 4$. We actually have $E_{0.1}/E_{0.05} = 4.22$.

## Exercises 9.7

**3.** The substitution $y' = u$ leads to the system

$$y' = u, \qquad u' = 4u - 4y.$$

Using formulas (5) and (6) in the text with $x$ corresponding to $t$, $y$ corresponding to $x$, and $u$ corresponding to $y$, we obtain

Runge-Kutta method with h=0.2

| m1 | m2 | m3 | m4 | k1 | k2 | k3 | k4 | x | y | u |
|----|----|----|----|----|----|----|----|----|----|----|
| | | | | | | | | 0.00 | -2.0000 | 1.0000 |
| 0.2000 | 0.4400 | 0.5280 | 0.9072 | 2.4000 | 3.2800 | 3.5360 | 4.8064 | 0.20 | -1.4928 | 4.4731 |

Runge-Kutta method with h=0.1

| m1 | m2 | m3 | m4 | k1 | k2 | k3 | k4 | x | y | u |
|----|----|----|----|----|----|----|----|----|----|----|
| | | | | | | | | 0.00 | -2.0000 | 1.0000 |
| 0.1000 | 0.1600 | 0.1710 | 0.2452 | 1.2000 | 1.4200 | 1.4520 | 1.7124 | 0.10 | -1.8321 | 2.4427 |
| 0.2443 | 0.3298 | 0.3444 | 0.4487 | 1.7099 | 2.0031 | 2.0446 | 2.3900 | 0.20 | -1.4919 | 4.4753 |

**6.**

Runge-Kutta method with h=0.1

| m1 | m2 | m3 | m4 | k1 | k2 | k3 | k4 | t | i1 | i2 |
|----|----|----|----|----|----|----|----|----|----|----|
| | | | | | | | | 0.00 | 0.0000 | 0.0000 |
| 10.0000 | 0.0000 | 12.5000 | -20.0000 | 0.0000 | 5.0000 | -5.0000 | 22.5000 | 0.10 | 2.5000 | 3.7500 |
| 8.7500 | -2.5000 | 13.4375 | -28.7500 | -5.0000 | 4.3750 | -10.6250 | 29.6875 | 0.20 | 2.8125 | 5.7813 |
| 10.1563 | -4.3750 | 17.0703 | -40.0000 | -8.7500 | 5.0781 | -16.0156 | 40.3516 | 0.30 | 2.0703 | 7.4023 |
| 13.2617 | -6.3672 | 22.9443 | -55.1758 | -12.7344 | 6.6309 | -22.5488 | 55.3076 | 0.40 | 0.6104 | 9.1919 |
| 17.9712 | -8.8867 | 31.3507 | -75.9326 | -17.7734 | 8.9856 | -31.2024 | 75.9821 | 0.50 | -1.5619 | 11.4877 |

**9.**

Runge-Kutta method with h=0.2

| m1 | m2 | m3 | m4 | k1 | k2 | k3 | k4 | t | x | y |
|----|----|----|----|----|----|----|----|----|----|----|
| | | | | | | | | 0.00 | -3.0000 | 5.0000 |
| -1.0000 | -0.9200 | -0.9080 | -0.8176 | -0.6000 | -0.7200 | -0.7120 | -0.8216 | 0.20 | -3.9123 | 4.2857 |

Runge-Kutta method with h=0.1

| m1 | m2 | m3 | m4 | k1 | k2 | k3 | k4 | t | x | y |
|----|----|----|----|----|----|----|----|----|----|----|
| | | | | | | | | 0.00 | -3.0000 | 5.0000 |
| -0.5000 | -0.4800 | -0.4785 | -0.4571 | -0.3000 | -0.3300 | -0.3290 | -0.3579 | 0.10 | -3.4790 | 4.6707 |
| -0.4571 | -0.4342 | -0.4328 | -0.4086 | -0.3579 | -0.3858 | -0.3846 | -0.4112 | 0.20 | -3.9123 | 4.2857 |

## Exercises 9.8

3. We identify $P(x) = 2$, $Q(x) = 1$, $f(x) = 5x$, and $h = (1 - 0)/5 = 0.2$. Then the finite difference equation is

$$1.2y_{i+1} - 1.96y_i + 0.8y_{i-1} = 0.04(5x_i).$$

The solution of the corresponding linear system gives

| x | 0.0 | 0.2 | 0.4 | 0.6 | 0.8 | 1.0 |
|---|------|---------|---------|---------|---------|--------|
| y | 0.0000 | -0.2259 | -0.3356 | -0.3308 | -0.2167 | 0.0000 |

6. We identify $P(x) = 5$, $Q(x) = 0$, $f(x) = 4\sqrt{x}$, and $h = (2 - 1)/6 = 0.1667$. Then the finite difference equation is

$$1.4167y_{i+1} - 2y_i + 0.5833y_{i-1} = 0.2778(4\sqrt{x_i}).$$

The solution of the corresponding linear system gives

| x | 1.0000 | 1.1667 | 1.3333 | 1.5000 | 1.6667 | 1.8333 | 2.0000 |
|---|--------|---------|---------|---------|---------|---------|---------|
| y | 1.0000 | -0.5918 | -1.1626 | -1.3070 | -1.2704 | -1.1541 | -1.0000 |

9. We identify $P(x) = 1 - x$, $Q(x) = x$, $f(x) = x$, and $h = (1 - 0)/10 = 0.1$. Then the finite difference equation is

$$[1 + 0.05(1 - x_i)]y_{i+1} + [-2 + 0.01x_i]y_i + [1 - 0.05(1 - x_i)]y_{i-1} = 0.01x_i.$$

The solution of the corresponding linear system gives

| x | 0.0 | 0.1 | 0.2 | 0.3 | 0.4 | 0.5 | 0.6 |
|---|--------|--------|--------|--------|--------|--------|--------|
| y | 0.0000 | 0.2660 | 0.5097 | 0.7357 | 0.9471 | 1.1465 | 1.3353 |

| 0.7 | 0.8 | 0.9 | 1.0 |
|--------|--------|--------|--------|
| 1.5149 | 1.6855 | 1.8474 | 2.0000 |

12. We identify $P(r) = 2/r$, $Q(r) = 0$, $f(r) = 0$, and $h = (4 - 1)/6 = 0.5$. Then the finite difference equation is

$$\left(1 + \frac{0.5}{r_i}\right)u_{i+1} - 2u_i + \left(1 - \frac{0.5}{r_i}\right)u_{i-1} = 0.$$

The solution of the corresponding linear system gives

| r | 1.0 | 1.5 | 2.0 | 2.5 | 3.0 | 3.5 | 4.0 |
|---|---------|---------|---------|---------|---------|---------|----------|
| u | 50.0000 | 72.2222 | 83.3333 | 90.0000 | 94.4444 | 97.6190 | 100.0000 |

# Chapter 9 Review Exercises

**3.**

| h=0.1<br>x(n) | EULER | IMPROVED<br>EULER | 3-TERM<br>TAYLOR | RUNGE<br>KUTTA |
|---|---|---|---|---|
| 1.00 | 2.0000 | 2.0000 | 2.0000 | 2.0000 |
| 1.10 | 2.1386 | 2.1549 | 2.1556 | 2.1556 |
| 1.20 | 2.3097 | 2.3439 | 2.3446 | 2.3454 |
| 1.30 | 2.5136 | 2.5672 | 2.5680 | 2.5695 |
| 1.40 | 2.7504 | 2.8246 | 2.8255 | 2.8278 |
| 1.50 | 3.0201 | 3.1157 | 3.1167 | 3.1197 |

| h=0.05<br>x(n) | EULER | IMPROVED<br>EULER | 3-TERM<br>TAYLOR | RUNGE<br>KUTTA |
|---|---|---|---|---|
| 1.00 | 2.0000 | 2.0000 | 2.0000 | 2.0000 |
| 1.05 | 2.0693 | 2.0735 | 2.0735 | 2.0736 |
| 1.10 | 2.1469 | 2.1554 | 2.1555 | 2.1556 |
| 1.15 | 2.2328 | 2.2459 | 2.2460 | 2.2462 |
| 1.20 | 2.3272 | 2.3450 | 2.3451 | 2.3454 |
| 1.25 | 2.4299 | 2.4527 | 2.4528 | 2.4532 |
| 1.30 | 2.5409 | 2.5689 | 2.5690 | 2.5695 |
| 1.35 | 2.6604 | 2.6937 | 2.6938 | 2.6944 |
| 1.40 | 2.7883 | 2.8269 | 2.8271 | 2.8278 |
| 1.45 | 2.9245 | 2.9686 | 2.9688 | 2.9696 |
| 1.50 | 3.0690 | 3.1187 | 3.1188 | 3.1197 |

**6.**

| h=0.1<br>x(n) | EULER | IMPROVED<br>EULER | 3-TERM<br>TAYLOR | RUNGE<br>KUTTA |
|---|---|---|---|---|
| 1.00 | 1.0000 | 1.0000 | 1.0000 | 1.0000 |
| 1.10 | 1.2000 | 1.2380 | 1.2350 | 1.2415 |
| 1.20 | 1.4760 | 1.5910 | 1.5866 | 1.6036 |
| 1.30 | 1.8710 | 2.1524 | 2.1453 | 2.1909 |
| 1.40 | 2.4643 | 3.1458 | 3.1329 | 3.2745 |
| 1.50 | 3.4165 | 5.2510 | 5.2208 | 5.8338 |

| h=0.05<br>x(n) | EULER | IMPROVED<br>EULER | 3-TERM<br>TAYLOR | RUNGE<br>KUTTA |
|---|---|---|---|---|
| 1.00 | 1.0000 | 1.0000 | 1.0000 | 1.0000 |
| 1.05 | 1.1000 | 1.1091 | 1.1088 | 1.1095 |
| 1.10 | 1.2183 | 1.2405 | 1.2401 | 1.2415 |
| 1.15 | 1.3595 | 1.4010 | 1.4004 | 1.4029 |
| 1.20 | 1.5300 | 1.6001 | 1.5994 | 1.6036 |
| 1.25 | 1.7389 | 1.8523 | 1.8515 | 1.8586 |
| 1.30 | 1.9988 | 2.1799 | 2.1789 | 2.1911 |
| 1.35 | 2.3284 | 2.6197 | 2.6182 | 2.6401 |
| 1.40 | 2.7567 | 3.2360 | 3.2340 | 3.2755 |
| 1.45 | 3.3296 | 4.1528 | 4.1497 | 4.2363 |
| 1.50 | 4.1253 | 5.6404 | 5.6350 | 5.8446 |

**9.** Using $x_0 = 1$, $y_0 = 2$, and $h = 0.1$ we have

$$x_1 = x_0 + h(x_0 + y_0) = 1 + 0.1(1 + 2) = 1.3$$
$$y_1 = y_0 + h(x_0 - y_0) = 2 + 0.1(1 - 2) = 1.9$$

and

$$x_2 = x_1 + h(x_1 + y_1) = 1.3 + 0.1(1.3 + 1.9) = 1.62$$
$$y_2 = y_1 + h(x_1 - y_1) = 1.9 + 0.1(1.3 - 1.9) = 1.84.$$

Thus, $x(0.2) \approx 1.62$ and $y(0.2) \approx 1.84$.

# 10 Plane Autonomous Systems and Stability

———————— **Exercises 10.1** ————————————————————

**3.** The corresponding plane autonomous system is

$$x' = y, \quad y' = x^2 - y(1 - x^3).$$

If $(x, y)$ is a critical point, $y = 0$ and so $x^2 - y(1 - x^3) = x^2 = 0$. Therefore $(0, 0)$ is the sole critical point.

**6.** The corresponding plane autonomous system is

$$x' = y, \quad y' = -x + \epsilon x|x|.$$

If $(x, y)$ is a critical point, $y = 0$ and $-x + \epsilon x|x| = x(-1 + \epsilon|x|) = 0$. Hence $x = 0$, $1/\epsilon$, $-1/\epsilon$. The critical points are $(0, 0)$, $(1/\epsilon, 0)$ and $(-1/\epsilon, 0)$.

**9.** From $x - y = 0$ we have $y = x$. Substituting into $3x^2 - 4y = 0$ we obtain $3x^2 - 4x = x(3x - 4) = 0$. It follows that $(0, 0)$ and $(4/3, 4/3)$ are the critical points of the system.

**12.** Adding the two equations we obtain $10 - 15 \dfrac{y}{y + 5} = 0$. It follows that $y = 10$, and from $-2x + y + 10 = 0$ we may conclude that $x = 10$. Therefore $(10, 10)$ is the sole critical point of the system.

**15.** From $x(1 - x^2 - 3y^2) = 0$ we have $x = 0$ or $x^2 + 3y^2 = 1$. If $x = 0$, then substituting into $y(3 - x^2 - 3y^2)$ gives $y(3 - 3y^2) = 0$. Therefore $y = 0$, $1$, $-1$. Likewise $x^2 = 1 - 3y^2$ yields $2y = 0$ so that $y = 0$ and $x^2 = 1 - 3(0)^2 = 1$. The critical points of the system are therefore $(0, 0)$, $(0, 1)$, $(0, -1)$, $(1, 0)$, and $(-1, 0)$.

**18. (a)** From Exercises 8.6, Problem 6, $x = c_1 + 2c_2 e^{-5t}$ and $y = 3c_1 + c_2 e^{-5t}$.

**(b)** From $\mathbf{X}(0) = (3, 4)$ it follows that $c_1 = c_2 = 1$. Therefore $x = 1 + 2e^{-5t}$ and $y = 3 + e^{-5t}$.

**(c)**

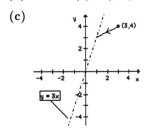

**21. (a)** From Exercises 8.6, Problem 17, $x = c_1(\sin t - \cos t)e^{4t} + c_2(-\sin t - \cos t)e^{4t}$ and
$y = 2c_1(\cos t)\,e^{4t} + 2c_2(\sin t)\,e^{4t}$. Because of the presence of $e^{4t}$, there are no periodic solutions.

**(b)** From $\mathbf{X}(0) = (-1, 2)$ it follows that $c_1 = 1$ and $c_2 = 0$. Therefore $x = (\sin t - \cos t)e^{4t}$ and
$y = 2(\cos t)\,e^{4t}$.

**(c)**

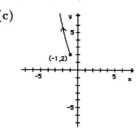

d the origin as $t$ increases.

**24.** Switching to polar coordinates,

$$\frac{dr}{dt} = \frac{1}{r}\left(x\frac{dx}{dt} + y\frac{dy}{dt}\right) = \frac{1}{r}(xy - x^2r^2 - xy + y^2r^2) = r^3$$

$$\frac{d\theta}{dt} = \frac{1}{r^2}\left(-y\frac{dx}{dt} + x\frac{dy}{dt}\right) = \frac{1}{r^2}(-y^2 - xyr^2 - x^2 + xyr^2) = -1.$$

If we use separation of variables, it follows that

$$r = \frac{1}{\sqrt{-2t + c_1}} \quad \text{and} \quad \theta = -t + c_2.$$

Since $\mathbf{X}(0) = (4, 0)$, $r = 4$ and $\theta = 0$ when $t = 0$. It follows that $c_2 = 0$ and $c_1 = 1/16$. The final
solution may be written as

$$r = \frac{4}{\sqrt{1 - 32t}}, \quad \theta = -t.$$

Note that $r \to \infty$ as $t \to (1/32)^-$. Because $0 \le t \le 1/32$, the curve is not a spiral.

**27.** If $\mathbf{X}(t) = (x(t), y(t))$ is a solution,

$$\frac{d}{dt} f(x(t), y(t)) = \frac{\partial f}{\partial x}\frac{dx}{dt} + \frac{\partial f}{\partial y}\frac{dy}{dt} = QP - PQ = 0,$$

using the chain rule. Therefore $f(x(t), y(t)) = c$ for some constant $c$, and the solution lies on a level
curve of $f$.

━━━━━━━━ **Exercises 10.2** ━━━━━━━━━━━

**3. (a)** All solutions are unstable spirals which become unbounded as $t$ increases.

**(b)**

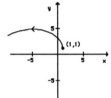

**6. (a)** All solutions become unbounded and $y = x/2$ serves as the asymptote.

**(b)**

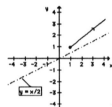

**9.** Since $\Delta = -41 < 0$, we may conclude from Figure 10.15 that $(0,0)$ is a saddle point.

**12.** Since $\Delta = 1$ and $\tau = -1$, $\tau^2 - 4\Delta = -3$ and so from Figure 10.15, $(0,0)$ is a stable spiral point.

**15.** Since $\Delta = 0.01$ and $\tau = -0.03$, $\tau^2 - 4\Delta < 0$ and so from Figure 10.15, $(0,0)$ is a stable spiral point.

**18.** Note that $\Delta = 1$ and $\tau = \mu$. Therefore we need both $\tau = \mu < 0$ and $\tau^2 - 4\Delta = \mu^2 - 4 < 0$ for $(0,0)$ to be a stable spiral point. These two conditions may be written as $-2 < \mu < 0$.

**21.** $\mathbf{AX_1 + F = 0}$ implies that $\mathbf{AX_1 = -F}$ or $\mathbf{X_1 = -A^{-1}F}$. Since $\mathbf{X_p}(t) = -\mathbf{A^{-1}F}$ is a particular solution, it follows from Theorem 8.7 that $\mathbf{X}(t) = \mathbf{X_c}(t) + \mathbf{X_1}$ is the general solution to $\mathbf{X' = AX + F}$. If $\tau < 0$ and $\Delta > 0$ then $\mathbf{X_c}(t)$ approaches $(0,0)$ by Theorem 10.1(i). It follows that $\mathbf{X}(t)$ approaches $\mathbf{X_1}$ as $t \to \infty$.

━━━━━━━━ **Exercises 10.3** ━━━━━━━━━━━

**3.** The critical points are $x = 0$ and $x = n + 1$. Since $g'(x) = k(n+1) - 2kx$, $g'(0) = k(n+1) > 0$ and $g'(n+1) = -k(n+1) < 0$. Therefore $x = 0$ is unstable while $x = n+1$ is asymptotically stable. See Theorem 10.2.

**6.** The only critical point is $v = mg/k$. Now $g(v) = g - (k/m)v$ and so $g'(v) = -k/m < 0$. Therefore $v = mg/k$ is an asymptotically stable critical point by Theorem 10.2.

**144**

**9.** Critical points occur at $P = a/b$, $c$ but not at $P = 0$. Since $g'(P) = (a - bP) + (P - c)(-b)$,

$$g'(a/b) = (a/b - c)(-b) = -a + bc \quad \text{and} \quad g'(c) = a - bc.$$

Since $a < bc$, $-a + bc > 0$ and $a - bc < 0$. Therefore $P = a/b$ is unstable while $P = c$ is asymptotically stable.

**12.** Critical points are $(1, 0)$ and $(-1, 0)$, and

$$g'(\mathbf{X}) = \begin{pmatrix} 2x & -2y \\ 0 & 2 \end{pmatrix}.$$

At $\mathbf{X} = (1, 0)$, $\tau = 4$, $\Delta = 4$, and so $\tau^2 - 4\Delta = 0$. We may conclude that $(1, 0)$ is unstable but we are unable to classify this critical point any further. At $\mathbf{X} = (-1, 0)$, $\Delta = -4 < 0$ and so $(-1, 0)$ is a saddle point.

**15.** Since $x^2 - y^2 = 0$, $y^2 = x^2$ and so $x^2 - 3x + 2 = (x - 1)(x - 2) = 0$. It follows that the critical points are $(1, 1)$, $(1, -1)$, $(2, 2)$, and $(2, -2)$. We next use the Jacobian

$$g'(\mathbf{X}) = \begin{pmatrix} -3 & 2y \\ 2x & -2y \end{pmatrix}$$

to classify these four critical points. For $\mathbf{X} = (1, 1)$, $\tau = -5$, $\Delta = 2$, and so $\tau^2 - 4\Delta = 17 > 0$. Therefore $(1, 1)$ is a stable node. For $\mathbf{X} = (1, -1)$, $\Delta = -2 < 0$ and so $(1, -1)$ is a saddle point. For $\mathbf{X} = (2, 2)$, $\Delta = -4 < 0$ and so we have another saddle point. Finally, if $\mathbf{X} = (2, -2)$, $\tau = 1$, $\Delta = 4$, and so $\tau^2 - 4\Delta = -15 < 0$. Therefore $(2, -2)$ is an unstable spiral point.

**18.** We found that $(0, 0)$, $(0, 1)$, $(0, -1)$, $(1, 0)$ and $(-1, 0)$ were the critical points in Exercise 15, Section 10.1. The Jacobian is

$$g'(\mathbf{X}) = \begin{pmatrix} 1 - 3x^2 - 3y^2 & -6xy \\ -2xy & 3 - x^2 - 9y^2 \end{pmatrix}.$$

For $\mathbf{X} = (0, 0)$, $\tau = 4$, $\Delta = 3$ and so $\tau^2 - 4\Delta = 4 > 0$. Therefore $(0, 0)$ is an unstable node. Both $(0, 1)$ and $(0, -1)$ give $\tau = -8$, $\Delta = 12$, and $\tau^2 - 4\Delta = 16 > 0$. These two critical points are therefore stable nodes. For $\mathbf{X} = (1, 0)$ or $(-1, 0)$, $\Delta = -4 < 0$ and so saddle points occur.

**21.** The corresponding plane autonomous system is

$$\theta' = y, \quad y' = (\cos \theta - \frac{1}{2}) \sin \theta.$$

Since $|\theta| < \pi$, it follows that critical points are $(0, 0)$, $(\pi/3, 0)$ and $(-\pi/3, 0)$. The Jacobian matrix is

$$g'(\mathbf{X}) = \begin{pmatrix} 0 & 1 \\ \cos 2\theta - \frac{1}{2} \cos \theta & 0 \end{pmatrix}$$

and so at $(0, 0)$, $\tau = 0$ and $\Delta = -1/2$. Therefore $(0, 0)$ is a saddle point. For $\mathbf{X} = (\pm \pi/3, 0)$, $\tau = 0$ and $\Delta = 3/4$. It is not possible to classify either critical point in this borderline case.

## Exercises 10.3

**24.** The corresponding plane autonomous system is

$$x' = y, \quad y' = -\frac{4x}{1+x^2} - 2y$$

and the only critical point is $(0,0)$. Since the Jacobian matrix is

$$\mathbf{g}'(\mathbf{X}) = \begin{pmatrix} 0 & 1 \\ -4\frac{1-x^2}{(1+x^2)^2} & -2 \end{pmatrix},$$

$\tau = -2$, $\Delta = 4$, $\tau^2 - 4\Delta = -12$, and so $(0,0)$ is a stable spiral point.

**27.** The corresponding plane autonomous system is

$$x' = y, \quad y' = -\frac{(\beta + \alpha^2 y^2)x}{1 + \alpha^2 x^2}$$

and the Jacobian matrix is

$$\mathbf{g}'(\mathbf{X}) = \begin{pmatrix} 0 & 1 \\ \frac{(\beta+\alpha y^2)(\alpha^2 x^2 - 1)}{(1+\alpha^2 x^2)^2} & \frac{-2\alpha^2 yx}{1+\alpha^2 x^2} \end{pmatrix}.$$

For $\mathbf{X} = (0,0)$, $\tau = 0$ and $\Delta = \beta$. Since $\beta < 0$, we may conclude that $(0,0)$ is a saddle point.

**30. (a)** The corresponding plane autonomous system is

$$x' = y, \quad y' = \epsilon\left(y - \frac{1}{3}y^3\right) - x$$

and so the only critical point is $(0,0)$. Since the Jacobian matrix is

$$\mathbf{g}'(\mathbf{X}) = \begin{pmatrix} 0 & 1 \\ -1 & \epsilon(1 - y^2) \end{pmatrix},$$

$\tau = \epsilon$, $\Delta = 1$, and so $\tau^2 - 4\Delta = \epsilon^2 - 4$ at the critical point $(0,0)$.

**(b)** When $\tau = \epsilon > 0$, $(0,0)$ is an unstable critical point.

**(c)** When $\epsilon < 0$ and $\tau^2 - 4\Delta = \epsilon^2 - 4 < 0$, $(0,0)$ is a stable spiral point. These two requirements can be written as $-2 < \epsilon < 0$.

**(d)** When $\epsilon = 0$, $x'' + x = 0$ and so $x = c_1 \cos t + c_2 \sin t$. Therefore all solutions are periodic (with period $2\pi$) and so $(0,0)$ is a center.

**33. (a)** $x' = 2xy = 0$ implies that either $x = 0$ or $y = 0$. If $x = 0$, then from $1 - x^2 + y^2 = 0$, $y^2 = -1$ and there are no real solutions. If $y = 0$, $1 - x^2 = 0$ and so $(1,0)$ and $(-1,0)$ are critical points. The Jacobian matrix is

$$\mathbf{g}'(\mathbf{X}) = \begin{pmatrix} 2y & 2x \\ -2x & 2y \end{pmatrix}$$

and so $\tau = 0$ and $\Delta = 4$ at either $\mathbf{X} = (1,0)$ or $(-1,0)$. We obtain no information about these critical points in this borderline case.

**(b)** $\dfrac{dy}{dx} = \dfrac{y'}{x'} = \dfrac{1 - x^2 + y^2}{2xy}$ or $2xy\dfrac{dy}{dx} = 1 - x^2 + y^2$. Letting $\mu = \dfrac{y^2}{x}$,

it follows that $\dfrac{d\mu}{dx} = \dfrac{1}{x^2} - 1$ and so $\mu = -\dfrac{1}{x} - x + 2c$. Therefore

$\dfrac{y^2}{x} = -\dfrac{1}{x} - x + 2c$ which can be put in the form

$$(x - c)^2 + y^2 = c^2 - 1.$$

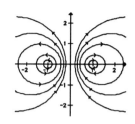

The solution curves are shown and so both $(1,0)$ and $(-1,0)$ are centers.

**36.** The corresponding plane autonomous system is

$$x' = y, \quad y' = \epsilon x^2 - x + 1$$

and so the critical points must satisfy $y = 0$ and

$$x = \frac{1 \pm \sqrt{1 - 4\epsilon}}{2\epsilon}.$$

Therefore we must require that $\epsilon \leq \frac{1}{4}$ for real solutions to exist. We will use the Jacobian matrix

$$g'(\mathbf{X}) = \begin{pmatrix} 0 & 1 \\ 2\epsilon x - 1 & 0 \end{pmatrix}$$

to attempt to classify $((1 \pm \sqrt{1 - 4\epsilon})/2\epsilon, 0)$ when $\epsilon \leq 1/4$. Note that $\tau = 0$ and $\Delta = \mp\sqrt{1 - 4\epsilon}$. For $\mathbf{X} = ((1 + \sqrt{1 - 4\epsilon})/2\epsilon, 0)$ and $\epsilon < 1/4$, $\Delta < 0$ and so a saddle point occurs. For $\mathbf{X} = ((1 - \sqrt{1 - 4\epsilon})/2\epsilon, 0)$ $\Delta \geq 0$ and we are not able to classify this critical point using linearization.

_____ **Exercises 10.4** _____

**3.** The corresponding plane autonomous system is

$$x' = y, \quad y' = -g\frac{f'(x)}{1 + [f'(x)]^2} - \frac{\beta}{m}y$$

and

$$\frac{\partial}{\partial x}\left(-g\frac{f'(x)}{1 + [f'(x)]^2} - \frac{\beta}{m}y\right) = -g\frac{(1 + [f'(x)]^2)f''(x) - f'(x)2f'(x)f''(x)}{(1 + [f'(x)]^2)^2}.$$

If $\mathbf{X}_1 = (x_1, y_1)$ is a critical point, $y_1 = 0$ and $f'(x_1) = 0$. The Jacobian at this critical point is therefore

$$g'(\mathbf{X}_1) = \begin{pmatrix} 0 & 1 \\ -gf''(x_1) & -\frac{\beta}{m} \end{pmatrix}.$$

**6. (a)** If $f(x) = \cosh x$, $f'(x) = \sinh x$ and $[f'(x)]^2 + 1 = \sinh^2 x + 1 = \cosh^2 x$. Therefore

$$\frac{dy}{dx} = \frac{y'}{x'} = -g\frac{\sinh x}{\cosh^2 x}\frac{1}{y}.$$

We may separate variables to show that $y^2 = \dfrac{2g}{\cosh x} + c$. But $x(0) = x_0$ and $y(0) = x'(0) = v_0$.

Therefore $c = v_0^2 - \dfrac{2g}{\cosh x_0}$ and so

$$y^2 = \frac{2g}{\cosh x} - \frac{2g}{\cosh x_0} + v_0^2.$$

Now

$$\frac{2g}{\cosh x} - \frac{2g}{\cosh x_0} + v_0^2 \geq 0 \quad \text{if and only if} \quad \cosh x \leq \frac{2g\cosh x_0}{2g - v_0^2\cosh x_0}$$

and the solution to this inequality is an interval $[-a, a]$. Therefore each $x$ in $(-a, a)$ has two corresponding values of $y$ and so the solution is periodic.

**(b)** Since $z = \cosh x$, the maximum height occurs at the largest value of $x$ on the cycle. From **(a)**, $x_{\max} = a$ where $\cosh a = \dfrac{2g\cosh x_0}{2g - v_0^2\cosh x_0}$. Therefore

$$z_{\max} = \frac{2g\cosh x_0}{2g - v_0^2\cosh x_0}.$$

**9. (a)** In the Lotka-Volterra Model the average number of predators is $d/c$ and the average number of prey is $a/b$. But

$$x' = -ax + bxy - \epsilon_1 x = -(a + \epsilon_1)x + bxy$$

$$y' = -cxy + dy - \epsilon_2 y = -cxy + (d - \epsilon_2)y$$

and so the new critical point in the first quadrant is $(d/c - \epsilon_2/c, a/b + \epsilon_1/b)$.

**(b)** The average number of predators $d/c - \epsilon_2/c$ has decreased while the average number of prey $a/b + \epsilon_1/b$ has increased. The fishery science model is consistent with Volterra's principle.

**12.** $(\hat{x}, \hat{y})$ is a stable node if and only if $\dfrac{K_1}{\alpha_{12}} > K_2$ and $\dfrac{K_2}{\alpha_{21}} > K_1$. [See Figure 10.31(a) in the text.]

From Problem 10, $(0,0)$ is an unstable node and from Problem 11, since $K_1 < \dfrac{K_2}{\alpha_{21}}$, $(K_1, 0)$ is a saddle point. Finally, when $K_2 < \dfrac{K_1}{\alpha_{12}}$, $(0, K_2)$ is a saddle point. This is Problem 10 with the roles of 1 and 2 interchanged. Therefore $(0,0)$, $(K_1,0)$, and $(0, K_2)$ are unstable.

**15. (a)** The corresponding plane autonomous system is

$$x = y, \quad y' = -\frac{\beta}{m}y|y| - \frac{k}{m}x$$

and so a critical point must satisfy both $y = 0$ and $x = 0$. Therefore $(0,0)$ is the unique critical point.

**(b)** The Jacobian matrix is

$$\begin{pmatrix} 0 & 1 \\ -\frac{k}{m} & -\frac{\beta}{m}2|y| \end{pmatrix}$$

and so $\tau = 0$ and $\Delta = \dfrac{k}{m} > 0$. Therefore $(0,0)$ is a center, stable spiral point, or an unstable spiral point. Physical considerations suggest that $(0,0)$ must be asymptotically stable and so $(0,0)$ must be a stable spiral point.

**18. (a)** $x' = x(-a + by) = 0$ implies that $x = 0$ or $y = a/b$. If $x = 0$, then, from

$$-cxy + \frac{r}{K}y(K - y) = 0,$$

$y = 0$ or $K$. Therefore $(0,0)$ and $(0,K)$ are critical points. If $\hat{y} = a/b$, then

$$\hat{y}\left[-cx + \frac{r}{K}(K - \hat{y})\right] = 0.$$

The corresponding value of $x$, $x = \hat{x}$, therefore satisfies the equation $c\hat{x} = \dfrac{r}{K}(K - \hat{y})$.

**(b)** The Jacobian matrix is

$$\mathbf{g'(X)} = \begin{pmatrix} -a + by & bx \\ -cy & -cx + \frac{r}{K}(K - 2y) \end{pmatrix}$$

and so at $\mathbf{X_1} = (0,0)$, $\Delta = -ar < 0$. For $\mathbf{X_1} = (0,K)$, $\Delta = n(Kb - a) = -rb\left(K - \dfrac{a}{b}\right)$. Since we are given that $K > \dfrac{a}{b}$, $\Delta < 0$ in this case. Therefore $(0,0)$ and $(0,K)$ are each saddle points. For $\mathbf{X_1} = (\hat{x}, \hat{y})$ where $\hat{y} = \dfrac{a}{b}$ and $c\hat{x} = \dfrac{r}{K}(K - \hat{y})$, we may write the Jacobian matrix as

$$\mathbf{g'}((\hat{x}, \hat{y})) = \begin{pmatrix} 0 & b\hat{x} \\ -c\hat{y} & -\frac{r}{K}\hat{y} \end{pmatrix}$$

and so $\tau = -\dfrac{r}{K}\hat{y} < 0$ and $\Delta = bc\hat{x}\hat{y} > 0$. Therefore $(\hat{x}, \hat{y})$ is a stable critical point and so it is either a stable node (perhaps degenerate) or a stable spiral point.

**(c)** Write

$$\tau^2 - 4\Delta = \frac{r^2}{K^2}\hat{y}^2 - 4bc\hat{x}\hat{y} = \hat{y}\left[\frac{r^2}{K^2}\hat{y} - 4bc\hat{x}\right] = \hat{y}\left[\frac{r^2}{K^2}\hat{y} - 4b\frac{r}{K}(K - \hat{y})\right]$$

using

$$c\hat{x} = \frac{r}{K}(K - \hat{y}) = \frac{r}{K}\hat{y}\left[\left(\frac{r}{K} + 4b\right)\hat{y} - 4bK\right].$$

Therefore $\tau^2 - 4\Delta < 0$ if and only if

$$\hat{y} < \frac{4bK}{\frac{r}{K} + 4b} = \frac{4bK^2}{r + 4bK}.$$

Note that

$$\frac{4bK^2}{r + 4bK} = \frac{4bK}{r + 4bK} \cdot K \approx K$$

where $K$ is large, and $\hat{y} = \dfrac{a}{b} < K$. Therefore $\tau^2 - 4\Delta < 0$ when $K$ is large and a stable spiral point will result.

# Chapter 10 Review Exercises

**3.** a center or a saddle point

**6.** True

**9.** Switching to polar coordinates,

$$\frac{dr}{dt} = \frac{1}{r}\left(x\frac{dx}{dt} + y\frac{dy}{dt}\right) = \frac{1}{r}(-xy - x^2r^3 + xy - y^2r^3) = -r^4$$

$$\frac{d\theta}{dt} = \frac{1}{r^2}\left(-y\frac{dx}{dt} + x\frac{dy}{dt}\right) = \frac{1}{r^2}(y^2 + xyr^3 + x^2 - xyr^3) = 1.$$

Using separation of variables it follows that $r = \dfrac{1}{\sqrt[3]{3t + c_1}}$ and $\theta = t + c_2$. Since $\mathbf{X}(0) = (1, 0)$, $r = 1$ and $\theta = 0$. It follows that $c_1 = 1$, $c_2 = 0$, and so

$$r = \frac{1}{\sqrt[3]{3t + 1}}, \quad \theta = t.$$

As $t \to \infty$, $r \to 0$ and the solution spirals toward the origin.

**12.** From $x' = x(1 + y - 3x) = 0$, either $x = 0$ or $1 + y - 3x = 0$. If $x = 0$, then, from $y(4 - 2x - y) = 0$ we obtain $y(4 - y) = 0$. It follows that $(0, 0)$ and $(0, 4)$ are critical points. If $1 + y - 3x = 0$, then $y(5 - 5x) = 0$. Therefore $(1/3, 0)$ and $(1, 2)$ are the remaining critical points. We will use the Jacobian matrix

$$g'(\mathbf{X}) = \begin{pmatrix} 1 + y - 6x & x \\ -2y & 4 - 2x - 2y \end{pmatrix}$$

to classify these four critical points. The results are as follows:

| X | $\tau$ | $\Delta$ | $\tau^2 - 4\Delta$ | Conclusion |
|---|---|---|---|---|
| $(0,0)$ | 5 | 4 | 9 | unstable node |
| $(0,4)$ | – | $-20$ | – | saddle point |
| $(\frac{1}{3},0)$ | – | $-\frac{10}{3}$ | – | saddle point |
| $(1,2)$ | $-5$ | 10 | $-15$ | stable spiral point |

**15.** $\dfrac{dy}{dx} = \dfrac{y'}{x'} = \dfrac{-2x\sqrt{y^2+1}}{y}$. We may separate variables to show that $\sqrt{y^2+1} = -x^2 + c$. But

$x(0) = x_0$ and $y(0) = x'(0) = 0$. It follows that $c = 1 + x_0^2$ so that

$$y^2 = (1 + x_0^2 - x^2)^2 - 1.$$

Note that $1 + x_0^2 - x^2 > 1$ for $-x_0 < x < x_0$ and $y = 0$ for $x = \pm x_0$. Each $x$ with $-x_0 < x < x_0$ has two corresponding values of $y$ and so the solution $\mathbf{X}(t)$ with $\mathbf{X}(0) = (x_0, 0)$ is periodic.

# 11 Orthogonal Functions and Fourier Series

————— **Exercises 11.1** —————

**3.** $\displaystyle\int_0^2 e^x(xe^{-x} - e^{-x})\,dx = \int_0^2 (x-1)\,dx = \left(\frac{1}{2}x^2 - x\right)\Big|_0^2 = 0$

**6.** $\displaystyle\int_{\pi/4}^{5\pi/4} e^x \sin x\,dx = \left(\frac{1}{2}e^x \sin x - \frac{1}{2}e^x \cos x\right)\Big|_{\pi/4}^{5\pi/4} = 0$

**9.** For $m \neq n$

$$\int_0^\pi \sin nx \sin mx\,dx = \frac{1}{2}\int_0^\pi [\cos(n-m)x - \cos(n+m)x]\,dx$$

$$= \frac{1}{2(n-m)}\sin(n-m)x\Big|_0^\pi - \frac{1}{2(n+m)}\sin 2(n+m)x\Big|_0^\pi$$

$$= 0.$$

For $m = n$

$$\int_0^\pi \sin^2 nx\,dx = \int_0^\pi \left[\frac{1}{2} - \frac{1}{2}\cos 2nx\right]\,dx = \frac{1}{2}x\Big|_0^\pi - \frac{1}{4n}\sin 2nx\Big|_0^\pi = \frac{\pi}{2}$$

so that

$$\|\sin nx\| = \sqrt{\frac{\pi}{2}}.$$

**12.** For $m \neq n$, we use Problems 11 and 10:

$$\int_{-p}^p \cos\frac{n\pi}{p}x \cos\frac{m\pi}{p}x\,dx = 2\int_0^p \cos\frac{n\pi}{p}x \cos\frac{m\pi}{p}x\,dx = 0$$

$$\int_{-p}^p \sin\frac{n\pi}{p}x \sin\frac{m\pi}{p}x\,dx = 2\int_0^p \sin\frac{n\pi}{p}x \sin\frac{m\pi}{p}x\,dx = 0.$$

Also

$$\int_{-p}^p \sin\frac{n\pi}{p}x \cos\frac{m\pi}{p}x\,dx = \frac{1}{2}\int_{-p}^p \left(\sin\frac{(n-m)\pi}{p}x + \sin\frac{(n+m)\pi}{p}x\right)\,dx = 0,$$

$$\int_{-p}^p 1\cdot\cos\frac{n\pi}{p}x\,dx = \frac{p}{n\pi}\sin\frac{n\pi}{p}x\Big|_{-p}^p = 0,$$

$$\int_{-p}^p 1\cdot\sin\frac{n\pi}{p}x\,dx = -\frac{p}{n\pi}\cos\frac{n\pi}{p}x\Big|_{-p}^p = 0,$$

and

$$\int_{-p}^{p} \sin\frac{n\pi}{p}x \, \cos\frac{n\pi}{p}x \, dx = \int_{-p}^{p} \frac{1}{2}\sin\frac{2n\pi}{p}x \, dx = -\frac{p}{4n\pi}\cos\frac{2n\pi}{p}x \, \Big|_{-p}^{p} = 0.$$

For $m = n$

$$\int_{-p}^{p} \cos^2\frac{n\pi}{p}x \, dx = \int_{-p}^{p}\left(\frac{1}{2}+\frac{1}{2}\cos\frac{2n\pi}{p}x\right) dx = p,$$

$$\int_{-p}^{p} \sin^2\frac{n\pi}{p}x \, dx = \int_{-p}^{p}\left(\frac{1}{2}-\frac{1}{2}\cos\frac{2n\pi}{p}x\right) dx = p,$$

and

$$\int_{-p}^{p} 1^2 dx = 2p$$

so that

$$\|1\| = \sqrt{2p}, \quad \left\|\cos\frac{n\pi}{p}x\right\| = \sqrt{p}, \quad \text{and} \quad \left\|\sin\frac{n\pi}{p}x\right\| = \sqrt{p}.$$

**15.** By orthogonality $\int_a^b \phi_0(x)\phi_n(x)dx = 0$ for $n = 1, 2, 3, \ldots$; that is, $\int_a^b \phi_n(x)dx = 0$ for $n = 1, 2, 3, \ldots$.

**18.** Setting

$$0 = \int_{-2}^{2} f_3(x)f_1(x)\,dx = \int_{-2}^{2}\left(x^2 + c_1x^3 + c_2x^4\right) dx = \frac{16}{3} + \frac{64}{5}c_2$$

and

$$0 = \int_{-2}^{2} f_3(x)f_2(x)\,dx = \int_{-2}^{2}\left(x^3 + c_1x^4 + c_2x^5\right) dx = \frac{64}{5}c_1$$

we obtain $c_1 = 0$ and $c_2 = -5/12$.

## Exercises 11.2

**3.** $a_0 = \int_{-1}^{1} f(x)\,dx = \int_{-1}^{0} 1\,dx + \int_{0}^{1} x\,dx = \frac{3}{2}.$

$a_n = \int_{-1}^{1} f(x)\cos n\pi x\,dx = \int_{-1}^{0} \cos n\pi x\,dx + \int_{0}^{1} x\cos n\pi x\,dx = \frac{1}{n^2\pi^2}[(-1)^n - 1]$

$b_n = \int_{-1}^{1} f(x)\sin n\pi x\,dx = \int_{-1}^{0} \sin n\pi x\,dx + \int_{0}^{1} x\sin n\pi x\,dx = -\frac{1}{n\pi}$

$f(x) = \frac{3}{4} + \sum_{n=1}^{\infty}\left[\frac{(-1)^n - 1}{n^2\pi^2}\cos n\pi x - \frac{1}{n\pi}\sin n\pi x\right]$

**6.** $a_0 = \frac{1}{\pi}\int_{-\pi}^{\pi} f(x)\,dx = \frac{1}{\pi}\int_{-\pi}^{0} \pi^2\,dx + \frac{1}{\pi}\int_{0}^{\pi}\left(\pi^2 - x^2\right) dx = \frac{5}{3}\pi^2$

**153**

$$a_n = \frac{1}{\pi} \int_{-\pi}^{\pi} f(x) \cos nx \, dx = \frac{1}{\pi} \int_{-\pi}^{0} \pi^2 \cos nx \, dx + \frac{1}{\pi} \int_{0}^{\pi} \left( \pi^2 - x^2 \right) \cos nx \, dx$$

$$= \frac{1}{\pi} \left( \frac{\pi^2 - x^2}{n} \sin nx \, \Big|_0^{\pi} + \frac{2}{n} \int_0^{\pi} x \sin nx \, dx \right) = \frac{2}{n^2}(-1)^{n+1}$$

$$b_n = \frac{1}{\pi} \int_{-\pi}^{\pi} f(x) \sin nx \, dx = \frac{1}{\pi} \int_{-\pi}^{0} \pi^2 \sin nx \, dx + \frac{1}{\pi} \int_{0}^{\pi} \left( \pi^2 - x^2 \right) \sin nx \, dx$$

$$= \frac{\pi}{n}[(-1)^n - 1] + \frac{1}{\pi} \left( \frac{x^2 - \pi^2}{n} \cos nx \, \Big|_0^{\pi} - \frac{2}{n} \int_0^{\pi} x \cos nx \, dx \right) = \frac{\pi}{n}(-1)^n + \frac{2}{n^3 \pi}[1 - (-1)^n]$$

$$f(x) = \frac{5\pi^2}{6} + \sum_{n=1}^{\infty} \left[ \frac{2}{n^2}(-1)^{n+1} \cos nx + \left( \frac{\pi}{n}(-1)^n + \frac{2[1 - (-1^n)]}{n^3 \pi} \right) \sin nx \right]$$

**9.** $a_0 = \dfrac{1}{\pi} \displaystyle\int_{-\pi}^{\pi} f(x) \, dx = \dfrac{1}{\pi} \int_{0}^{\pi} \sin x \, dx = \dfrac{2}{\pi}$

$$a_n = \frac{1}{\pi} \int_{-\pi}^{\pi} f(x) \cos nx \, dx = \frac{1}{\pi} \int_0^{\pi} \sin x \, \cos nx \, dx = \frac{1}{2\pi} \int_0^{\pi} [\sin(n+1)x + \sin(1-n)x] \, dx$$

$$= \frac{1 + (-1)^n}{\pi(1 - n^2)} \quad \text{for } n = 2, 3, 4, \ldots$$

$$a_1 = \frac{1}{2\pi} \int_0^{\pi} \sin 2x \, dx = 0$$

$$b_n = \frac{1}{\pi} \int_{-\pi}^{\pi} f(x) \sin nx \, dx = \frac{1}{\pi} \int_0^{\pi} \sin x \, \sin nx \, dx$$

$$= \frac{1}{2\pi} \int_0^{\pi} [\cos(1-n)x - \cos(1+n)x] \, dx = 0 \quad \text{for } n = 2, 3, 4, \ldots$$

$$b_1 = \frac{1}{2\pi} \int_0^{\pi} (1 - \cos 2x) \, dx = \frac{1}{2}$$

$$f(x) = \frac{1}{\pi} + \frac{1}{2} \sin x + \sum_{n=2}^{\infty} \frac{1 + (-1)^n}{\pi(1 - n^2)} \cos nx$$

**12.** $a_0 = \dfrac{1}{2} \displaystyle\int_{-2}^{2} f(x) \, dx = \dfrac{1}{2} \left( \int_0^1 x \, dx + \int_1^2 1 \, dx \right) = \dfrac{3}{4}$

$$a_n = \frac{1}{2} \int_{-2}^{2} f(x) \cos \frac{n\pi}{2} x \, dx = \frac{1}{2} \left( \int_0^1 x \cos \frac{n\pi}{2} x \, dx + \int_1^2 \cos \frac{n\pi}{2} x \, dx \right) = \frac{2}{n^2 \pi^2} \left( \cos \frac{n\pi}{2} - 1 \right)$$

$$b_n = \frac{1}{2} \int_{-2}^{2} f(x) \sin \frac{n\pi}{2} x \, dx = \frac{1}{2} \left( \int_0^1 x \sin \frac{n\pi}{2} x \, dx + \int_1^2 \sin \frac{n\pi}{2} x \, dx \right)$$

$$= \frac{2}{n^2 \pi^2} \left( \sin \frac{n\pi}{2} + \frac{n\pi}{2}(-1)^{n+1} \right)$$

$$f(x) = \frac{3}{8} + \sum_{n=1}^{\infty} \left[\frac{2}{n^2\pi^2}\left(\cos\frac{n\pi}{2} - 1\right)\cos\frac{n\pi}{2}x + \frac{2}{n^2\pi^2}\left(\sin\frac{n\pi}{2} + \frac{n\pi}{2}(-1)^{n+1}\right)\sin\frac{n\pi}{2}x\right]$$

**15.** $a_0 = \dfrac{1}{\pi}\displaystyle\int_{-\pi}^{\pi} f(x)\,dx = \dfrac{1}{\pi}\displaystyle\int_{-\pi}^{\pi} e^x\,dx = \dfrac{1}{\pi}(e^\pi - e^{-\pi})$

$a_n = \dfrac{1}{\pi}\displaystyle\int_{-\pi}^{\pi} f(x)\cos nx\,dx = \dfrac{(-1)^n(e^\pi - e^{-\pi})}{\pi(1+n^2)}$

$b_n = \dfrac{1}{\pi}\displaystyle\int_{-\pi}^{\pi} f(x)\sin nx\,dx = \dfrac{1}{\pi}\displaystyle\int_{-\pi}^{\pi} e^x\sin nx\,dx = \dfrac{(-1)^n n(e^{-\pi} - e^{\pi})}{\pi(1+n^2)}$

$f(x) = \dfrac{e^\pi - e^{-\pi}}{2\pi} + \displaystyle\sum_{n=1}^{\infty}\left[\dfrac{(-1)^n(e^\pi - e^{-\pi})}{\pi(1+n^2)}\cos nx + \dfrac{(-1)^n n(e^{-\pi} - e^\pi)}{\pi(1+n^2)}\sin nx\right]$

**18.** From Problem 17

$$\frac{\pi^2}{8} = \frac{1}{2}\left(\frac{\pi^2}{6} + \frac{\pi^2}{12}\right) = \frac{1}{2}\left(2 + \frac{2}{3^2} + \frac{2}{5^2} + \cdots\right) = 1 + \frac{1}{3^2} + \frac{1}{5^2} + \cdots.$$

**21. (a)** Letting $c_0 = a_0/2$, $c_n = (a_n - ib_n)$, and $c_{-n} = (a_n + ib_n)/2$ we have

$$f(x) = \frac{a_0}{2} + \sum_{n=1}^{\infty}\left(a_n\cos\frac{n\pi}{p}x + b_n\sin\frac{n\pi}{p}x\right)$$

$$= c_0 + \sum_{n=1}^{\infty}\left(a_n\frac{e^{in\pi x/p} + e^{-in\pi x/p}}{2} + b_n\frac{e^{in\pi x/p} - e^{-in\pi x/p}}{2i}\right)$$

$$= c_0 + \sum_{n=1}^{\infty}\left(a_n\frac{e^{in\pi x/p} + e^{-in\pi x/p}}{2} - b_n\frac{ie^{in\pi x/p} - ie^{-in\pi x/p}}{2}\right)$$

$$= c_0 + \sum_{n=1}^{\infty}\left(\frac{a_n - ib_n}{2}e^{in\pi x/p} + \frac{a_n + ib_n}{2}e^{-in\pi x/p}\right)$$

$$= c_0 + \sum_{n=1}^{\infty}\left(c_n e^{in\pi x/p} + c_{-n}e^{i(-n)\pi x/p}\right) = \sum_{n=-\infty}^{\infty} c_n e^{in\pi x/p}.$$

**155**

**(b)** Multiplying both sides of the expression in **(a)** by $e^{-im\pi x/p}$ and integrating we obtain

$$\int_{-p}^{p} f(x)e^{-im\pi x/p}dx = \int_{-p}^{p} \left( \sum_{n=-\infty}^{\infty} c_n e^{in\pi x/p} e^{-im\pi x/p} \right) dx$$

$$= \sum_{n=-\infty}^{\infty} c_n \int_{-p}^{p} e^{i(n-m)\pi x/p}dx$$

$$= \sum_{n\neq m} c_n \int_{-p}^{p} e^{i(n-m)\pi x/p}dx + c_m \int_{-p}^{p} e^{i(m-n)\pi x/p}dx$$

$$= \sum_{n\neq m} c_n \int_{-p}^{p} e^{i(n-m)\pi x/p}dx + c_m \int_{-p}^{p} dx$$

$$= \sum_{n\neq m} c_n \int_{-p}^{p} e^{i(n-m)\pi x/p}dx + 2pc_m.$$

Recalling that

$$e^{iy} = \cos y + i \sin y \quad \text{and} \quad e^{-iy} = \cos y - i \sin y$$

we have for $n - m$ an integer and $n \neq m$

$$\int_{-p}^{p} e^{i(n-m)\pi x/p}dx = \frac{p}{i(n-m)\pi} e^{i(n-m)\pi x/p} \Big|_{-p}^{p}$$

$$= \frac{p}{i(n-m)\pi} \left( e^{i(n-m)\pi} - e^{-i(n-m)\pi} \right)$$

$$= \frac{p}{i(n-m)\pi} [\cos(n-m)\pi + i\sin(n-m)\pi - \cos(n-m)\pi + i\sin(n-m)\pi]$$

$$= 0.$$

Thus

$$\int_{-p}^{p} f(x)e^{-im\pi x/p}dx = 2pc_m$$

and

$$c_m = \frac{1}{2p} \int_{-p}^{p} f(x)e^{-im\pi x/p}dx.$$

**3.** Since

$$f(-x) = (-x)^2 - x = x^2 - x,$$

$f(x)$ is neither even nor odd.

**6.** Since

$$f(-x) = \left|(-x)^5\right| = \left|x^5\right| = f(x),$$

$f(x)$ is an even function.

**9.** Since $f(x)$ is not defined for $x < 0$, it is neither even nor odd.

**12.** Since $f(x)$ is an even function, we expand in a cosine series:

$$a_0 = \int_1^2 1 \, dx = 1$$

$$a_n = \int_1^2 \cos \frac{n\pi}{2} x \, dx = -\frac{2}{n\pi} \sin \frac{n\pi}{2}.$$

Thus

$$f(x) = \frac{1}{2} + \sum_{n=1}^{\infty} \frac{-2}{n\pi} \sin \frac{n\pi}{2} \cos \frac{n\pi}{2} x.$$

**15.** Since $f(x)$ is an even function, we expand in a cosine series:

$$a_0 = 2 \int_0^1 x^2 \, dx = \frac{2}{3}$$

$$a_n = 2 \int_0^1 x^2 \cos n\pi x \, dx = 2 \left( \frac{x^2}{n\pi} \sin n\pi x \bigg|_0^1 - \frac{2}{n\pi} \int_0^1 x \sin n\pi x \, dx \right) = \frac{4}{n^2 \pi^2} (-1)^n.$$

Thus

$$f(x) = \frac{1}{3} + \sum_{n=1}^{\infty} \frac{4}{n^2 \pi^2} (-1)^n \cos n\pi x.$$

**18.** Since $f(x)$ is an odd function, we expand in a sine series:

$$b_n = \frac{2}{\pi} \int_0^\pi x^3 \sin nx \, dx = \frac{2}{\pi} \left( -\frac{x^3}{n} \cos nx \bigg|_0^\pi + \frac{3}{n} \int_0^\pi x^2 \cos nx \, dx \right)$$

$$= \frac{2\pi^2}{n} (-1)^{n+1} - \frac{12}{n^2 \pi} \int_0^\pi x \sin nx \, dx$$

$$= \frac{2\pi^2}{n} (-1)^{n+1} - \frac{12}{n^2 \pi} \left( -\frac{x}{n} \cos nx \bigg|_0^\pi + \frac{1}{n} \int_0^\pi \cos nx \, dx \right) = \frac{2\pi^2}{n} (-1)^{n+1} + \frac{12}{n^3} (-1)^n.$$

Thus

$$f(x) = \sum_{n=1}^{\infty}\left(\frac{2\pi^2}{n}(-1)^{n+1} + \frac{12}{n^3}(-1)^n\right)\sin nx.$$

**21.** Since $f(x)$ is an even function, we expand in a cosine series:

$$a_0 = \int_0^1 x\,dx + \int_1^2 1\,dx = \frac{3}{2}$$

$$a_n = \int_0^1 x\cos\frac{n\pi}{2}x\,dx + \int_1^2 \cos\frac{n\pi}{2}x\,dx = \frac{4}{n^2\pi^2}\left(\cos\frac{n\pi}{2} - 1\right).$$

Thus

$$f(x) = \frac{3}{4} + \sum_{n=1}^{\infty}\frac{4}{n^2\pi^2}\left(\cos\frac{n\pi}{2} - 1\right)\cos\frac{n\pi}{2}x.$$

**24.** Since $f(x)$ is an even function, we expand in a cosine series:

$$a_0 = \frac{2}{\pi/2}\int_0^{\pi/2}\cos x\,dx = \frac{4}{\pi}$$

$$a_n = \frac{2}{\pi/2}\int_0^{\pi/2}\cos x\cos\frac{n\pi}{\pi/2}x\,dx = \frac{4}{\pi}\int_0^{\pi/2}\cos x\cos 2nx\,dx$$

$$= \frac{2}{\pi}\int_0^{\pi/2}[\cos(2n-1)x + \cos(2n+1)x]\,dx = \frac{4(-1)^{n+1}}{\pi(4n^2 - 1)}.$$

Thus

$$f(x) = \frac{2}{\pi} + \sum_{n=1}^{\infty}\frac{4(-1)^{n+1}}{\pi(4n^2 - 1)}\cos 2nx.$$

**27.** $a_0 = \frac{4}{\pi}\int_0^{\pi/2}\cos x\,dx = \frac{4}{\pi}$

$$a_n = \frac{4}{\pi}\int_0^{\pi/2}\cos x\cos 2nx\,dx = \frac{2}{\pi}\int_0^{\pi/2}[\cos(2n+1)x + \cos(2n-1)x]\,dx = \frac{4(-1)^n}{\pi(1 - 4n^2)}$$

$$b_n = \frac{4}{\pi}\int_0^{\pi/2}\cos x\sin 2nx\,dx = \frac{2}{\pi}\int_0^{\pi/2}[\sin(2n+1)x + \sin(2n-1)x]\,dx = \frac{8n}{\pi(4n^2 - 1)}$$

$$f(x) = \frac{2}{\pi} + \sum_{n=1}^{\infty}\frac{4(-1)^n}{\pi(1 - 4n^2)}\cos 2nx$$

$$f(x) = \sum_{n=1}^{\infty}\frac{8n}{\pi(4n^2 - 1)}\sin 2nx$$

**30.** $a_0 = \frac{1}{\pi}\int_\pi^{2\pi}(x - \pi)\,dx = \frac{\pi}{2}$

$$a_n = \frac{1}{\pi} \int_\pi^{2\pi} (x - \pi) \cos \frac{n}{2} x \, dx = \frac{4}{n^2 \pi} \left[ (-1)^n - \cos \frac{n\pi}{2} \right]$$

$$b_n = \frac{1}{\pi} \int_\pi^{2\pi} (x - \pi) \sin \frac{n}{2} x \, dx = \frac{2}{n}(-1)^{n+1} - \frac{4}{n^2 \pi} \sin \frac{n\pi}{2}$$

$$f(x) = \frac{\pi}{4} + \sum_{n=1}^\infty \frac{4}{n^2 \pi} \left[ (-1)^n - \cos \frac{n\pi}{2} \right] \cos \frac{n}{2} x$$

$$f(x) = \sum_{n=1}^\infty \left( \frac{2}{n}(-1)^{n+1} - \frac{4}{n^2 \pi} \sin \frac{n\pi}{2} \right) \sin \frac{n}{2} x$$

**33.** $a_0 = 2 \int_0^1 (x^2 + x) \, dx = \frac{5}{3}$

$$a_n = 2 \int_0^1 (x^2 + x) \cos n\pi x \, dx = \frac{2(x^2 + x)}{n\pi} \sin n\pi x \Big|_0^1 - \frac{2}{n\pi} \int_0^1 (2x+1) \sin n\pi x \, dx = \frac{2}{n^2 \pi^2}[3(-1)^n - 1]$$

$$b_n = 2 \int_0^1 (x^2 + x) \sin n\pi x \, dx = -\frac{2(x^2 + x)}{n\pi} \cos n\pi x \Big|_0^1 + \frac{2}{n\pi} \int_0^1 (2x + 1) \cos n\pi x \, dx$$

$$= \frac{4}{n\pi}(-1)^{n+1} + \frac{4}{n^3 \pi^3}[(-1)^n - 1]$$

$$f(x) = \frac{5}{6} + \sum_{n=1}^\infty \frac{2}{n^2 \pi^2}[3(-1)^n - 1] \cos n\pi x$$

$$f(x) = \sum_{n=1}^\infty \left( \frac{4}{n\pi}(-1)^{n+1} + \frac{4}{n^3 \pi^3}[(-1)^n - 1] \right) \sin n\pi x$$

**36.** $a_0 = \frac{2}{\pi} \int_0^\pi x \, dx = \pi$

$$a_n = \frac{2}{\pi} \int_0^\pi x \cos 2nx \, dx = 0$$

$$b_n = \frac{2}{\pi} \int_0^\pi x \sin 2nx \, dx = -\frac{1}{n}$$

$$f(x) = \frac{\pi}{2} + \sum_{n=1}^\infty \left( -\frac{1}{n} \sin 2nx \right)$$

**39.** We have

$$b_n = \frac{2}{\pi} \int_0^\pi 5 \sin nt \, dt = \frac{10}{n\pi}[1 - (-1)^n]$$

so that

$$f(t) = \sum_{n=1}^\infty \frac{10[1 - (-1)^n]}{n\pi} \sin nt.$$

Substituting the assumption $x_p(t) = \sum\limits_{n=1}^{\infty} B_n \sin nt$ into the differential equation then gives

$$x_p'' + 10x_p = \sum_{n=1}^{\infty} B_n(10 - n^2) \sin nt = \sum_{n=1}^{\infty} \frac{10[1 - (-1)^n]}{n\pi} \sin nt$$

and so $B_n = \dfrac{10[1 - (-1)^n]}{n\pi(10 - n^2)}$. Thus

$$x_p(t) = \frac{10}{\pi} \sum_{n=1}^{\infty} \frac{1 - (-1)^n}{n(10 - n^2)} \sin nt.$$

**42.** We have

$$a_0 = \frac{2}{(1/2)} \int_0^{1/2} t\, dt = \frac{1}{2}$$

$$a_n = \frac{2}{(1/2)} \int_0^{1/2} t \cos 2n\pi t\, dt = \frac{1}{n^2\pi^2}[(-1)^n - 1]$$

so that

$$f(t) = \frac{1}{4} + \sum_{n=1}^{\infty} \frac{(-1)^n - 1}{n^2\pi^2} \cos 2n\pi t.$$

Substituting the assumption

$$x_p(t) = \frac{A_0}{2} + \sum_{n=1}^{\infty} A_n \cos 2n\pi t$$

into the differential equation then gives

$$\frac{1}{4} x_p'' + 12x_p = 6A_0 + \sum_{n=1}^{\infty} A_n(12 - n^2\pi^2)\cos 2n\pi t = \frac{1}{4} + \sum_{n=1}^{\infty} \frac{(-1)^n - 1}{n^2\pi^2} \cos 2n\pi t$$

and $A_0 = \dfrac{1}{24}$, $A_n = \dfrac{(-1)^n - 1}{n^2\pi^2(12 - n^2\pi^2)}$. Thus

$$x_p(t) = \frac{1}{48} + \frac{1}{\pi^2} \sum_{n=1}^{\infty} \frac{(-1)^n - 1}{n^2(12 - n^2\pi^2)} \cos 2n\pi t.$$

**45.** If $f$ and $g$ are even and $h(x) = f(x)g(x)$ then

$$h(-x) = f(-x)g(-x) = f(x)g(x) = h(x)$$

and $h$ is even.

**48.** If $f$ is even then

$$\int_{-a}^{a} f(x)\, dx = -\int_a^0 f(-u)\, du + \int_0^a f(x)\, dx = \int_0^a f(u)\, du + \int_0^a f(x)\, dx = 2\int_0^a f(x)\, dx.$$

**51.** Using Problems 13 and 14 we obtain

$$f(x) = \frac{1}{2}|x| + \frac{1}{2}x = \frac{\pi}{4} + \sum_{n=1}^{\infty} \left( \frac{(-1)^n - 1}{\pi n^2} \cos nx + \frac{(-1)^{n+1}}{n} \sin nx \right).$$

# Exercises 11.4

Recall that a homogeneous linear system of algebraic equations has a nontrivial solution if and only if the determinant of coefficients is 0.

**3.** For $\lambda \leq 0$ the only solution of the boundary-value problem is $y = 0$. For $\lambda > 0$ we have

$$y = c_1 \cos \sqrt{\lambda}\, x + c_2 \sin \sqrt{\lambda}\, x.$$

Now

$$y'(x) = -c_1 \sqrt{\lambda} \sin \sqrt{\lambda}\, x + c_2 \sqrt{\lambda} \cos \sqrt{\lambda}\, x$$

and $y'(0) = 0$ implies $c_2 = 0$, so

$$y(L) = c_1 \cos \sqrt{\lambda}\, L = 0$$

gives

$$\sqrt{\lambda}\, L = \frac{(2n-1)\pi}{2} \quad \text{or} \quad \lambda = \frac{(2n-1)^2 \pi^2}{4L^2}, \quad n = 1, 2, 3, \dots.$$

The eigenvalues $(2n-1)^2 \pi^2 / 4L^2$ correspond to the eigenfunctions $\cos \dfrac{(2n-1)\pi}{2L} x$ for $n = 1, 2, 3, \dots.$

**6.** For $\lambda \leq 0$ the only solution of the boundary-value problem is $y = 0$. For $\lambda > 0$ we have

$$y = c_1 \cos \sqrt{\lambda}\, x + c_2 \sin \sqrt{\lambda}\, x.$$

Now $y(-\pi) = y(\pi) = 0$ implies

$$c_1 \cos \sqrt{\lambda}\, \pi - c_2 \sin \sqrt{\lambda}\, \pi = 0$$

$$\tag{1}$$

$$c_1 \cos \sqrt{\lambda}\, \pi + c_2 \sin \sqrt{\lambda}\, \pi = 0.$$

This homogeneous system will have a nontrivial solution when

$$\begin{vmatrix} \cos \sqrt{\lambda}\, \pi & -\sin \sqrt{\lambda}\, \pi \\ \cos \sqrt{\lambda}\, \pi & \sin \sqrt{\lambda}\, \pi \end{vmatrix} = 2 \sin \sqrt{\lambda}\, \pi \cos \sqrt{\lambda}\, \pi = \sin 2\sqrt{\lambda}\, \pi = 0.$$

Then

$$2\sqrt{\lambda}\, \pi = n\pi \quad \text{or} \quad \lambda = \frac{n^2}{4}; \quad n = 1, 2, 3, \dots.$$

When $n = 2k - 1$ is odd, the eigenvalues are $(2k-1)^2/4$. Since $\cos(2k-1)\pi/2 = 0$ and $\sin(2k-1)\pi/2 \neq 0$, we see from either equation in (1) that $c_2 = 0$. Thus, the eigenfunctions corresponding to the eigenvalues $(2k-1)^2/4$ are $y = \cos(2k-1)x/2$ for $k = 1, 2, 3, \ldots$ . Similarly, when $n = 2k$ is even, the eigenvalues are $k^2$ with corresponding eigenfunctions $y = \sin kx$ for $k = 1, 2, 3, \ldots$ .

9. The auxiliary equation has solutions

$$m = \frac{1}{2}\left(-2 \pm \sqrt{4 - 4(\lambda + 1)}\right) = -1 \pm \sqrt{-\lambda}.$$

For $\lambda < 0$ we have

$$y = e^{-x}\left(c_1 \cosh \sqrt{-\lambda}\, x + c_2 \sinh \sqrt{-\lambda}\, x\right).$$

The boundary conditions imply

$$y(0) = c_1 = 0$$

$$y(5) = c_2 e^{-5} \sinh 5\sqrt{-\lambda} = 0$$

so $c_1 = c_2 = 0$ and the only solution of the boundary-value problem is $y = 0$.

For $\lambda = 0$ we have

$$y = c_1 e^{-x} + c_2 x e^{-x}$$

and the only solution of the boundary-value problem is $y = 0$.

For $\lambda > 0$ we have

$$y = e^{-x}\left(c_1 \cos \sqrt{\lambda}\, x + c_2 \sin \sqrt{\lambda}\, x\right).$$

Now $y(0) = 0$ implies $c_1 = 0$, so

$$y(5) = c_2 e^{-5} \sin 5\sqrt{\lambda} = 0$$

gives

$$5\sqrt{\lambda} = n\pi \quad \text{or} \quad \lambda = \frac{n^2\pi^2}{25}, \quad n = 1, 2, 3, \ldots .$$

The eigenvalues $n^2\pi^2/25$ correspond to the eigenfunctions $e^{-x} \sin \dfrac{n\pi}{5} x$ for $n = 1, 2, 3, \ldots$ .

12. For $\lambda = 0$ the only solution of the boundary-value problem is $y = 0$. For $\lambda \neq 0$ we have

$$y = c_1 \cos \lambda x + c_2 \sin \lambda x.$$

Now $y(0) = 0$ implies $c_1 = 0$, so

$$y'(3\pi) = c_2 \lambda \cos 3\pi\lambda = 0$$

gives

$$3\pi\lambda = \frac{(2n-1)\pi}{2} \quad \text{or} \quad \lambda = \frac{2n-1}{6}, \quad n = 1, 2, 3, \ldots .$$

The eigenvalues $(2n-1)/6$ correspond to the eigenfunctions $\sin \dfrac{2n-1}{6} x$ for $n = 1, 2, 3, \ldots$ .

**15.** For $\lambda > 0$ a general solution of the given differential equation is

$$y = c_1 \cos(\sqrt{\lambda} \ln x) + c_2 \sin(\sqrt{\lambda} \ln x).$$

Since $\ln 1 = 0$, the boundary condition $y(1) = 0$ implies $c_1 = 0$. Therefore

$$y = c_2 \sin(\sqrt{\lambda} \ln x).$$

Using $\ln e^{\pi} = \pi$ we find that $y(e^{\pi}) = 0$ implies

$$c_2 \sin \sqrt{\lambda}\, \pi = 0$$

or $\sqrt{\lambda}\, \pi = n\pi$, $n = 1, 2, 3, \ldots$ . The eigenvalues and eigenfunctions are, in turn,

$$\lambda = n^2, \quad n = 1, 2, 3, \ldots \quad \text{and} \quad y = \sin(n \ln x).$$

For $\lambda \leq 0$ the only solution of the boundary-value problem is $y = 0$.

To obtain the self-adjoint form we note that the integrating factor is $(1/x^2)e^{\int dx/x} = 1/x$. That is, the self-adjoint form is

$$\frac{d}{dx}[xy'] + \frac{\lambda}{x}y = 0.$$

Identifying the weight function $p(x) = 1/x$ we can then write the orthogonality relation

$$\int_1^{e^{\pi}} \frac{1}{x} \sin(n \ln x) \sin(m \ln x)\, dx = 0, \quad m \neq n.$$

**18.** The eigenfunctions are $\sin \sqrt{\lambda_n}\, x$ where $\tan \sqrt{\lambda_n} = -\lambda_n$. Thus

$$\| \sin \sqrt{\lambda_n}\, x \|^2 = \int_0^1 \sin^2 \sqrt{\lambda_n}\, x\, dx = \frac{1}{2} \int_0^1 \left( 1 - \cos 2\sqrt{\lambda_n}\, x \right) dx$$

$$= \frac{1}{2} \left( x - \frac{1}{2\sqrt{\lambda_n}} \sin 2\sqrt{\lambda_n}\, x \right) \Big|_0^1 = \frac{1}{2} \left( 1 - \frac{1}{2\sqrt{\lambda_n}} \sin 2\sqrt{\lambda_n} \right)$$

$$= \frac{1}{2} \left[ 1 - \frac{1}{2\sqrt{\lambda_n}} \left( 2 \sin \sqrt{\lambda_n} \cos \sqrt{\lambda_n} \right) \right]$$

$$= \frac{1}{2} \left[ 1 - \frac{1}{\sqrt{\lambda_n}} \tan \sqrt{\lambda_n} \cos \sqrt{\lambda_n} \cos \sqrt{\lambda_n} \right]$$

$$= \frac{1}{2} \left[ 1 - \frac{1}{\sqrt{\lambda_n}} \left( -\sqrt{\lambda_n} \cos^2 \sqrt{\lambda_n} \right) \right] = \frac{1}{2} \left( 1 + \cos^2 \sqrt{\lambda_n} \right).$$

**21.** Identifying $f(x) = \tan x + x$ we have $f'(x) = \sec^2 x + 1$. Then Newton's method involves

$$x_{n+1} = x_n - \frac{\tan x_n + x_n}{\sec^2 x_n + 1}.$$

## Exercises 11.4

Using initial values slightly to the right of the asymptotes, we take $x_0 = 1.6, 4.8, 7.9$, and $11.1$. The resulting roots are $2.0288, 4.9132, 7.9787$, and $11.0855$. Squaring, we obtain the eigenvalues $4.1159$, $24.1393, 63.6591$, and $122.8892$.

**24.** For $\lambda < 0$ we have

$$y = c_1 \cosh \sqrt{-\lambda}\, x + c_2 \sinh \sqrt{-\lambda}\, x$$

$$y' = c_1 \sqrt{-\lambda} \sinh \sqrt{-\lambda}\, x + c_2 \sqrt{-\lambda} \cosh \sqrt{-\lambda}\, x.$$

Using the fact that $\cosh x$ is an even function and $\sinh x$ is odd we have

$$y(-L) = c_1 \cosh(-\sqrt{-\lambda}\, L) + c_2 \sinh(-\sqrt{-\lambda}\, L)$$

$$= c_1 \cosh \sqrt{-\lambda}\, L - c_2 \sinh \sqrt{-\lambda}\, L$$

and

$$y'(-L) = c_1 \sqrt{-\lambda} \sinh(-\sqrt{-\lambda}\, L) + c_2 \sqrt{-\lambda} \cosh(-\sqrt{-\lambda}\, L)$$

$$= -c_1 \sqrt{-\lambda} \sinh \sqrt{-\lambda}\, L + c_2 \sqrt{-\lambda} \cosh \sqrt{-\lambda}\, L.$$

The boundary conditions imply

$$c_1 \cosh \sqrt{-\lambda}\, L - c_2 \sinh \sqrt{-\lambda}\, L = c_1 \cosh \sqrt{-\lambda}\, L + c_2 \sinh \sqrt{-\lambda}\, L$$

or

$$2 c_2 \sinh \sqrt{-\lambda}\, L = 0$$

and

$$-c_1 \sqrt{-\lambda} \sinh \sqrt{-\lambda}\, L + c_2 \sqrt{-\lambda} \cosh \sqrt{-\lambda}\, L = c_1 \sqrt{-\lambda} \sinh \sqrt{-\lambda}\, L + c_2 \sqrt{-\lambda} \cosh \sqrt{-\lambda}\, L$$

or

$$2 c_1 \sqrt{-\lambda} \sinh \sqrt{-\lambda}\, L = 0.$$

Since $\sqrt{-\lambda}\, L \neq 0$, $c_1 = c_2 = 0$ and the only solution of the boundary-value problem in this case is $y = 0$.

For $\lambda = 0$ we have

$$y = c_1 x + c_2$$

$$y' = c_1.$$

From $y(-L) = y(L)$ we obtain

$$-c_1 L + c_2 = c_1 L + c_2.$$

Then $c_1 = 0$ and $y = 1$ is an eigenfunction corresponding to the eigenvalue $\lambda = 0$.

For $\lambda > 0$ we have

$$y = c_1 \cos \sqrt{\lambda}\, x + c_2 \sin \sqrt{\lambda}\, x$$

$$y' = -c_1 \sqrt{\lambda} \sin \sqrt{\lambda}\, x + c_2 \sqrt{\lambda} \cos \sqrt{\lambda}\, x.$$

**164**

The first boundary condition implies

$$c_1 \cos \sqrt{\lambda}\, L - c_2 \sin \sqrt{\lambda}\, L = c_1 \cos \sqrt{\lambda}\, L + c_2 \sin \sqrt{\lambda}\, L$$

or

$$2c_2 \sin \sqrt{\lambda}\, L = 0.$$

Thus, if $c_1 = 0$ and $c_2 \neq 0$,

$$\sqrt{\lambda}\, L = n\pi \quad \text{or} \quad \lambda = \frac{n^2 \pi^2}{L^2}, \ n = 1, 2, 3, \ldots .$$

The corresponding eigenfunctions are $\sin \dfrac{n\pi}{L}x$, for $n = 1, 2, 3, \ldots$ . Similarly, the second boundary condition implies

$$2c_1 \sqrt{\lambda} \sin \sqrt{\lambda}\, L = 0.$$

If $c_2 = 0$ and $c_1 \neq 0$,

$$\sqrt{\lambda}\, L = n\pi \quad \text{or} \quad \lambda = \frac{n^2 \pi^2}{L^2}, \ n = 1, 2, 3, \ldots,$$

and the corresponding eigenfunctions are $\cos \dfrac{n\pi}{L}x$, for $n = 1, 2, 3, \ldots$ .

# Exercises 11.5

**3.** The boundary condition indicates that we use (15) and (16) of Section 11.5. With $b = 2$ we obtain

$$c_i = \frac{2}{4J_1^2(2\lambda_i)} \int_0^2 x J_0(\lambda_i x)\, dx$$

$$\boxed{t = \lambda_i x \qquad dt = \lambda_i\, dx}$$

$$= \frac{1}{2J_1^2(2\lambda_i)} \cdot \frac{1}{\lambda_i^2} \int_0^{2\lambda_i} t J_0(t)\, dt$$

$$= \frac{1}{2\lambda_i^2 J_1^2(2\lambda_i)} \int_0^{2\lambda_i} \frac{d}{dt}[t J_1(t)]\, dt \qquad \text{[From (4) in the text]}$$

$$= \frac{1}{2\lambda_i^2 J_1^2(2\lambda_i)} t J_1(t) \Big|_0^{2\lambda_i}$$

$$= \frac{1}{\lambda_i J_1(2\lambda_i)}.$$

Thus

$$f(x) = \sum_{i=1}^{\infty} \frac{1}{\lambda_i J_1(2\lambda_i)} J_0(\lambda_i x).$$

**165**

**6.** Writing the boundary condition in the form

$$2J_0(2\lambda) + 2\lambda J_0'(2\lambda) = 0$$

we identify $b = 2$ and $h = 2$. Using (17) and (18) of Section 11.5 we obtain

$$c_i = \frac{2\lambda_i^2}{(4\lambda_i^2 + 4)J_0^2(2\lambda_i)} \int_0^2 x J_0(\lambda_i x)\, dx$$

$$\boxed{\; t = \lambda_i x \qquad dt = \lambda_i\, dx \;}$$

$$= \frac{\lambda_i^2}{2(\lambda_i^2 + 1)J_0^2(2\lambda_i)} \cdot \frac{1}{\lambda_i^2} \int_0^{2\lambda_i} t J_0(t)\, dt$$

$$= \frac{1}{2(\lambda_i^2 + 1)J_0^2(2\lambda_i)} \int_0^{2\lambda_i} \frac{d}{dt}[t J_1(t)]\, dt \qquad \text{[From (4) in the text]}$$

$$= \frac{1}{2(\lambda_i^2 + 1)J_0^2(2\lambda_i)} t J_1(t) \Big|_0^{2\lambda_i}$$

$$= \frac{\lambda_i J_1(2\lambda_i)}{(\lambda_i^2 + 1)J_0^2(2\lambda_i)} .$$

Thus

$$f(x) = \sum_{i=1}^{\infty} \frac{\lambda_i J_1(2\lambda_i)}{(\lambda_i^2 + 1)J_0^2(2\lambda_i)} J_0(\lambda_i x).$$

**9.** The boundary condition indicates that we use (19) and (20) of Section 11.5. With $b = 3$ we obtain

$$c_1 = \frac{2}{9}\int_0^3 x x^2\, dx = \frac{2}{9}\frac{x^4}{4}\Big|_0^3 = \frac{9}{2},$$

$$c_i = \frac{2}{9 J_0^2(3\lambda_i)} \int_0^3 x J_0(\lambda_i x) x^2\, dx$$

$$\boxed{\; t = \lambda_i x \qquad dt = \lambda_i\, dx \;}$$

$$= \frac{2}{9 J_0^2(3\lambda_i)} \cdot \frac{1}{\lambda_i^4} \int_0^{3\lambda_i} t^3 J_0(t)\, dt$$

$$= \frac{2}{9\lambda_i^4 J_0^2(3\lambda_i)} \int_0^{3\lambda_i} t^2 \frac{d}{dt}[t J_1(t)]\, dt$$

$$\boxed{\begin{array}{ll} u = t^2 & dv = \frac{d}{dt}[t J_1(t)]\, dt \\ du = 2t\, dt & v = t J_1(t) \end{array}}$$

$$= \frac{2}{9\lambda_i^4 J_0^2(3\lambda_i)} \left( t^3 J_1(t)\Big|_0^{3\lambda_i} - 2\int_0^{3\lambda_i} t^2 J_1(t)\, dt \right)$$

**166**

With $n = 0$ in equation (5) in Section 11.5 in the text we have $J_0'(x) = -J_1(x)$, so the boundary condition $J_0'(3\lambda_i) = 0$ implies $J_1(3\lambda_i) = 0$. Then

$$c_i = \frac{2}{9\lambda_i^4 J_0^2(3\lambda_i)} \left( -2 \int_0^{3\lambda_i} \frac{d}{dt} \left[ t^2 J_2(t) \right] \, dt \right) = \frac{2}{9\lambda_i^4 J_0^2(3\lambda_i)} \left( -2t^2 J_2(t) \Big|_0^{3\lambda_i} \right)$$

$$= \frac{2}{9\lambda_i^4 J_0^2(3\lambda_i)} \left[ -18\lambda_i^2 J_2(3\lambda_i) \right] = \frac{-4 J_2(3\lambda_i)}{\lambda_i^2 J_0^2(3\lambda_i)}.$$

Thus,

$$f(x) = \frac{9}{2} - 4 \sum_{i=1}^{\infty} \frac{J_2(3\lambda_i)}{\lambda_i^2 J_0^2(3\lambda_i)} J_0(\lambda_i x).$$

**12.** Since $f(x) = x^3$ is a polynomial in $x$, an expansion of $f$ in polynomials in $x$ must terminate with the term having the same degree as $f$. We have

$$c_0 = \frac{1}{2} \int_{-1}^{1} x^3 P_0(x) \, dx = \frac{1}{2} \int_{-1}^{1} x^3 \, dx = 0,$$

$$c_1 = \frac{3}{2} \int_{-1}^{1} x^3 P_1(x) \, dx = \frac{3}{2} \int_{-1}^{1} x^4 \, dx = \frac{3}{5},$$

$$c_2 = \frac{5}{2} \int_{-1}^{1} x^3 P_2(x) \, dx = \frac{5}{2} \int_{-1}^{1} x^3 \frac{1}{2} \left( 3x^2 - 1 \right) \, dx = 0,$$

$$c_3 = \frac{7}{2} \int_{-1}^{1} x^3 P_3(x) \, dx = \frac{7}{2} \int_{-1}^{1} x^3 \frac{1}{2} \left( 5x^3 - 3x \right) \, dx = \frac{2}{5}.$$

Thus

$$f(x) = c_0 P_0(x) + c_1 P_1(x) + c_2 P_2(x) + c_3 P_3(x)$$

$$= \frac{3}{5} P_1(x) + \frac{2}{5} P_3(x).$$

**15.** Using $\cos^2 \theta = \frac{1}{2}(\cos 2\theta + 1)$ we have

$$P_2(\cos \theta) = \frac{1}{2}(3 \cos^2 \theta - 1) = \frac{3}{2} \cos^2 \theta - \frac{1}{2}$$

$$= \frac{3}{4}(\cos 2\theta + 1) - \frac{1}{2} = \frac{3}{4} \cos 2\theta + \frac{1}{4} = \frac{1}{4}(3 \cos 2\theta + 1).$$

**18.** If $f$ is an odd function on $(-1, 1)$ then

$$\int_{-1}^{1} f(x) P_{2n}(x) \, dx = 0$$

and

$$\int_{-1}^{1} f(x) P_{2n+1}(x) \, dx = 2 \int_0^1 f(x) P_{2n+1}(x) \, dx.$$

**167**

Thus

$$c_{2n+1} = \frac{2(2n+1)+1}{2} \int_{-1}^{1} f(x)P_{2n+1}(x)\,dx = \frac{4n+3}{2}\left(2\int_{0}^{1} f(x)P_{2n+1}(x)\,dx\right)$$

$$= (4n+1)\int_{0}^{1} f(x)P_{2n+1}(x)\,dx,$$

$c_{2n} = 0$, and

$$f(x) = \sum_{n=0}^{\infty} c_{2n+1}P_{2n+1}(x).$$

# —————— Chapter 11 Review Exercises ——————

**3.** cosine, since $f$ is even.

**6.** True

**9.** Since the coefficient of $y$ in the differential equation is $n^2$, the weight function is the integrating factor

$$\frac{1}{a(x)}e^{\int (b/a)dx} = \frac{1}{1-x^2}e^{\int -\frac{x}{1-x^2}\,dx} = \frac{1}{1-x^2}e^{\frac{1}{2}\ln(1-x^2)} = \frac{\sqrt{1-x^2}}{1-x^2} = \frac{1}{\sqrt{1-x^2}}$$

on the interval $[-1,1]$.

**12.** From

$$\int_{0}^{L} \sin^2 \frac{(2n+1)\pi}{2L}x\,dx = \int_{0}^{L}\left(\frac{1}{2} - \frac{1}{2}\cos\frac{(2n+1)\pi}{2L}x\right)dx = \frac{L}{2}$$

we see that

$$\left\|\sin\frac{(2n+1)\pi}{2L}x\right\| = \sqrt{\frac{L}{2}}.$$

**15.** Since

$$A_0 = 2\int_{0}^{1} e^{-x}\,dx$$

and

$$A_n = 2\int_{-1}^{1} e^{-x}\cos n\pi x\,dx = \frac{2}{1+n^2\pi^2}[(1-(-1)^n e^{-1}]$$

for $n = 1, 2, 3, \ldots$ we have

$$f(x) = 1 - e^{-1} + 2\sum_{n=1}^{\infty} \frac{1-(-1)^n e^{-1}}{1+n^2\pi^2}\cos n\pi x.$$

**18.** To obtain the self-adjoint form of the differential equation in Problem 17 we note that an integrating factor is $(1/x^2)e^{\int dx/x} = 1/x$. Thus the weight function is $9/x$ and an orthogonality relation is

$$\int_{1}^{e} \frac{9}{x}\cos\left(\frac{2n-1}{2}\pi\ln x\right)\cos\left(\frac{2m-1}{2}\pi\ln x\right)dx = 0, \quad m \neq n.$$

# 12 Boundary-Value Problems in Rectangular Coordinate Systems

**3.** If $u = XY$ then

$$u_x = X'Y,$$

$$u_y = XY',$$

$$X'Y = X(Y - Y'),$$

and

$$\frac{X'}{X} = \frac{Y - Y'}{Y} = \pm\lambda^2.$$

Then

$$X' \mp \lambda^2 X = 0 \quad \text{and} \quad Y' - (1 \mp \lambda^2)Y = 0$$

so that

$$X = A_1 e^{\pm\lambda^2 x},$$

$$Y = A_2 e^{(1 \mp \lambda^2)y},$$

and

$$u = XY = c_1 e^{y + c_2(x - y)}.$$

**6.** If $u = XY$ then

$$u_x = X'Y,$$

$$u_y = XY',$$

$$yX'Y = xXY',$$

and

$$\frac{X'}{xX} = \frac{Y'}{-yY} = \pm\lambda^2.$$

Then

$$X \mp \lambda^2 xX = 0 \quad \text{and} \quad Y' \pm \lambda^2 yY = 0$$

**169**

so that

$$X = A_1 e^{\pm \lambda^2 x^2/2},$$

$$Y = A_2 e^{\mp \lambda^2 y^2/2},$$

and

$$u = XY = c_1 e^{c_2(x^2-y^2)}.$$

**9.** If $u = XT$ then

$$u_t = XT',$$

$$u_{xx} = X''T,$$

$$kX''T - XT = XT',$$

and we choose

$$\frac{T'}{T} = \frac{kX'' - X}{X} = -1 \pm k\lambda^2$$

so that

$$T' - (-1 \pm k\lambda^2)T = 0 \quad \text{and} \quad X'' - (\pm\lambda^2)X = 0.$$

For $\lambda^2 > 0$ we obtain

$$X = A_1 \cosh \lambda x + A_2 \sinh \lambda x \quad \text{and} \quad T = A_3 e^{(-1+k\lambda^2)t}$$

so that

$$u = XT = e^{(-1+k\lambda^2)t} \left( c_1 \cosh \lambda x + c_2 \sinh \lambda x \right).$$

For $-\lambda^2 < 0$ we obtain

$$X = A_1 \cos \lambda x + A_2 \sin \lambda x \quad \text{and} \quad T = A_3 e^{(-1-k\lambda^2)t}$$

so that

$$u = XT = e^{(-1-k\lambda^2)t}(c_3 \cos \lambda x + c_4 \sin \lambda x).$$

If $\lambda^2 = 0$ then

$$X'' = 0 \quad \text{and} \quad T' + T = 0,$$

and we obtain

$$X = A_1 x + A_2 \quad \text{and} \quad T = A_3 e^{-t}.$$

In this case

$$u = XT = e^{-t}(c_5 x + c_6)$$

**12.** If $u = XT$ then

$$u_t = XT',$$

$$u_{tt} = XT'',$$

$$u_{xx} = X''T,$$

$$a^2 X''T = XT'' + 2kXT',$$

and

$$\frac{X''}{X} = \frac{T'' + 2kT'}{a^2 T} = \pm\lambda^2$$

so that

$$X'' \mp \lambda^2 X = 0 \quad \text{and} \quad T'' + 2kT' \mp a^2\lambda^2 T = 0.$$

For $\lambda^2 > 0$ we obtain

$$X = A_1 e^{\lambda x} + A_2 e^{-\lambda x},$$

$$T = A_3 e^{(-k+\sqrt{k^2+a^2\lambda^2})t} + A_4 e^{(-k-\sqrt{k^2+a^2\lambda^2})t},$$

and

$$u = XT = \left(A_1 e^{\lambda x} + A_2 e^{-\lambda x}\right)\left(A_3 e^{(-k+\sqrt{k^2+a^2\lambda^2})t} + A_4 e^{(-k-\sqrt{k^2+a^2\lambda^2})t}\right).$$

For $-\lambda^2 < 0$ we obtain

$$X = A_1 \cos\lambda x + A_2 \sin\lambda x.$$

If $k^2 - a^2\lambda^2 > 0$ then

$$T = A_3 e^{(-k+\sqrt{k^2-a^2\lambda^2})t} + A_4 e^{(-k-\sqrt{k^2-a^2\lambda^2})t}.$$

If $k^2 - a^2\lambda^2 < 0$ then

$$T = e^{-kt}\left(A_3 \cos\sqrt{a^2\lambda^2 - k^2}\,t + A_4 \sin\sqrt{a^2\lambda^2 - k^2}\,t\right).$$

If $k^2 - a^2\lambda^2 = 0$ then

$$T = A_3 e^{-kt} + A_4 t e^{-kt}$$

so that

$$u = XT = (A_1 \cos\lambda x + A_2 \sin\lambda x)\left(A_3 e^{(-k+\sqrt{k^2-a^2\lambda^2})t} + A_4 e^{(-k-\sqrt{k^2-a^2\lambda^2})t}\right)$$

$$= (A_1 \cos\lambda x + A_2 \sin\lambda x)e^{-kt}\left(A_3 \cos\sqrt{a^2\lambda^2 - k^2}\,t + A_4 \sin\sqrt{a^2\lambda^2 - k^2}\,t\right)$$

$$= \left(A_1 \cos\frac{k}{a}x + A_2 \sin\frac{k}{a}x\right)\left(A_3 e^{-kt} + A_4 t e^{-kt}\right).$$

For $\lambda^2 = 0$ we obtain

$$X A_1 x + A_2,$$

$$T = A_3 + A_4 e^{-2kt},$$

and

**171**

$$u = XT = (A_1 x + A_2)(A_3 + A_4 e^{-2kt}).$$

**15.** If $u = XY$ then

$$u_{xx} = X''Y,$$

$$u_{yy} = XY'',$$

$$X''Y + XY'' = XY,$$

and

$$\frac{X''}{X} = \frac{Y - Y''}{Y} = \pm\lambda^2$$

so that

$$X'' \mp \lambda^2 X = 0 \quad \text{and} \quad Y'' + (\pm\lambda^2 - 1)Y = 0.$$

For $\lambda^2 > 0$ we obtain

$$X = A_1 e^{\lambda x} + A_2 e^{-\lambda x}.$$

If $\lambda^2 - 1 > 0$ then

$$Y = A_3 \cos \sqrt{\lambda^2 - 1}\, y + A_4 \sin \sqrt{\lambda^2 - 1}\, y.$$

If $\lambda^2 - 1 < 0$ then

$$Y = A_3 e^{\sqrt{1-\lambda^2}\, y} + A_4 e^{-\sqrt{1-\lambda^2}\, y}.$$

If $\lambda^2 - 1 = 0$ then $Y = A_3 y + A_4$ so that

$$u = XY = \left(A_1 e^{\lambda x} + A_2 e^{-\lambda x}\right)\left(A_3 \cos \sqrt{\lambda^2 - 1}\, y + A_4 \sin \sqrt{\lambda^2 - 1}\, y\right),$$

$$= \left(A_1 e^{\lambda x} + A_2 e^{-\lambda x}\right)\left(A_3 e^{\sqrt{1-\lambda^2}\, y} + A_4 e^{-\sqrt{1-\lambda^2}\, y}\right)$$

$$= (A_1 e^x + A_2 e^{-x})(A_3 y + A_4).$$

For $-\lambda^2 < 0$ we obtain

$$X = A_1 \cos \lambda x + A_2 \sin \lambda x,$$

$$Y = A_3 e^{\sqrt{1+\lambda^2}\, y} + A_4 e^{-\sqrt{1+\lambda^2}\, y},$$

and

$$u = XY = (A_1 \cos \lambda x + A_2 \sin \lambda x)\left(A_3 e^{\sqrt{1+\lambda^2}\, y} + A_4 e^{-\sqrt{1+\lambda^2}\, y}\right).$$

For $\lambda^2 = 0$ we obtain

$$X = A_1 x + A_2,$$

$$Y = A_3 e^y + A_4 e^{-y},$$

and

$$u = XY = (A_1 x + A_2)(A_3 e^y + A_4 e^{-y}).$$

**18.** Identifying $A = 3$, $B = 5$, and $C = 1$, we compute $B^2 - 4AC = 13 > 0$. The equation is hyperbolic.

**21.** Identifying $A = 1$, $B = -9$, and $C = 0$, we compute $B^2 - 4AC = 81 > 0$. The equation is hyperbolic.

**24.** Identifying $A = 1$, $B = 0$, and $C = 1$, we compute $B^2 - 4AC = -4 < 0$. The equation is elliptic.

**27.** If $u = RT$ then

$$u_r = R'T,$$

$$u_{rr} = R''T,$$

$$u_t = RT',$$

$$RT' = k\left(R''T + \frac{1}{r}R'T\right),$$

and

$$\frac{r^2 R'' + rR'}{r^2 R} = \frac{T'}{kT} = \pm\lambda^2.$$

If we use $-\lambda^2 < 0$ then

$$r^2 R'' + rR' + \lambda^2 r^2 R = 0 \quad \text{and} \quad T'' \mp \lambda^2 kT = 0$$

so that

$$T = A_1 e^{-k\lambda^2 t},$$

$$R = A_2 J_0(\lambda r) + A_3 Y_0(\lambda r),$$

and

$$u = RT = e^{-k\lambda^2 t}[c_1 J_0(\lambda r) + c_2 Y_0(\lambda r)]$$

**30.** For $u = A_1 e^{\lambda^2 y} \cosh 2\lambda x + B_1 e^{\lambda^2 y} \sinh 2\lambda x$ we compute

$$\frac{\partial^2 u}{\partial x^2} = 4\lambda^2 A_1 e^{\lambda^2 y} \cosh 2\lambda x + 4\lambda^2 B_1 e^{\lambda^2 y} \sinh 2\lambda x$$

and

$$\frac{\partial u}{\partial y} = \lambda^2 A_1 e^{\lambda^2 y} \cosh 2\lambda x + \lambda^2 B_1 e^{\lambda^2 y} \sinh 2\lambda x.$$

Then $\partial^2 u/\partial x^2 = 4\partial u/\partial y$.

For $u = A_2 e^{-\lambda^2 y} \cos 2\lambda x + B_2 e^{-\lambda^2 y} \sin 2\lambda x$ we compute

$$\frac{\partial^2 u}{\partial x^2} = -4\lambda^2 A_2 e^{-\lambda^2 y} \cos 2\lambda x - 4\lambda^2 B_2 e^{-\lambda^2 y} \sin 2\lambda x$$

and

$$\frac{\partial u}{\partial y} = -\lambda^2 A_2 \cos 2\lambda x - \lambda^2 B_2 \sin 2\lambda x.$$

**173**

Then $\partial^2 u/\partial x^2 = 4\partial u/\partial y$.

For $u = A_3 x + B_3$ we compute $\partial^2 u/\partial x^2 = \partial u/\partial y = 0$. Then $\partial^2 u/\partial x^2 = 4\partial u/\partial y$.

## Exercises 12.2

**3.** $k\dfrac{\partial^2 u}{\partial x^2} = \dfrac{\partial u}{\partial t}$,   $0 < x < L$, $t > 0$

$u(0,t) = 100$,   $\dfrac{\partial u}{\partial x}\Big|_{x=L} = -hu(L,t)$,   $t > 0$

$u(x,0) = f(x)$,   $0 < x < L$

**6.** $a^2\dfrac{\partial^2 u}{\partial x^2} = \dfrac{\partial^2 u}{\partial t^2}$,   $0 < x < L$, $t > 0$

$u(0,t) = 0$,   $u(L,t) = 0$,   $t > 0$

$u(x,0) = 0$,   $\dfrac{\partial u}{\partial x}\Big|_{t=0} = \sin\dfrac{\pi x}{L}$,   $0 < x < L$

**9.** $\dfrac{\partial^2 u}{\partial x^2} + \dfrac{\partial^2 u}{\partial y^2} = 0$,   $0 < x < 4$, $0 < y < 2$

$\dfrac{\partial u}{\partial x}\Big|_{x=0} = 0$,   $u(4,y) = f(y)$,   $0 < y < 2$

$\dfrac{\partial u}{\partial y}\Big|_{y=0} = 0$,   $u(x,2) = 0$,   $0 < x < 4$

## Exercises 12.3

**3.** Using $u = XT$ and $-\lambda^2$ as a separation constant leads to

$$X'' + \lambda^2 X = 0,$$
$$X'(0) = 0,$$
$$X'(L) = 0,$$

and

$$T' + k\lambda^2 T = 0.$$

Then

$$X = c_1 \cos\frac{n\pi}{L}x \quad \text{and} \quad T = c_2 e^{-\frac{kn^2\pi^2}{L^2}t}$$

for $n = 0, 1, 2, \ldots$ so that

$$u = \sum_{n=0}^{\infty} A_n e^{-\frac{kn^2\pi^2}{L^2}t} \cos \frac{n\pi}{L}x.$$

Imposing

$$u(x,0) = f(x) = \sum_{n=0}^{\infty} A_n \cos \frac{n\pi}{L}x$$

gives

$$u(x,t) = \frac{1}{L}\int_0^L f(x)\,dx + \frac{2}{L}\sum_{n=1}^{\infty}\left(\int_0^L f(x)\cos\frac{n\pi}{L}x\,dx\right)e^{-\frac{kn^2\pi^2}{L^2}t}\cos\frac{n\pi}{L}x.$$

**6.** Using $u = XT$ and $-\lambda^2$ as a separation constant leads to

$$X'' + \lambda^2 X = 0,$$

$$X(0) = 0,$$

$$X(L) = 0,$$

and

$$T' + (h + k\lambda^2)T = 0.$$

Then

$$X = c_1 \sin \frac{n\pi}{L}x \quad \text{and} \quad T = c_2 e^{-\left(h + \frac{kn^2\pi^2}{L^2}\right)t}$$

for $n = 1, 2, 3, \ldots$ so that

$$u = \sum_{n=1}^{\infty} A_n e^{-\left(h + \frac{kn^2\pi^2}{L^2}\right)t} \sin \frac{n\pi}{L}x.$$

Imposing

$$u(x,0) = f(x) = \sum_{n=1}^{\infty} A_n \sin \frac{n\pi}{L}x$$

gives

$$u = \frac{2}{L}\sum_{n=1}^{\infty}\left(\int_0^L f(x)\sin\frac{n\pi}{L}x\,dx\right)e^{-\left(h + \frac{kn^2\pi^2}{L^2}\right)t}\sin\frac{n\pi}{L}x.$$

**3.** Using $u = XT$ and $-\lambda^2$ as a separation constant leads to

$$X'' + \lambda^2 X = 0,$$
$$X(0) = 0,$$
$$X(L) = 0,$$

and

$$T'' + \lambda^2 a^2 T = 0.$$

Then

$$X = c_1 \sin \frac{n\pi}{L} x \quad \text{and} \quad T = c_2 \cos \frac{n\pi a}{L} t + c_3 \sin \frac{n\pi a}{L} t$$

for $n = 1, 2, 3, \ldots$ so that

$$u = \sum_{n=1}^{\infty} \left( A_n \cos \frac{n\pi a}{L} t + B_n \sin \frac{n\pi a}{L} t \right) \sin \frac{n\pi}{L} x.$$

Imposing

$$u(x,0) = \sum_{n=1}^{\infty} A_n \sin \frac{n\pi}{L} x$$

gives

$$A_n = \frac{2}{L} \left( \int_0^{L/3} \frac{3}{L} x \sin \frac{n\pi}{L} x \, dx + \int_{L/3}^{2L/3} \sin \frac{n\pi}{L} x \, dx + \int_{2L/3}^{L} \left( 3 - \frac{3}{L} x \right) \sin \frac{n\pi}{L} x \, dx \right)$$

so that

$$A_1 = \frac{6\sqrt{3}}{\pi^2},$$

$$A_2 = A_3 = A_4 = 0,$$

$$A_5 = -\frac{6\sqrt{3}}{5^2 \pi^2},$$

$$A_6 = 0,$$

$$a_7 = \frac{6\sqrt{3}}{7^2 \pi^2} \cdots.$$

Imposing

$$u_t(x,0) = 0 = \sum_{n=1}^{\infty} B_n \frac{n\pi a}{L} \sin \frac{n\pi}{L} x$$

gives $B_n = 0$ for $n = 1, 2, 3, \ldots$ so that

$$u(x,t) = \frac{6\sqrt{3}}{\pi^2} \left( \cos \frac{\pi a}{L} t \, \sin \frac{\pi}{L} x - \frac{1}{5^2} \cos \frac{5\pi a}{L} t \, \sin \frac{5\pi}{L} x + \frac{1}{7^2} \cos \frac{7\pi a}{L} t \, \sin \frac{7\pi}{L} x - \cdots \right).$$

**6.** Using $u = XT$ and $-\lambda^2$ as a separation constant leads to

$$X'' + \lambda^2 X = 0,$$

$$X(0) = 0,$$

$$X(1) = 0,$$

and

$$T'' + \lambda^2 a^2 T = 0.$$

Then

$$X = c_1 \sin n\pi x \quad \text{and} \quad T = c_2 \cos n\pi at + c_3 \sin n\pi at$$

for $n = 1, 2, 3, \ldots$ so that

$$u = \sum_{n=1}^{\infty} \left( A_n \cos n\pi at + B_n \sin n\pi at \right) \sin n\pi x.$$

Imposing

$$u(x,0) = 0.01 \sin 3\pi x = \sum_{n=1}^{\infty} A_n \sin n\pi x$$

and

$$u_t(x,0) = 0 = \sum_{n=1}^{\infty} B_n n\pi a \, \sin n\pi x$$

gives $B_n = 0$ for $n = 1, 2, 3, \ldots$, $A_3 = 0.01$, and $A_n = 0$ for $n = 1, 2, 4, 5, 6, \ldots$ so that

$$u(x,t) = 0.01 \sin 3\pi x \, \cos 3\pi at.$$

**9.** Using $u = XT$ and $-\lambda^2$ as a separation constant leads to

$$X'' + \lambda^2 X = 0,$$

$$X(0) = 0,$$

$$X(\pi) = 0,$$

and

$$T'' + 2\beta T' + \lambda^2 T = 0.$$

Then

$$X = c_1 \sin nx \quad \text{and} \quad T = e^{-\beta t} \left( c_2 \cos \sqrt{n^2 - \beta^2} \, t + c_3 \sin \sqrt{n^2 - \beta^2} \, t \right)$$

**177**

so that

$$u = \sum_{n=1}^{\infty} e^{-\beta t} \left( A_n \cos \sqrt{n^2 - \beta^2}\, t + B_n \sin \sqrt{n^2 - \beta^2}\, t \right) \sin nx.$$

Imposing

$$u(x,0) = f(x) = \sum_{n=1}^{\infty} A_n \sin nx$$

and

$$u_t(x,0) = 0 = \sum_{n=1}^{\infty} \left( B_n \sqrt{n^2 - \beta^2} - \beta A_n \right) \sin nx$$

gives

$$u(x,t) = e^{-\beta t} \sum_{n=1}^{\infty} A_n \left( \cos \sqrt{n^2 - \beta^2}\, t + \frac{\beta}{\sqrt{n^2 - \beta^2}} \sin \sqrt{n^2 - \beta^2}\, t \right) \sin nx,$$

where

$$A_n = \frac{2}{\pi} \int_0^{\pi} f(x) \sin nx \, dx.$$

**12.** In this case the boundary conditions become, for $t > 0$,

$$u(0,t) = 0, \qquad u(L,t) = 0,$$

$$\frac{\partial u}{\partial x} \bigg|_{x=0} = 0, \qquad \frac{\partial u}{\partial x} \bigg|_{x=L} = 0.$$

**15.** $u(x,t) = \frac{1}{2}[\sin(x+at) + \sin(x-at)] + \frac{1}{2a} \int_{x-at}^{x+at} ds$

$= \frac{1}{2}[\sin x \cos at + \cos x \sin at + \sin x \cos at - \cos x \sin at] + \frac{1}{2a} s \Big|_{x-at}^{x+at} = \sin x \cos at + t$

**18.**

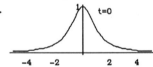 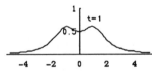 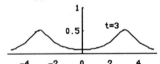

## Exercises 12.5

**3.** Using $u = XY$ and $-\lambda^2$ as a separation constant leads to

$$X'' + \lambda^2 X = 0,$$

$$X(0) = 0,$$

$$X(a) = 0,$$

and

$$Y'' - \lambda^2 Y = 0,$$

$$Y(b) = 0.$$

Then

$$X = c_1 \sin \frac{n\pi}{a}x \quad \text{and} \quad Y = c_2 \cosh \frac{n\pi}{a}y - c_2 \frac{\cosh \frac{n\pi b}{a}}{\sinh \frac{n\pi b}{a}} \sinh \frac{n\pi}{a}y$$

for $n = 1, 2, 3, \ldots$ so that

$$u = \sum_{n=1}^{\infty} A_n \sin \frac{n\pi}{a}x \left( \cosh \frac{n\pi}{a}y - \frac{\cosh \frac{n\pi b}{a}}{\sinh \frac{n\pi b}{a}} \sinh \frac{n\pi}{a}y \right).$$

Imposing

$$u(x,0) = f(x) = \sum_{n=1}^{\infty} A_n \sin \frac{n\pi}{a}x$$

gives

$$A_n = \frac{2}{a} \int_0^a f(x) \sin \frac{n\pi}{a}x \, dx$$

so that

$$u(x,y) = \frac{2}{a} \sum_{n=1}^{\infty} \left( \int_0^a f(x) \sin \frac{n\pi}{a}x \, dx \right) \sin \frac{n\pi}{a}x \left( \cosh \frac{n\pi}{a}y - \frac{\cosh \frac{n\pi b}{a}}{\sinh \frac{n\pi b}{a}} \sinh \frac{n\pi}{a}y \right).$$

**6.** Using $u = XY$ and $\lambda^2$ as a separation constant leads to

$$X'' - \lambda^2 X = 0,$$

$$X'(1) = 0,$$

and

$$Y'' + \lambda^2 Y = 0,$$

$$Y'(0) = 0,$$

$$Y'(\pi) = 0.$$

Then

$$Y = c_1 \cos ny$$

for $n = 0, 1, 2, \ldots$ and

$$X = c_2 \cosh nx - c_2 \frac{\sinh n}{\cosh n} \sinh nx$$

for $n = 0, 1, 2, \ldots$ so that

$$u = A_0 + \sum_{n=1}^{\infty} A_n \left( \cosh nx - \frac{\sinh n}{\cosh n} \sinh nx \right) \cos ny.$$

**179**

## Exercises 12.5

Imposing

$$u(0, y) = g(y) = A_0 + \sum_{n=1}^{\infty} A_n \cos ny$$

gives

$$A_0 = \frac{1}{\pi} \int_0^{\pi} g(y)\, dy \quad \text{and} \quad A_n = \frac{2}{\pi} \int_0^{\pi} g(y) \cos ny\, dy$$

for $n = 1, 2, 3, \ldots$ so that

$$u(x, y) = \frac{1}{\pi} \int_0^{\pi} g(y)\, dy + \sum_{n=1}^{\infty} \left( \frac{2}{\pi} \int_0^{\pi} g(y) \cos ny\, dy \right) \left( \cosh nx - \frac{\sinh n}{\cosh n} \sinh nx \right) \cos ny.$$

9. Using $u = XY$ and $-\lambda^2$ as a separation constant leads to

$$X'' + \lambda^2 X = 0,$$

$$X(0) = 0,$$

$$X(\pi) = 0,$$

and

$$Y'' - \lambda^2 Y = 0.$$

Then the boundedness of $u$ as $y \to \infty$ gives $Y = c_1 e^{-ny}$ and $X = c_2 \sin nx$ for $n = 1, 2, 3, \ldots$ so that

$$u = \sum_{n=1}^{\infty} A_n e^{-ny} \sin nx.$$

Imposing

$$u(x, 0) = f(x) = \sum_{n=1}^{\infty} A_n \sin nx$$

gives

$$A_n = \frac{2}{\pi} \int_0^{\pi} f(x) \sin nx\, dx$$

so that

$$u(x, y) = \sum_{n=1}^{\infty} \left( \frac{2}{\pi} \int_0^{\pi} f(x) \sin nx\, dx \right) e^{-ny} \sin nx.$$

12. Since the boundary conditions at $x = 0$ and $x = a$ are functions of $y$ we choose to separate Laplace's equation as

$$\frac{X''}{X} = -\frac{Y''}{Y} = \lambda^2$$

so that

$$X'' - \lambda^2 X = 0$$

$$Y'' + \lambda^2 Y = 0$$

and

$$X(x) = c_1 \cosh \lambda x + c_2 \sinh \lambda x$$

$$Y(y) = c_3 \cos \lambda y + c_2 \sin \lambda y.$$

Now $Y(0) = 0$ gives $c_3 = 0$ and $Y(b) = 0$ implies $\sin \lambda b = 0$ or $\lambda = n\pi/b$ for $n = 1, 2, 3, \ldots$ . Thus

$$u_n(x, y) = XY = \left( A_n \cosh \frac{n\pi}{b} x + B_n \sinh \frac{n\pi}{b} x \right) \sin \frac{n\pi}{b} y$$

and

$$u(x, y) = \sum_{n=1}^{\infty} \left( A_n \cosh \frac{n\pi}{b} x + B_n \sinh \frac{n\pi}{b} x \right) \sin \frac{n\pi}{b} y. \tag{1}$$

At $x = 0$ we then have

$$F(y) = \sum_{n=1}^{\infty} A_n \sin \frac{n\pi}{b} y$$

and consequently

$$A_n = \frac{2}{b} \int_0^b F(y) \sin \frac{n\pi}{b} y \, dy. \tag{2}$$

At $x = a$,

$$G(y) = \sum_{n=1}^{\infty} \left( A_n \cosh \frac{n\pi}{b} a + B_n \sinh \frac{n\pi}{b} a \right) \sin \frac{n\pi}{b} y$$

indicates that the entire expression in the parentheses is given by

$$A_n \cosh \frac{n\pi}{b} a + B_n \sinh \frac{n\pi}{b} a = \frac{2}{b} \int_0^b G(y) \sin \frac{n\pi}{b} y \, dy.$$

We can now solve for $B_n$:

$$B_n \sinh \frac{n\pi}{b} a = \frac{2}{b} \int_0^b G(y) \sin \frac{n\pi}{b} y \, dy - A_n \cosh \frac{n\pi}{b} a$$

$$B_n = \frac{1}{\sinh \frac{n\pi}{b} a} \left( \frac{2}{b} \int_0^b G(y) \sin \frac{n\pi}{b} y \, dy - A_n \cosh \frac{n\pi}{b} a \right). \tag{3}$$

A solution to the given boundary-value problem consists of the series (1) with coefficients $A_n$ and $B_n$ given in (2) and (3), respectively.

**3.** If we let $u(x, t) = v(x, t) + \psi(x)$, then we obtain as in Example 1 in the text

$$k\psi'' + r = 0$$

or

$$\psi(x) = -\frac{r}{2k}x^2 + c_1 x + c_2.$$

The boundary conditions become

$$u(0, t) = v(0, t) + \psi(0) = u_0$$

$$u(1, t) = v(1, t) + \psi(1) = u_0.$$

Letting $\psi(0) = \psi(1) = u_0$ we obtain homogeneous boundary conditions in $v$:

$$v(0, t) = 0 \quad \text{and} \quad v(1, t) = 0.$$

Now $\psi(0) = \psi(1) = u_0$ implies $c_2 = u_0$ and $c_1 = r/2k$. Thus

$$\psi(x) = -\frac{r}{2k}x^2 + \frac{r}{2k}x + u_0 = u_0 - \frac{r}{2k}x(x - 1).$$

To determine $v(x, t)$ we solve

$$k\frac{\partial^2 v}{\partial x^2} = \frac{\partial v}{\partial t}, \quad 0 < x < 1, \ t > 0$$

$$v(0, t) = 0, \quad v(1, t) = 0,$$

$$v(x, 0) = \frac{r}{2k}x(x - 1) - u_0.$$

Separating variables, we find

$$v(x, t) = \sum_{n=1}^{\infty} A_n e^{-kn^2\pi^2 t} \sin n\pi x,$$

where

$$A_n = 2\int_0^1 \left[\frac{r}{2k}x(x - 1) - u_0\right] \sin n\pi x \, dx = 2\left[\frac{u_0}{n\pi} + \frac{r}{kn^3\pi^3}\right][(-1)^n - 1]. \tag{4}$$

Hence, a solution of the original problem is

$$u(x, t) = \psi(x) + v(x, t)$$

$$= u_0 - \frac{r}{2k}x(x - 1) + \sum_{n=1}^{\infty} A_n e^{-kn^2\pi^2 t} \sin n\pi x,$$

where $A_n$ is defined in (4).

**6.** Substituting $u(x,t) = v(x,t) + \psi(x)$ into the partial differential equation gives

$$k\frac{\partial^2 v}{\partial x^2} + k\psi'' - hv - h\psi = \frac{\partial v}{\partial t}.$$

This equation will be homogeneous provided $\psi$ satisfies

$$k\psi'' - h\psi = 0.$$

Since $k$ and $h$ are positive, the general solution of this latter equation is

$$\psi(x) = c_1 \cosh\sqrt{\frac{h}{k}}\,x + c_2 \sinh\sqrt{\frac{h}{k}}\,x.$$

From $\psi(0) = 0$ and $\psi(\pi) = u_0$ we find $c_1 = 0$ and $c_2 = u_0/\sinh\sqrt{h/k}\,\pi$. Hence

$$\psi(x) = u_0\frac{\sinh\sqrt{h/k}\,x}{\sinh\sqrt{h/k}\,\pi}.$$

Now the new problem is

$$k\frac{\partial^2 v}{\partial x^2} - hv = \frac{\partial v}{\partial t}, \quad 0 < x < \pi,\ t > 0$$

$$v(0,t) = 0, \quad v(\pi,t) = 0, \quad t > 0$$

$$v(x,0) = -\psi(x), \quad 0 < x < \pi.$$

If we let $v = XT$ then

$$\frac{X''}{X} = \frac{T' + hT}{kT} = -\lambda^2$$

gives the separated differential equations

$$X'' + \lambda^2 X = 0 \quad \text{and} \quad T' + \left(h + k\lambda^2\right)T = 0.$$

The respective solutions are

$$X(x) = c_3 \cos\lambda x + c_4 \sin\lambda x$$

$$T(t) = c_5 e^{-\left(h + k\lambda^2\right)t}.$$

From $X(0) = 0$ we get $c_3 = 0$ and from $X(\pi) = 0$ we find $\lambda = n$ for $n = 1, 2, 3, \ldots$. Consequently, it follows that

$$v(x,t) = \sum_{n=1}^{\infty} A_n e^{-\left(h + kn^2\right)t}\sin nx$$

where

$$A_n = -\frac{2}{\pi}\int_0^{\pi}\psi(x)\sin nx\,dx.$$

Hence a solution of the original problem is

$$u(x,t) = u_0 \frac{\sinh\sqrt{h/k}\,x}{\sinh\sqrt{h/k}\,\pi} + e^{-ht}\sum_{n=1}^{\infty} A_n e^{-kn^2 t}\sin nx$$

where

$$A_n = -\frac{2}{\pi}\int_0^{\pi} u_0 \frac{\sinh\sqrt{h/k}\,x}{\sinh\sqrt{h/k}\,\pi}\sin nx\, dx.$$

Using the exponential definition of the hyperbolic sine and integration by parts we find

$$A_n = \frac{2u_0 nk(-1)^n}{\pi\,(h+kn^2)}.$$

9. Substituting $u(x,t) = v(x,t) + \psi(x)$ into the partial differential equation gives

$$a^2 \frac{\partial^2 v}{\partial x^2} + a^2\psi'' + Ax = \frac{\partial^2 v}{\partial t^2}.$$

This equation will be homogeneous provided $\psi$ satisfies

$$a^2\psi'' + Ax = 0.$$

The general solution of this differential equation is

$$\psi(x) = -\frac{A}{6a^2}x^3 + c_1 x + c_2.$$

From $\psi(0) = 0$ we obtain $c_2 = 0$, and from $\psi(1) = 0$ we obtain $c_1 = A/6a^2$. Hence

$$\psi(x) = \frac{A}{6a^2}(x - x^3).$$

Now the new problem is

$$a^2 \frac{\partial^2 v}{\partial x^2} = \frac{\partial^2 v}{\partial t^2}$$

$$v(0,t) = 0, \quad v(1,t) = 0, \quad t > 0,$$

$$v(x,0) = -\psi(x), \quad v_t(x,0) = 0, \quad 0 < x < 1.$$

Identifying this as the wave equation solved in Section 12.4 in the text with $L = 1$, $f(x) = -\psi(x)$, and $g(x) = 0$ we obtain

$$v(x,t) = \sum_{n=1}^{\infty} A_n \cos n\pi at \sin n\pi x$$

where

$$A_n = 2\int_0^1 [-\psi(x)]\sin n\pi x\, dx = \frac{A}{3a^2}\int_0^1 (x^3 - x)\sin n\pi x\, dx = \frac{2A(-1)^n}{a^2\pi^3 n^3}.$$

Thus

$$u(x,t) = \frac{A}{6a^2}(x - x^3) + \frac{2A}{a^2\pi^3}\sum_{n=1}^{\infty}\frac{(-1)^n}{n^3}\cos n\pi at \sin n\pi x.$$

**12.** Substituting $u(x,y) = v(x,y) + \psi(x)$ into Poisson's equation we obtain

$$\frac{\partial^2 v}{\partial x^2} + \psi''(x) + h + \frac{\partial^2 v}{\partial y^2} = 0.$$

The equation will be homogeneous provided $\psi$ satisfies $\psi''(x) + h = 0$ or $\psi(x) = -\frac{h}{2}x^2 + c_1 x + c_2$. From $\psi(0) = 0$ we obtain $c_2 = 0$. From $\psi(\pi) = 1$ we obtain

$$c_1 = \frac{1}{\pi} + \frac{h\pi}{2}.$$

Then

$$\psi(x) = \left(\frac{1}{\pi} + \frac{h\pi}{2}\right)x - \frac{h}{2}x^2.$$

The new boundary-value problem is

$$\frac{\partial^2 v}{\partial x^2} + \frac{\partial^2 v}{\partial y^2} = 0$$

$$v(0,y) = 0, \quad v(\pi,y) = 0,$$

$$v(x,0) = -\psi(x), \quad 0 < x < \pi.$$

This is Problem 9 in Section 12.5. The solution is

$$v(x,y) = \sum_{n=1}^{\infty}A_n e^{-ny}\sin nx$$

where

$$A_n = \frac{2}{\pi}\int_0^{\pi}[-\psi(x)\sin nx]\,dx$$

$$= \frac{2(-1)^n}{m}\left(\frac{1}{\pi} + \frac{h\pi}{2}\right) - h(-1)^n\left(\frac{\pi}{n} + \frac{2}{n^2}\right).$$

Thus

$$u(x,y) = v(x,y) + \psi(x) = \left(\frac{1}{\pi} + \frac{h\pi}{2}\right)x - \frac{h}{2}x^2 + \sum_{n=1}^{\infty}A_n e^{-ny}\sin nx.$$

**185**

**3.** Separating variables in Laplace's equation gives

$$X'' + \lambda^2 X = 0$$

$$Y'' - \lambda^2 Y = 0$$

and

$$X(x) = c_1 \cos \lambda x + c_2 \sin \lambda x$$

$$Y(y) = c_3 \cosh \lambda y + c_4 \sinh \lambda y.$$

From $u(0, y) = 0$ we obtain $X(0) = 0$ and $c_1 = 0$. From $u_x(a, y) = -hu(a, y)$ we obtain $X'(a) = -hX(a)$ and

$$\lambda \cos \lambda a = -h \sin \lambda a \quad \text{or} \quad \tan \lambda a = -\frac{\lambda}{h}.$$

Let $\lambda_n$, where $n = 1, 2, 3, \ldots$, be the cosecutive positive roots of this equation. From $u(x, 0) = 0$ we obtain $Y(0) = 0$ and $c_3 = 0$. Thus

$$u(x, y) = \sum_{n=1}^{\infty} A_n \sinh \lambda_n y \sin \lambda_n x.$$

Now

$$f(x) = \sum_{n=1}^{\infty} A_n \sinh \lambda_n b \sin \lambda_n x$$

and

$$A_n \sinh \lambda_n b = \frac{\int_0^a f(x) \sin \lambda_n x \, dx}{\int_0^a \sin^2 \lambda_n x \, dx}.$$

Since

$$\int_0^a \sin^2 \lambda_n x \, dx = \frac{1}{2}\left[a - \frac{1}{2\lambda_n} \sin 2\lambda_n a\right] = \frac{1}{2}\left[a - \frac{1}{\lambda_n} \sin \lambda_n a \cos \lambda_n a\right]$$

$$= \frac{1}{2}\left[a - \frac{1}{h\lambda_n}(h \sin \lambda_n a) \cos \lambda_n a\right]$$

$$= \frac{1}{2}\left[a - \frac{1}{h\lambda_n}(-\lambda_n \cos \lambda_n a) \cos \lambda_n a\right] = \frac{1}{2h}\left[ah + \cos^2 \lambda_n a\right],$$

we have

$$A_n = \frac{2h}{\sinh \lambda_n b[ah + \cos^2 \lambda_n a]} \int_0^a f(x) \sin \lambda_n x \, dx.$$

**6.** Substituting $u(x, t) = v(x, t) + \psi(x)$ into the partial differential equation gives

$$a^2 \frac{\partial^2 v}{\partial x^2} + \psi''(x) = \frac{\partial^2 v}{\partial t^2}.$$

**186**

This equation will be homogeneous if $\psi''(x) = 0$ or $\psi(x) = c_1 x + c_2$. The boundary condition $u(0,t) = 0$ implies $\psi(0) = 0$ which implies $c_2 = 0$. Thus $\psi(x) = c_1 x$. Using the second boundary condition, we obtain

$$E\left(\frac{\partial v}{\partial x} + \psi'\right)\Big|_{x=L} = F_0,$$

which will be homogeneous when

$$E\psi'(L) = F_0.$$

Since $\psi'(x) = c_1$ we conclude that $c_1 = F_0/E$ and

$$\psi(x) = \frac{F_0}{E} x.$$

The new boundary-value problem is

$$a^2 \frac{\partial^2 v}{\partial x^2} = \frac{\partial^2 v}{\partial t^2}, \quad 0 < x < L, \quad t > 0$$

$$v(0,t) = 0, \quad \frac{\partial v}{\partial x}\Big|_{x=L} = 0, \quad t > 0,$$

$$v(x,0) = -\frac{F_0}{E} x, \quad \frac{\partial v}{\partial t}\Big|_{t=0} = 0, \quad 0 < x < L.$$

Referring to Example 2 in the text we see that

$$v(x,t) = \sum_{n=1}^{\infty} A_n \cos a\left(\frac{2n-1}{2L}\right)\pi t \sin\left(\frac{2n-1}{2L}\right)\pi x$$

where

$$-\frac{F_0}{E} x = \sum_{n=1}^{\infty} A_n \sin\left(\frac{2n-1}{2L}\right)\pi x$$

and

$$A_n = \frac{-F_0 \int_0^L x \sin\left(\frac{2n-1}{2L}\right)\pi x\, dx}{E \int_0^L \sin^2\left(\frac{2n-1}{2L}\right)\pi x\, dx} = \frac{8F_0 L(-1)^n}{E\pi^2(2n-1)^2}.$$

Thus

$$u(x,t) = v(x,t) + \psi(x)$$

$$= \frac{F_0}{E} x + \frac{8F_0 L}{E\pi^2} \sum_{n=1}^{\infty} \frac{(-1)^n}{(2n-1)^2} \cos a\left(\frac{2n-1}{2L}\right)\pi t \sin\left(\frac{2n-1}{2L}\right)\pi x.$$

9. (a) Using $u = XT$ and separation constant $\lambda^4$ we find

$$X^{(4)} - \lambda^4 X = 0$$

and

$$X(x) = c_1 \cos \lambda x + c_2 \sin \lambda x + c_3 \cosh \lambda x + c_4 \sinh \lambda x.$$

Since $u = XT$ the boundary conditions become

$$X(0) = 0, \quad X'(0) = 0, \quad X''(1) = 0, \quad X'''(1) = 0.$$

Now $X(0) = 0$ implies $c_1 + c_3 = 0$, while $X'(0) = 0$ implies $c_2 + c_4 = 0$. Thus

$$X(x) = c_1 \cos \lambda x + c_2 \sin \lambda x - c_1 \cosh \lambda x - c_2 \sinh \lambda x.$$

The boundary condition $X''(1) = 0$ implies

$$-c_1 \cos \lambda - c_2 \sin \lambda - c_1 \cosh \lambda - c_2 \sinh \lambda = 0$$

while the boundary condition $X'''(1) = 0$ implies

$$c_1 \sin \lambda - c_2 \cos \lambda - c_1 \sinh \lambda - c_2 \cosh \lambda = 0.$$

We then have the system of two equations in two unknowns

$$(\cos \lambda + \cosh \lambda)c_1 + (\sin \lambda + \sinh \lambda)c_2 = 0$$

$$(\sin \lambda - \sinh \lambda)c_1 - (\cos \lambda + \cosh \lambda)c_2 = 0.$$

This homogeneous system will have nontrivial solutions for $c_1$ and $c_2$ provided

$$\begin{vmatrix} \cos \lambda + \cosh \lambda & \sin \lambda + \sinh \lambda \\ \sin \lambda - \sinh \lambda & - \cos \lambda - \cosh \lambda \end{vmatrix} = 0$$

or

$$-2 - 2\cos \lambda \cosh \lambda = 0.$$

Thus, the eigenvalues are determined by the equation $\cos \lambda \cosh \lambda = -1$.

(b) Using a computer to graph $\cosh \lambda$ and $-1/\cos \lambda = -\sec \lambda$ we see that the first two positive eigenvalues occur near 1.9 and 4.7. Applying Newton's method with these initial values we find that the eigenvalues are $\lambda_1 = 1.8751$ and $\lambda_2 = 4.6941$.

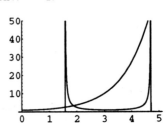

———————— **Exercises 12.8** ————————

**3.** In this problem we need to solve the partial differential equation

$$a^2 \left( \frac{\partial^2 u}{\partial x^2} + \frac{\partial^2 u}{\partial y^2} \right) = \frac{\partial^2 u}{\partial t^2}.$$

To separate this equation we try $u(x, y, t) = X(x)Y(y)T(t)$:

$$a^2(X''YT + XY''T) = XYT''$$

$$\frac{X''}{X} = -\frac{Y''}{Y} + \frac{T''}{a^2 T} = -\lambda^2.$$

Then

$$X'' + \lambda^2 X = 0 \tag{5}$$

$$\frac{Y''}{Y} = \frac{T''}{a^2 T} + \lambda^2 = -\mu^2$$

$$Y'' + \mu^2 Y = 0 \tag{6}$$

$$T'' + a^2 \left( \lambda^2 + \mu^2 \right) T = 0. \tag{7}$$

The general solutions of equations (5), (6), and (7) are, respectively,

$$X(x) = c_1 \cos \lambda x + c_2 \sin \lambda x$$

$$Y(y) = c_3 \cos \mu y + c_4 \sin \mu y$$

$$T(t) = c_5 \cos a\sqrt{\lambda^2 + \mu^2}\, t + c_6 \sin a\sqrt{\lambda^2 + \mu^2}\, t.$$

The conditions $X(0) = 0$ and $Y(0) = 0$ give $c_1 = 0$ and $c_3 = 0$. The conditions $X(\pi) = 0$ and $Y(\pi) = 0$ yield two sets of eigenvalues:

$$\lambda = m, \ m = 1, 2, 3, \ldots \quad \text{and} \quad \mu = n, \ n = 1, 2, 3, \ldots .$$

A product solution of the partial differential equation that satisfies the boundary conditions is

$$u_{mn}(x, y, t) = \left( A_{mn} \cos a\sqrt{m^2 + n^2}\, t + B_{mn} \sin a\sqrt{m^2 + n^2}\, t \right) \sin mx \sin ny.$$

To satisfy the initial conditions we use the superposition principle:

$$u(x, y, t) = \sum_{m=1}^{\infty} \sum_{n=1}^{\infty} \left( A_{mn} \cos a\sqrt{m^2 + n^2}\, t + B_{mn} \sin a\sqrt{m^2 + n^2}\, t \right) \sin mx \sin ny.$$

The initial condition $u_t(x, y, 0) = 0$ implies $B_{mn} = 0$ and

$$u(x, y, t) = \sum_{m=1}^{\infty} \sum_{n=1}^{\infty} A_{mn} \cos a\sqrt{m^2 + n^2}\, t \sin mx \sin ny.$$

**189**

## Exercises 12.8

At $t = 0$ we have

$$xy(x - \pi)(y - \pi) = \sum_{m=1}^{\infty} \sum_{n=1}^{\infty} A_{mn} \sin mx \sin ny.$$

It follows that (See Problem 52, Exercises 11.3)

$$A_{mn} = \frac{4}{\pi^2} \int_0^{\pi} \int_0^{\pi} xy(x - \pi)(y - \pi) \sin mx \sin ny \, dx \, dy$$

$$= \frac{4}{\pi^2} \int_0^{\pi} x(x - \pi) \sin mx \, dx \int_0^{\pi} y(y - \pi) \sin ny \, dy$$

$$= \frac{16}{m^3 n^3 \pi^2} [(-1)^m - 1][(-1)^n - 1].$$

**6.** The boundary and initial conditions are

$$u(0, y, z) = 0, \qquad\qquad u(a, y, z) = 0,$$

$$u(x, 0, z) = 0, \qquad\qquad u(x, b, z) = 0,$$

$$u(x, y, 0) = f(x, y), \qquad u(x, y, c) = 0.$$

The conditions $X(0) = Y(0) = 0$ give $c_1 = c_3 = 0$. The conditions $X(a) = Y(b) = 0$ yield two sets of eigenvalues:

$$\lambda = \frac{m\pi}{a}, \; m = 1, 2, 3, \ldots \quad \text{and} \quad \mu = \frac{n\pi}{b}, \; n = 1, 2, 3, \ldots .$$

Let

$$\omega_{mn}^2 = \frac{m^2 \pi^2}{a^2} + \frac{n^2 \pi^2}{b^2}.$$

Then the boundary condition $Z(c) = 0$ gives

$$c_5 \cosh c\omega_{mn} + c_6 \sinh c\omega_{mn} = 0$$

from which we obtain

$$Z(z) = c_5 \left( \cosh \omega_{mn} z - \frac{\cosh c\omega_{mn}}{\sinh c\omega_{mn}} \sinh \omega z \right)$$

$$= \frac{c_5}{\sinh c\omega_{mn}} (\sinh c\omega_{mn} \cosh \omega_{mn} z - \cosh c\omega_{mn} \sinh \omega_{mn} z)$$

$$= c_{mn} \sinh \omega_{mn}(c - z).$$

By the superposition principle

$$u(x, y, t) = \sum_{m=1}^{\infty} \sum_{n=1}^{\infty} A_{mn} \sinh \omega_{mn}(c - z) \sin \frac{m\pi}{a} x \sin \frac{n\pi}{b} y$$

**190**

where

$$A_{mn} = \frac{4}{ab\sinh c\omega_{mn}} \int_0^b \int_0^a f(x,y) \sin\frac{m\pi}{a}x \sin\frac{n\pi}{b}y\, dx\, dy.$$

# Chapter 12 Review Exercises

**3.** Substituting $u(x,t) = v(x,t) + \psi(x)$ into the partial differential equation we obtain

$$k\frac{\partial^2 v}{\partial x^2} + k\psi''(x) = \frac{\partial v}{\partial t}.$$

This equation will be homogeneous provided $\psi$ satisfies

$$k\psi'' = 0 \quad \text{or} \quad \psi = c_1 x + c_2.$$

Considering

$$u(0,t) = v(0,t) + \psi(0) = u_0$$

we set $\psi(0) = u_0$ so that $\psi(x) = c_1 x + u_0$. Now

$$-\frac{\partial u}{\partial x}\bigg|_{x=\pi} = -\frac{\partial v}{\partial x}\bigg|_{x=\pi} - \psi'(x) = v(\pi,t) + \psi(\pi) - u_1$$

is equivalent to

$$\frac{\partial v}{\partial x}\bigg|_{x=\pi} + v(\pi,t) = u_1 - \psi'(x) - \psi(\pi) = u_1 - c_1 - (c_1\pi + u_0),$$

which will be homogeneous when

$$u_1 - c_1 - c_1\pi - u_0 = 0 \quad \text{or} \quad c_1 = \frac{u_1 - u_0}{1 + \pi}.$$

The steady-state solution is

$$\psi(x) = \left(\frac{u_1 - u_0}{1 + \pi}\right)x + u_0.$$

**6.** The boundary-value problem is

$$\frac{\partial^2 u}{\partial x^2} + x^2 = \frac{\partial^2 u}{\partial t^2}, \quad 0 < x < 1, \quad t > 0,$$

$$u(0,t) = 1, \quad u(1,t) = 0, \quad t > 0,$$

$$u(x,0) = f(x), \quad u_t(x,0) = 0, \quad 0 < x < 1.$$

Substituting $u(x,t) = v(x,t) + \psi(x)$ into the partial differential equation gives

$$\frac{\partial^2 v}{\partial x^2} + \psi''(x) + x^2 = \frac{\partial^2 v}{\partial t^2}.$$

**191**

This equation will be homogeneous provided $\psi''(x) + x^2 = 0$ or

$$\psi(x) = -\frac{1}{12}x^4 + c_1 x + c_2.$$

From $\psi(0) = 1$ and $\psi(1) = 0$ we obtain $c_1 = -11/12$ and $c_2 = 1$. The new problem is

$$\frac{\partial^2 v}{\partial x^2} = \frac{\partial^2 v}{\partial t^2}, \quad 0 < x < 1, \quad t > 0,$$

$$v(0,t) = 0, \quad v(1,t) = 0, \quad t > 0,$$

$$v(x,0) = f(x) - \psi(x), \quad v_t(x,0) = 0, \quad 0 < x < 1.$$

From Section 12.4 in the text we see that $B_n = 0$,

$$A_n = 2\int_0^1 [f(x) - \psi(x)]\sin n\pi x \, dx = 2\int_0^1 \left[ f(x) + \frac{1}{12}x^4 + \frac{11}{12}x - 1 \right]\sin n\pi x \, dx,$$

and

$$v(x,t) = \sum_{n=1}^{\infty} A_n \cos n\pi t \sin n\pi x.$$

Thus

$$u(x,t) = v(x,t) + \psi(x) = -\frac{1}{12}x^4 - \frac{11}{12}x + 1 + \sum_{n=1}^{\infty} A_n \cos n\pi t \sin n\pi x.$$

9. Using $u = XY$ and $\lambda^2$ as a separation constant leads to

$$X'' - \lambda^2 X = 0,$$

and

$$Y'' + \lambda^2 Y = 0,$$

$$Y(0) = 0,$$

$$Y(\pi) = 0.$$

Then

$$Y = c_1 \sin ny \quad \text{and} \quad X = c_2 e^{-nx}$$

for $n = 1, 2, 3, \ldots$(since $u$ must be bounded as $x \to \infty$) so that

$$u = \sum_{n=1}^{\infty} A_n e^{-nx} \sin ny.$$

Imposing

$$u(0,y) = 50 = \sum_{n=1}^{\infty} A_n \sin ny$$

gives

$$A_n = \frac{2}{\pi}\int_0^\pi 50 \sin ny \, dy = \frac{100}{n\pi}[1 - (-1)^n]$$

so that

$$u(x, y) = \sum_{n=1}^{\infty} \frac{100}{n\pi}[1 - (-1)^n]e^{-nx} \sin ny.$$

**12.** Substituting $u(x, t) = v(x, t) + \psi(x)$ into the partial differential equation gives

$$k\frac{\partial^2 v}{\partial x^2} + k\psi'' + \sin 2\pi x = \frac{\partial v}{\partial t}.$$

This equation will be homogeneous provided $\psi$ satisfies

$$k\psi'' + \sin 2\pi x = 0.$$

The general solution of this equation is

$$\psi(x) = \frac{1}{4k\pi^2} \sin 2\pi x + c_1 x + c_2.$$

From $\psi(0) = \psi(1) = 0$ we find that $c_1 = c_2 = 0$ and

$$\psi(x) = \frac{1}{4k\pi^2} \sin 2\pi x.$$

Now the new problem is

$$k\frac{\partial^2 v}{\partial x^2} = \frac{\partial v}{\partial t}, \quad 0 < x < 1, \quad t > 0$$

$$v(0, t) = 0, \quad v(1, t) = 0, \quad t > 0$$

$$v(x, 0) = \sin \pi x - \psi(x), \quad 0 < x < 1.$$

If we let $v = XT$ then

$$\frac{X''}{X} = \frac{T'}{kT} = -\lambda^2$$

gives the separated differential equations

$$X'' + \lambda^2 X = 0 \quad \text{and} \quad T' + k\lambda^2 T = 0.$$

The respective solutions are

$$X(x) = c_3 \cos \lambda x + c_4 \sin \lambda x$$

$$T(t) = c_5 e^{-k\lambda^2 t}.$$

From $X(0) = 0$ we get $c_3 = 0$ and from $X(1) = 0$ we find $\lambda = n\pi$ for $n = 1, 2, 3, \ldots$ . Consequently, it follows that

$$v(x, t) = \sum_{n=1}^{\infty} A_n e^{-kn^2\pi^2 t} \sin n\pi x$$

where

$$v(x, 0) = \sin \pi x - \frac{1}{4k\pi^2} \sin 2\pi x = 0$$

**193**

implies

$$A_n = 2 \int_0^1 \left( \sin \pi x - \frac{1}{4k\pi^2} \sin 2\pi x \right) \sin n\pi x \, dx.$$

By orthogonality $A_n = 0$ for $n = 3, 4, 5, \ldots$, and only $A_1$ and $A_2$ can be nonzero. We have

$$A_1 = 2 \left[ \int_0^1 \sin^2 \pi x \, dx - \frac{1}{4k\pi^2} \int_0^1 \sin 2\pi x \sin \pi x \, dx \right] = 2 \int_0^1 \frac{1}{2}(1 - \cos 2\pi x) \, dx = 1$$

and

$$A_2 = 2 \left[ \int_0^1 \sin \pi x \sin 2\pi x \, dx - \frac{1}{4k\pi^2} \int_0^1 \sin^2 2\pi x \, dx \right]$$

$$= -\frac{1}{2k\pi^2} \int_0^1 \frac{1}{2}(1 - \cos 4\pi x) \, dx = -\frac{1}{4k\pi^2}.$$

Therefore

$$v(x, t) = A_1 e^{-k\pi^2 t} \sin \pi x + A_2 e^{-k4\pi^2 t} \sin 2\pi x$$

$$= e^{-k\pi^2 t} \sin \pi x - \frac{1}{4k\pi^2} e^{-4k\pi^2 t} \sin 2\pi x$$

and

$$u(x, t) = v(x, t) + \psi(x) = e^{-k\pi^2 t} \sin \pi x + \frac{1}{4k\pi^2}(1 - e^{-4k\pi^2 t}) \sin 2\pi x.$$

# 13 Boundary-Value Problems in Other Coordinate Systems

━━━━━━━━ **Exercises 13.1** ━━━━━━━━━━━━━━━━━━━━━━━━━━━━━━

**3.** We have

$$A_0 = \frac{1}{2\pi} \int_0^{2\pi} (2\pi\theta - \theta^2)\, d\theta = \frac{2\pi^2}{3}$$

$$A_n = \frac{1}{\pi} \int_0^{2\pi} (2\pi\theta - \theta^2) \cos n\theta\, d\theta = -\frac{4}{n^2}$$

$$B_n = \frac{1}{\pi} \int_0^{2\pi} (2\pi\theta - \theta^2) \sin n\theta\, d\theta = 0$$

and so

$$u(r, \theta) = \frac{2\pi^2}{3} - 4 \sum_{n=1}^{\infty} \frac{r^n}{n^2} \cos n\theta.$$

**6.** We solve

$$\frac{\partial^2 u}{\partial r^2} + \frac{1}{r} \frac{\partial u}{\partial r} + \frac{1}{r^2} \frac{\partial^2 u}{\partial \theta^2} = 0, \quad 0 < \theta < \frac{\pi}{2}, \quad 0 < r < c,$$

$$u(c, \theta) = f(\theta), \quad 0 < \theta < \frac{\pi}{2},$$

$$u(r, 0) = 0, \quad u(r, \pi/2) = 0, \quad 0 < r < c.$$

Proceeding as in Example 1 of Section 13.1 in the text we obtain the separated differential equations

$$r^2 R'' + rR' - \lambda^2 R = 0$$

$$\Theta'' + \lambda^2 \Theta = 0$$

with solutions

$$\Theta(\theta) = c_1 \cos \lambda\theta + c_2 \sin \lambda\theta$$

$$R(r) = c_3 r^\lambda + c_4 r^{-\lambda}.$$

Since we want $R(r)$ to be bounded as $r \to 0$ we require $c_4 = 0$. Applying the boundary conditions $\Theta(0) = 0$ and $\Theta(\pi/2) = 0$ we find that $c_1 = 0$ and $\lambda = 2n$ for $n = 1, 2, 3, \ldots$. Therefore

$$u(r, \theta) = \sum_{n=1}^{\infty} A_n r^{2n} \sin 2n\theta.$$

From
$$u(c, \theta) = f(\theta) = \sum_{n=1}^{\infty} A_n c^n \sin 2n\theta$$

we find
$$A_n = \frac{4}{\pi c^{2n}} \int_0^{\pi/2} f(\theta) \sin 2n\theta \, d\theta.$$

9. Proceeding as in Example 1 of Section 13.1 and using the periodicity of $u(r, \theta)$, we have
$$\Theta(\theta) = c_1 \cos \lambda\theta + c_2 \lambda\theta$$

where $\lambda = n$ for $n = 0, 1, 2, \ldots$ . Then
$$R(r) = c_3 r^n + c_4 r^{-n}.$$

[We do not have $c_4 = 0$ in this case since $0 < a \le r$.] Since $u(b, \theta) = 0$ we have
$$u(r, \theta) = A_0 \ln \frac{r}{b} + \sum_{n=1}^{\infty} \left[ \left( \frac{b}{r} \right)^n - \left( \frac{r}{b} \right)^n \right] [A_n \cos n\theta + B_n \sin n\theta].$$

From
$$u(a, \theta) = f(\theta) = A_0 \ln \frac{a}{b} + \sum_{n=1}^{\infty} \left[ \left( \frac{b}{a} \right)^n - \left( \frac{a}{b} \right)^n \right] [A_n \cos n\theta + B_n \sin n\theta]$$

we find
$$A_0 \ln \frac{a}{b} = \frac{1}{2\pi} \int_0^{2\pi} f(\theta) \, d\theta,$$

$$\left[ \left( \frac{b}{a} \right)^n - \left( \frac{a}{b} \right)^n \right] A_n = \frac{1}{\pi} \int_0^{2\pi} f(\theta) \cos n\theta \, d\theta,$$

and
$$\left[ \left( \frac{b}{a} \right)^n - \left( \frac{a}{b} \right)^n \right] B_n = \frac{1}{\pi} \int_0^{2\pi} f(\theta) \sin n\theta \, d\theta.$$

12. Letting $u(r, \theta) = v(r, \theta) + \psi(\theta)$ we obtain $\psi''(\theta) = 0$ and so $\psi(\theta) = c_1\theta + c_2$. From $\psi(0) = 0$ and $\psi(\pi) = u_0$ we find, in turn, $c_2 = 0$ and $c_1 = u_0/\pi$. Therefore $\psi(\theta) = \frac{u_0}{\pi} \theta$.

Now $u(1, \theta) = v(1, \theta) + \psi(\theta)$ so that $v(1, \theta) = u_0 - \frac{u_0}{\pi} \theta$.

From
$$v(r, \theta) = \sum_{n=1}^{\infty} A_n r^n \sin n\theta \quad \text{and} \quad v(1, \theta) = \sum_{n=1}^{\infty} A_n \sin n\theta$$

we obtain
$$A_n = \frac{2}{\pi} \int_0^{\pi} \left( u_0 - \frac{u_0}{\pi} \theta \right) \sin n\theta \, d\theta = \frac{2u_0}{\pi n}.$$

Thus
$$u(r, \theta) = \frac{u_0}{\pi} \theta + \frac{2u_0}{\pi} \sum_{n=1}^{\infty} \frac{r^n}{n} \sin n\theta.$$

## Exercises 13.2

**3.** Referring to Example 2 in the text we have

$$R(r) = c_1 J_0(\lambda r) + c_2 Y_0(\lambda r)$$

$$Z(z) = c_3 \cosh \lambda z + c_4 \sinh \lambda z$$

where $c_2 = 0$ and $J_0(2\lambda) = 0$ defines the positive eigenvalues $\lambda_n$. From $Z(4) = 0$ we obtain

$$c_3 \cosh 4\lambda_n + c_4 \sinh 4\lambda_n = 0 \quad \text{or} \quad c_4 = -c_3 \frac{\cosh 4\lambda_n}{\sinh 4\lambda_n}.$$

Then

$$Z(z) = c_3 \left[ \cosh \lambda_n z - \frac{\cosh 4\lambda_n}{\sinh 4\lambda_n} \sinh \lambda_n z \right] = c_3 \frac{\sinh 4\lambda_n \cosh \lambda_n z - \cosh 4\lambda_n \sinh \lambda_n z}{\sinh 4\lambda_n}$$

$$= c_3 \frac{\sinh \lambda_n (4 - z)}{\sinh 4\lambda_n}$$

and

$$u(r, z) = \sum_{n=1}^{\infty} A_n \frac{\sinh \lambda_n (4 - z)}{\sinh 4\lambda_n} J_0(\lambda_n r).$$

From

$$u(r, 0) = u_0 = \sum_{n=1}^{\infty} A_n J_0(\lambda_n r)$$

we obtain

$$A_n = \frac{2u_0}{4 J_1^2(2\lambda_n)} \int_0^2 r J_0(\lambda_n r) \, dr = \frac{u_0}{\lambda_n J_1(2\lambda_n)}.$$

Thus the temperature in the cylinder is

$$u(r, z) = u_0 \sum_{n=1}^{\infty} \frac{\sinh \lambda_n (4 - z) J_0(\lambda_n r)}{\lambda_n \sinh 4\lambda_n J_1(2\lambda_n)}.$$

**6.** If the edge $r = c$ is insulated we have the boundary condition $u_r(c, t) = 0$. Letting $u(r, t) = R(r)T(t)$ and separating variables we obtain

$$\frac{R'' + \frac{1}{r} R'}{R} = \frac{T'}{kT} = \mu \quad \text{and} \quad R'' + \frac{1}{r} R' - \mu R = 0, \quad T' - \mu kT = 0.$$

From the second equation we find $T(t) = e^{\mu k t}$. If $\mu > 0$, $T(t)$ increases without bound as $t \to \infty$. Thus we assume $\mu = -\lambda^2 \leq 0$. Now

$$R'' + \frac{1}{r} R' + \lambda^2 R = 0$$

**197**

is a parametric Bessel equation with solution

$$R(r) = c_1 J_0(\lambda r) + c_2 Y_0(\lambda r).$$

Since $Y_0$ is unbounded as $r \to 0$ we take $c_2 = 0$. Then $R(r) = c_1 J_0(\lambda r)$ and the boundary condition $u_r(c, t) = R'(c)T(t) = 0$ implies

$$R'(c) = \lambda c_1 J_0'(\lambda c) = 0.$$

This defines an eigenvalue $\lambda = 0$ and positive eigenvalues $\lambda_n$. Thus

$$u(r, t) = A_0 + \sum_{n=1}^{\infty} A_n J_0(\lambda_n r) e^{-\lambda_n^2 kt}.$$

From

$$u(r, 0) = f(r) = A_0 + \sum_{n=1}^{\infty} A_n J_0(\lambda_n r)$$

we find

$$A_0 = \frac{2}{c^2} \int_0^c r f(r)\, dr$$

$$A_n = \frac{2}{c^2 J_0^2(\lambda_n c)} \int_0^c r J_0(\lambda_n r) f(r)\, dr.$$

**9.** Substituting $u(r, t) = v(r, t) + \psi(r)$ into the partial differential equation gives

$$\frac{\partial^2 v}{\partial r^2} + \frac{1}{r}\frac{\partial v}{\partial r} + \psi'' + \frac{1}{r}\psi' = \frac{\partial v}{\partial t}.$$

This equation will be homogeneous provided $\psi'' + \frac{1}{r}\psi' = 0$ or

$$\psi(r) = c_1 \ln r + c_2.$$

Since $\ln r$ is unbounded as $r \to 0$ we take $c_1 = 0$. Then $\psi(r) = c_2$ and using

$$u(2, t) = v(2, t) + \psi(2) = 100$$

we set $c_2 = \psi(r) = 100$. Referring to Problem 6 above, the solution of

$$\frac{\partial^2 v}{\partial r^2} + \frac{1}{r}\frac{\partial v}{\partial r} = \frac{\partial v}{\partial t}, \quad 0 < r < 2, \quad t > 0$$

is

$$v(r, t) = c_1 J_0(\lambda r) e^{\mu t}.$$

The boundary conditions

$$v(2, t) = 0, \quad t > 0,$$

$$v(r, 0) = u(r, 0) - \psi(r)$$

then give

$$v(r,t) = \sum_{n=1}^{\infty} A_n J_0(\lambda_n r) e^{-\lambda_n^2 t}$$

where

$$A_n = \frac{2}{2^2 J_1^2(2\lambda_n)} \int_0^2 r J_0(\lambda_n r)[u(r,0) - \psi(r)]\,dr$$

$$= \frac{1}{2 J_1^2(2\lambda_n)}\left[\int_0^1 r J_0(\lambda_n r)[200 - 100]\,dr + \int_1^2 r J_0(\lambda_n r)[100 - 100]\,dr\right]$$

$$= \frac{50}{J_1^2(2\lambda_n)} \int_0^1 r J_0(\lambda_n r)\,dr \qquad \boxed{x = \lambda_n r,\ dx = \lambda_n\,dr}$$

$$= \frac{50}{J_1^2(2\lambda_n)} \int_0^{\lambda_n} \frac{1}{\lambda_n^2} x J_0(x)\,dx$$

$$= \frac{50}{\lambda_n^2 J_1^2(2\lambda_n)} \int_0^{\lambda_n} \frac{d}{dx}[x J_1(x)]\,dx \qquad \boxed{\text{see (4) of Section 11.5 in text}}$$

$$= \frac{50}{\lambda_n^2 J_1^2(2\lambda_n)} (x J_1(x))\Big|_0^{\lambda_n} = \frac{50 J_1(\lambda_n)}{\lambda_n J_1^2(2\lambda_n)}.$$

Thus

$$u(r,t) = v(r,t) + \psi(r) = 100 + 50 \sum_{n=1}^{\infty} \frac{J_1(\lambda_n) J_0(\lambda_n r)}{\lambda_n J_1^2(2\lambda_n)} e^{-\lambda_n^2 t}.$$

**12.** **(a)** First we see that

$$\frac{R''\Theta + \frac{1}{r}R'\Theta + \frac{1}{r^2}R\Theta''}{R\Theta} = \frac{T''}{a^2 T} = -\lambda^2.$$

This gives $T'' + a^2\lambda^2 T = 0$. Then from

$$\frac{R'' + \frac{1}{r}R' + \lambda^2 R}{-R/r^2} = \frac{\Theta''}{\Theta} = -\nu^2$$

we get $\Theta'' + \nu^2\Theta = 0$ and $r^2 R'' + r R' + (\lambda^2 r^2 - \nu^2)R = 0$.

**(b)** The general solutions of the differential equations in part (a) are

$$T = c_1 \cos a\lambda t + c_2 \sin a\lambda t$$

$$\Theta = c_3 \cos \nu\theta + c_4 \cos \nu\theta$$

$$R = c_5 J_\nu(\lambda r) + c_6 Y_\nu(\lambda r).$$

**(c)** Implicitly we expect $u(r,\theta,t) = u(r,\theta+2\pi,t)$ and so $\Theta$ must be $2\pi$-periodic. Therefore $\nu = n$, $n = 0, 1, 2, \ldots$. The corresponding eigenfunctions are $1$, $\cos\theta$, $\cos 2\theta$, $\ldots$, $\sin\theta$, $\sin 2\theta$, $\ldots$.

Arguing that $u(r,\theta,t)$ is bounded as $r \to 0$ we then define $c_6 = 0$ and so $R = c_3 J_n(\lambda r)$. But $R(c) = 0$ gives $J_n(\lambda c) = 0$; this equation defines the eigenvalues $\lambda_n$. For each $n$, $\lambda_{ni} = x_{ni}/c$, $i = 1, 2, 3, \ldots$.

**(d)** $u(r,\theta,t) = \displaystyle\sum_{i=1}^{n}(A_{0i} \cos a\lambda_{0i}t + B_{0i} \sin a\lambda_{0i}t)J_0(\lambda_{0i}r)$

$$+ \sum_{n=1}^{\infty}\sum_{i=1}^{\infty}\Big[(A_{ni} \cos a\lambda_{ni}t + B_{ni} \sin a\lambda_{ni}t) \cos n\theta$$

$$+ (C_{ni} \cos a\lambda_{ni}t + D_{ni} \sin a\lambda_{ni}t) \sin n\theta\Big] J_n(\lambda_{ni}r)$$

## Exercises 13.3

**3.** The coefficients are given by

$$A_n = \frac{2n+1}{2c^n}\int_0^{\pi} \cos\theta\, P_n(\cos\theta)\sin\theta\, d\theta = \frac{2n+1}{2c^n}\int_0^{\pi} P_1(\cos\theta)P_n(\cos\theta)\sin\theta\, d\theta$$

$$\boxed{x = \cos\theta,\ dx = -\sin\theta\, d\theta}$$

$$= \frac{2n+1}{2c^n}\int_{-1}^{1} P_1(x)P_n(x)\, dx.$$

Since $P_n(x)$ and $P_m(x)$ are orthogonal for $m \neq n$, $A_n = 0$ for $n \neq 1$ and

$$A_1 = \frac{2(1)+1}{2c^1}\int_{-1}^{1} P_1(x)P_1(x)\, dx = \frac{3}{2c}\int_{-1}^{1} x^2\, dx = \frac{1}{c}.$$

Thus

$$u(r,\theta) = \frac{r}{c}P_1(\cos\theta) = \frac{r}{c}\cos\theta.$$

**6.** Referring to Example 1 in the text we have

$$R(r) = c_1 r^n \quad \text{and} \quad \Theta(\theta) = P_n(\cos\theta).$$

Now $\Theta(\pi/2) = 0$ implies that $n$ is odd, so

$$u(r,\theta) = \sum_{n=0}^{\infty} A_{2n+1}r^{2n+1}P_{2n+1}(\cos\theta).$$

From

$$u(c,\theta) = f(\theta) = \sum_{n=0}^{\infty} A_{2n+1}c^{2n+1}P_{2n+1}(\cos\theta)$$

we see that

$$A_{2n+1}c^{2n+1} = (4n+3)\int_0^{\pi/2} f(\theta)\sin\theta\, P_{2n+1}(\cos\theta)\, d\theta.$$

Thus

$$u(r, \theta) = \sum_{n=0}^{\infty} A_{2n+1} r^{2n+1} P_{2n+1}(\cos \theta)$$

where

$$A_{2n+1} = \frac{4n + 3}{c^{2n+1}} \int_0^{\pi/2} f(\theta) \sin \theta \, P_{2n+1}(\cos \theta) \, d\theta.$$

9. Checking the hint, we find

$$\frac{1}{r} \frac{\partial^2}{\partial r^2}(ru) = \frac{1}{r} \frac{\partial}{\partial r}\left[r\frac{\partial u}{\partial r} + u\right] = \frac{1}{r}\left[r\frac{\partial^2 u}{\partial r^2} + \frac{\partial u}{\partial r} + \frac{\partial u}{\partial r}\right] = \frac{\partial^2 u}{\partial r^2} + \frac{2}{r} \frac{\partial u}{\partial r}.$$

The partial differential equation then becomes

$$\frac{\partial^2}{\partial r^2}(ru) = r\frac{\partial u}{\partial t}.$$

Now, letting $ru(r, t) = v(r, t) + \psi(r)$, since the boundary condition is nonhomogeneous, we obtain

$$\frac{\partial^2}{\partial r^2}[v(r, t) + \psi(r)] = r\frac{\partial}{\partial t}\left[\frac{1}{r}v(r, t) + \psi(r)\right]$$

or

$$\frac{\partial^2 v}{\partial r^2} + \psi''(r) = \frac{\partial v}{\partial t}.$$

This differential equation will be homogeneous if $\psi''(r) = 0$ or $\psi(r) = c_1 r + c_2$. Now

$$u(r, t) = \frac{1}{r}v(r, t) + \frac{1}{r}\psi(r) \quad \text{and} \quad \frac{1}{r}\psi(r) = c_1 + \frac{c_2}{r}.$$

Since we want $u(r, t)$ to be bounded as $r$ approaches 0, we require $c_2 = 0$. Then $\psi(r) = c_1 r$. When $r = 1$

$$u(1, t) = v(1, t) + \psi(1) = v(1, t) + c_1 = 100,$$

and we will have the homogeneous boundary condition $v(1, t) = 0$ when $c_1 = 100$. Consequently, $\psi(r) = 100r$. The initial condition

$$u(r, 0) = \frac{1}{r}v(r, 0) + \frac{1}{r}\psi(r) = \frac{1}{r}v(r, 0) + 100 = 0$$

implies $v(r, 0) = -100r$. We are thus led to solve the new boundary-value problem

$$\frac{\partial^2 v}{\partial r^2} = \frac{\partial v}{\partial t}, \quad 0 < r < 1, \quad t > 0,$$

$$v(1, t) = 0, \quad \lim_{r \to 0} \frac{1}{r}v(r, t) < \infty,$$

$$v(r, 0) = -100r.$$

**201**

Letting $v(r,t) = R(r)T(t)$ and separating variables leads to

$$R'' + \lambda^2 R = 0 \quad \text{and} \quad T' + \lambda^2 T = 0$$

with solutions

$$R(r) = c_3 \cos \lambda r + c_4 \sin \lambda r \quad \text{and} \quad T(t) = c_5 e^{-\lambda^2 t}.$$

The boundary conditions are equivalent to $R(1) = 0$ and $\lim_{r \to 0} \frac{1}{r} R(r) < \infty$. Since

$$\lim_{r \to 0} \frac{1}{r} R(r) = \lim_{r \to 0} \frac{c_3 \cos \lambda r}{r} + \lim_{r \to 0} \frac{c_4 \sin \lambda r}{r} = \lim_{r \to 0} \frac{c_3 \cos \lambda r}{r} + c_4 \lambda < \infty$$

we must have $c_3 = 0$. Then $R(r) = c_4 \sin \lambda r$, and $R(1) = 0$ implies $\lambda = n\pi$ for $n = 1, 2, 3, \ldots$. Thus

$$v_n(r,t) = A_n e^{-n^2 \pi^2 t} \sin n\pi r$$

for $n = 1, 2, 3, \ldots$. Using the condition $\lim_{r \to 0} \frac{1}{r} R(r) < \infty$ it is easily shown that there are no eigenvalues for $\lambda = 0$, nor does setting the common constant to $+\lambda^2$ when separating variables lead to any solutions. Now, by the superposition principle,

$$v(r,t) = \sum_{n=1}^{\infty} A_n e^{-n^2 \pi^2 t} \sin n\pi r.$$

The initial condition $v(r,0) = -100r$ implies

$$-100r = \sum_{n=1}^{\infty} A_n \sin n\pi r.$$

This is a Fourier sine series and so

$$A_n = 2 \int_0^1 (-100r \sin n\pi r) \, dr = -200 \left[ -\frac{r}{n\pi} \cos n\pi r \Big|_0^1 + \int_0^1 \frac{1}{n\pi} \cos n\pi r \, dr \right]$$

$$= -200 \left[ -\frac{\cos n\pi}{n\pi} + \frac{1}{n^2 \pi^2} \sin n\pi r \Big|_0^1 \right] = -200 \left[ -\frac{(-1)^n}{n\pi} \right] = \frac{(-1)^n 200}{n\pi}.$$

A solution of the problem is thus

$$u(r,t) = \frac{1}{r} v(r,t) + \frac{1}{r} \psi(r) = \frac{1}{r} \sum_{n=1}^{\infty} (-1)^n \frac{20}{n\pi} e^{-n^2 \pi^2 t} \sin n\pi r + \frac{1}{r} (100r)$$

$$= \frac{200}{\pi r} \sum_{n=1}^{\infty} \frac{(-1)^n}{n} e^{-n^2 \pi^2 t} \sin n\pi r + 100.$$

**12.** Proceeding as in Example 1 we obtain

$$\Theta(\theta) = P_n(\cos \theta) \quad \text{and} \quad R(r) = c_1 r^n + c_2 r^{-(n+1)}$$

so that

$$u(r, \theta) = \sum_{n=0}^{\infty} (A_n r^n + B_n r^{-(n+1)}) P_n(\cos \theta).$$

To satisfy $\lim_{r \to \infty} u(r, \theta) = -Er \cos \theta$ we must have $A_n = 0$ for $n = 2, 3, 4, \ldots$ . Then

$$\lim_{r \to \infty} u(r, \theta) = -Er \cos \theta = A_0 \cdot 1 + A_1 r \cos \theta,$$

so $A_0 = 0$ and $A_1 = -E$. Thus

$$u(r, \theta) = -Er \cos \theta + \sum_{n=0}^{\infty} B_n r^{-(n+1)} P_n(\cos \theta).$$

Now

$$u(c, \theta) = 0 = -Ec \cos \theta + \sum_{n=0}^{\infty} B_n c^{-(n+1)} P_n(\cos \theta)$$

so

$$\sum_{n=0}^{\infty} B_n c^{-(n+1)} P_n(\cos \theta) = Ec \cos \theta$$

and

$$B_n c^{-(n+1)} = \frac{2n+1}{2} \int_0^{\pi} Ec \cos \theta \, P_n(\cos \theta) \sin \theta \, d\theta.$$

Now $\cos \theta = P_1(\cos \theta)$ so, for $n \neq 1$,

$$\int_0^{\pi} \cos \theta \, P_n(\cos \theta) \sin \theta \, d\theta = 0$$

by orthogonality. Thus $B_n = 0$ for $n \neq 1$ and

$$B_1 = \frac{3}{2} Ec^3 \int_0^{\pi} \cos^2 \theta \sin \theta \, d\theta = Ec^3.$$

Therefore,

$$u(r, \theta) = -Er \cos \theta + Ec^3 r^{-2} \cos \theta.$$

## Chapter 13 Review Exercises

**3.** The conditions $\Theta(0) = 0$ and $\Theta(\pi) = 0$ applied to $\Theta = c_1 \cos \lambda \theta + c_2 \sin \lambda \theta$ give $c_1 = 0$ and $\lambda = n$, $n = 1, 2, 3, \ldots$, respectively. Thus we have the Fourier sine-series coefficients

$$A_n = \frac{2}{\pi} \int_0^{\pi} u_0(\pi \theta - \theta^2) \sin n\theta \, d\theta = \frac{4u_0}{n^3 \pi}[1 - (-1)^n].$$

Thus

$$u(r, \theta) = \frac{4u_0}{\pi} \sum_{n=1}^{\infty} \frac{1 - (-1)^n}{n^3} r^n \sin n\theta.$$

**6.** We solve

$$\frac{\partial^2 u}{\partial r^2} + \frac{1}{r}\frac{\partial u}{\partial r} + \frac{1}{r^2}\frac{\partial^2 u}{\partial \theta^2} = 0, \quad r > 1, \quad 0 < \theta < \pi,$$

$$u(r,0) = 0, \quad u(r,\pi) = 0, \quad r > 1,$$

$$u(1,\theta) = f(\theta), \quad 0 < \theta < \pi.$$

Separating variables we obtain

$$\Theta(\theta) = c_1 \cos \lambda\theta + c_2 \sin \lambda\theta$$

$$R(r) = c_3 r^\lambda + c_4 r^{-\lambda}.$$

Applying the boundary conditions $\Theta(0) = 0$, and $\Theta(\pi) = 0$ gives $c_1 = 0$ and $\lambda = n$ for $n = 1, 2, 3, \ldots$ . Assuming $f(\theta)$ to be bounded, we expect the solution $u(r,\theta)$ to also be bounded as $r \to \infty$. This requires that $c_3 = 0$. Therefore

$$u(r,\theta) = \sum_{n=1}^{\infty} A_n r^{-n} \sin n\theta.$$

From

$$u(1,\theta) = f(\theta) = \sum_{n=1}^{\infty} A_n \sin n\theta$$

we obtain

$$A_n = \frac{2}{\pi}\int_0^{\pi} f(\theta)\sin n\theta \, d\theta.$$

**9.** Referring to Example 2 in Section 13.2 we have

$$R(r) = c_1 J_0(\lambda r) + c_2 Y_0(\lambda r)$$

$$Z(z) = c_3 \cosh \lambda z + c_4 \sinh \lambda z$$

where $c_2 = 0$ and $J_0(2\lambda) = 0$ defines the positive eigenvalues $\lambda_n$. From $Z'(0) = 0$ we obtain $c_4 = 0$. Then

$$u(r,z) = \sum_{n=1}^{\infty} A_n \cosh \lambda_n z J_0(\lambda_n r).$$

From

$$u(r,4) = 50 = \sum_{n=1}^{\infty} A_n \cosh 4\lambda_n J_0(\lambda_n r)$$

we obtain (as in Example 1 of Section 13.1)

$$A_n \cosh 4\lambda_n = \frac{2(50)}{4J_1^2(2\lambda_n)}\int_0^2 r J_0(\lambda_n r)\, dr = \frac{50}{\lambda_n J_1(2\lambda_n)}.$$

Thus the temperature in the cylinder is

$$u(r, z) = 50 \sum_{n=1}^{\infty} \frac{\cosh \lambda_n z J_0(\lambda_n r)}{\lambda_n \cosh 4\lambda_n J_1(2\lambda_n)}.$$

**12.** Since

$$\frac{1}{r}\frac{\partial^2}{\partial r^2}(ru) = \frac{1}{r}\frac{\partial}{\partial r}\left[r\frac{\partial u}{\partial r} + u\right] = \frac{1}{r}\left[r\frac{\partial^2 u}{\partial r^2} + \frac{\partial u}{\partial r} + \frac{\partial u}{\partial r}\right] = \frac{\partial^2 u}{\partial r^2} + \frac{2}{r}\frac{\partial u}{r}$$

the differential equation becomes

$$\frac{1}{r}\frac{\partial^2}{\partial r^2}(ru) = \frac{\partial^2 u}{\partial t^2} \quad \text{or} \quad \frac{\partial^2}{\partial r^2}(ru) = r\frac{\partial^2 u}{\partial t^2}.$$

Letting $v(r, t) = ru(r, t)$ we obtain the boundary-value problem

$$\frac{\partial^2 v}{\partial r^2} = \frac{\partial^2 v}{\partial t^2}, \quad 0 < r < 1, \quad t > 0$$

$$\left.\frac{\partial v}{\partial r}\right|_{r=1} - v(1, t) = 0, \quad t > 0$$

$$v(r, 0) = rf(r), \quad \left.\frac{\partial v}{\partial t}\right|_{t=0} = rg(r), \quad 0 < r < 1.$$

If we separate variables using $v(r, t) = R(r)T(t)$ then we obtain

$$R(r) = c_1 \cos \lambda r + c_2 \sin \lambda r$$

$$T(t) = c_3 \cos \lambda t + c_4 \sin \lambda t.$$

Since $u(r, t) = v(r, t)/r$, in order to insure boundedness at $r = 0$ we define $c_1 = 0$. Then $R(r) = c_2 \sin \lambda r$. Now the boundary condition $R'(1) - R(1) = 0$ implies $\lambda \cos \lambda - \sin \lambda = 0$. Thus, the eigenvalues $\lambda_n$ are the positive solutions of $\tan \lambda = \lambda$. We now have

$$v_n(r, t) = (A_n \cos \lambda_n t + B_n \sin \lambda_n t) \sin \lambda_n r.$$

For the eigenvalue $\lambda = 0$,

$$R(r) = c_1 r + c_2 \quad \text{and} \quad T(t) = c_3 t + c_4,$$

and boundedness at $r = 0$ implies $c_2 = 0$. We then take

$$v_0(r, t) = A_0 tr + B_0 r$$

so that

$$v(r, t) = A_0 tr + B_0 r + \sum_{n=1}^{\infty}(a_n \cos \lambda_n t + B_n \sin \lambda_n t) \sin \lambda_n r.$$

Now

$$v(r, 0) = rf(r) = B_0 r + \sum_{n=1}^{\infty} A_n \sin \lambda_n r.$$

**205**

Since $\{r, \sin \lambda_n r\}$ is an orthogonal set on $[0, 1]$,

$$\int_0^1 r \sin \lambda_n r \, dr = 0 \quad \text{and} \quad \int_0^1 \sin \lambda_n r \sin \lambda_n r \, dr = 0$$

for $m \neq n$. Therefore

$$\int_0^1 r^2 f(r) \, dr = B_0 \int_0^1 r^2 \, dr = \frac{1}{3} B_0$$

and

$$B_0 = 3 \int_0^1 r^2 f(r) \, dr.$$

Also

$$\int_0^1 r f(r) \sin \lambda_n r \, dr = A_n \int_0^1 \sin^2 \lambda_n r \, dr$$

and

$$A_n = \frac{\int_0^1 r f(r) \sin \lambda_n r \, dr}{\int_0^1 \sin^2 \lambda_n r \, dr}.$$

Now

$$\int_0^1 \sin^2 \lambda_n r \, dr = \frac{1}{2} \int_0^1 (1 - \cos 2\lambda_n r) \, dr = \frac{1}{2}\left[1 - \frac{\sin 2\lambda_n}{2\lambda_n}\right] = \frac{1}{2}[1 - \cos^2 \lambda_n].$$

Since $\tan \lambda_n = \lambda_n$,

$$1 + \lambda_n^2 = 1 + \tan^2 \lambda_n = \sec^2 \lambda_n = \frac{1}{\cos^2 \lambda_n}$$

and

$$\cos^2 \lambda_n = \frac{1}{1 + \lambda_n^2}.$$

Then

$$\int_0^1 \sin^2 \lambda_n r \, dr = \frac{1}{2}\left[1 - \frac{1}{1 + \lambda_n^2}\right] = \frac{\lambda_n^2}{2(1 + \lambda_n^2)}$$

and

$$A_n = \frac{2(1 + \lambda_n^2)}{\lambda_n^2} \int_0^1 r f(r) \sin \lambda_n r \, dr.$$

Similarly, setting

$$\frac{\partial v}{\partial t}\bigg|_{t=0} = rg(r) = A_0 r + \sum_{n=1}^{\infty} B_n \lambda_n \sin \lambda_n r$$

we obtain

$$A_0 = 3 \int_0^1 r^2 g(r) \, dr$$

and

$$B_n = \frac{2(1 + \lambda_n^2)}{\lambda_n^3} \int_0^1 r g(r) \sin \lambda_n r \, dr.$$

Therefore, since $v(r,t) = ru(r,t)$ we have

$$u(r,t) = A_0 t + B_0 + \sum_{n=1}^{\infty} (A_n \cos \lambda_n t + B_n \sin \lambda_n t) \frac{\sin \lambda_n r}{r},$$

where the $\lambda_n$ are solutions of $\tan \lambda = \lambda$ and

$$A_0 = 3 \int_0^1 r^2 g(r) \, dr$$

$$B_0 = 3 \int_0^1 r^2 f(r) \, dr$$

$$A_n = \frac{2(1 + \lambda_n^2)}{\lambda_n^2} \int_0^1 rf(r) \sin \lambda_n r \, dr$$

$$B_n = \frac{2(1 + \lambda_n^2)}{\lambda_n^3} \int_0^1 rg(r) \sin \lambda_n r \, dr$$

for $n = 1, 2, 3, \ldots$ .

# 14 Integral Transform Method

━━━━━━━━ **Exercises 14.1** ━━━━━━━━━━━━━━━━━━━━━━━━━

**3.** By the first translation theorem,

$$\mathscr{L}\{e^t \operatorname{erf}(\sqrt{t})\} = \mathscr{L}\{\operatorname{erf}(\sqrt{t})\}\Big|_{s\to s-1} = \frac{1}{s\sqrt{s+1}}\Big|_{s\to s-1} = \frac{1}{\sqrt{s}(s-1)}.$$

**6.** We first compute

$$\frac{\sinh a\sqrt{s}}{s\sinh\sqrt{s}} = \frac{e^{a\sqrt{s}} - e^{-a\sqrt{s}}}{s(e^{\sqrt{s}} - e^{-\sqrt{s}})} = \frac{e^{(a-1)\sqrt{s}} - e^{-(a+1)\sqrt{s}}}{s(1 - e^{-2\sqrt{s}})}$$

$$= \frac{e^{(a-1)\sqrt{s}}}{s}\left[1 + e^{-2\sqrt{s}} + e^{-4\sqrt{s}} + \cdots\right] - \frac{e^{-(a+1)\sqrt{s}}}{s}\left[1 + e^{-2\sqrt{s}} + e^{-4\sqrt{s}} + \cdots\right]$$

$$= \left[\frac{e^{-(1-a)\sqrt{s}}}{s} + \frac{e^{-(3-a)\sqrt{s}}}{s} + \frac{e^{-(5-a)\sqrt{s}}}{s} + \cdots\right]$$

$$- \left[\frac{e^{-(1+a)\sqrt{s}}}{s} + \frac{e^{-(3+a)\sqrt{s}}}{s} + \frac{e^{-(5+a)\sqrt{s}}}{s} + \cdots\right]$$

$$= \sum_{n=0}^{\infty}\left[\frac{e^{-(2n+1-a)\sqrt{s}}}{s} - \frac{e^{-(2n+1+a)\sqrt{s}}}{s}\right].$$

Then

$$\mathscr{L}\left\{\frac{\sinh a\sqrt{s}}{s\sinh\sqrt{s}}\right\} = \sum_{n=0}^{\infty}\left[\mathscr{L}\left\{\frac{e^{-(2n+1-a)\sqrt{s}}}{s}\right\} - \mathscr{L}\left\{-\frac{e^{-(2n+1+a)\sqrt{s}}}{s}\right\}\right]$$

$$= \sum_{n=0}^{\infty}\left[\operatorname{erfc}\left(\frac{2n+1-a}{2\sqrt{t}}\right) - \operatorname{erfc}\left(\frac{2n+1+a}{2\sqrt{t}}\right)\right]$$

$$= \sum_{n=0}^{\infty}\left(\left[1 - \operatorname{erf}\left(\frac{2n+1-a}{2\sqrt{t}}\right)\right] - \left[1 - \operatorname{erf}\left(\frac{2n+1+a}{2\sqrt{t}}\right)\right]\right)$$

$$= \sum_{n=0}^{\infty}\left[\operatorname{erf}\left(\frac{2n+1+a}{2\sqrt{t}}\right) - \operatorname{erf}\left(\frac{2n+1-a}{2\sqrt{t}}\right)\right].$$

**9.** $\displaystyle\int_{a}^{b} e^{-u^2}\,du = \int_{a}^{0} e^{-u^2}\,du + \int_{0}^{b} e^{-u^2}\,du = \int_{0}^{b} e^{-u^2}\,du - \int_{0}^{a} e^{-u^2}\,du$

$$= \frac{\sqrt{\pi}}{2}\operatorname{erf}(b) - \frac{\sqrt{\pi}}{2}\operatorname{erf}(a) = \frac{\sqrt{\pi}}{2}[\operatorname{erf}(b) - \operatorname{erf}(a)]$$

3. The solution of

$$a^2 \frac{d^2U}{dx^2} - s^2U = 0$$

is in this case

$$U(x, s) = c_1 e^{-(x/a)s} + c_2 e^{(x/a)s}.$$

Since $\lim_{x \to \infty} u(x, t) = 0$ we have $\lim_{x \to \infty} U(x, s) = 0$. Thus $c_2 = 0$ and

$$U(x, s) = c_1 e^{-(x/a)s}.$$

If $\mathscr{L}\{u(0, t)\} = \mathscr{L}\{f(t)\} = F(s)$ then $U(0, s) = F(s)$. From this we have $c_1 = F(s)$ and

$$U(x, s) = F(s)e^{-(x/a)s}.$$

Hence, by the second translation theorem,

$$u(x, t) = f\left(t - \frac{x}{a}\right)\mathscr{U}\left(t - \frac{x}{a}\right).$$

6. Transforming the partial differential equation gives

$$\frac{d^2U}{dx^2} - s^2U = -\frac{\omega}{s^2 + \omega^2} \sin \pi x.$$

Using undetermined coefficients we obtain

$$U(x, s) = c_1 \cosh sx + c_2 \sinh sx + \frac{\omega}{(s^2 + \pi^2)(s^2 + \omega^2)} \sin \pi x.$$

The transformed boundary conditions $U(0, s) = 0$ and $U(1, s) = 0$ give, in turn, $c_1 = 0$ and $c_2 = 0$. Therefore

$$U(x, s) = \frac{\omega}{(s^2 + \pi^2)(s^2 + \omega^2)} \sin \pi x$$

and

$$u(x, t) = \omega \sin \pi x \, \mathscr{L}^{-1}\left\{\frac{1}{(s^2 + \pi^2)(s^2 + \omega^2)}\right\}$$

$$= \frac{\omega}{\omega^2 - \pi^2} \sin \pi x \, \mathscr{L}^{-1}\left\{\frac{1}{\pi} \frac{\pi}{s^2 + \pi^2} - \frac{1}{\omega} \frac{\omega}{s^2 + \omega^2}\right\}$$

$$= \frac{\omega}{\pi(\omega^2 - \pi^2)} \sin \pi t \sin \pi x - \frac{1}{\omega^2 - \pi^2} \sin \omega t \sin \pi x.$$

9. Transforming the partial differential equation gives

$$\frac{d^2U}{dx^2} - s^2U = -sxe^{-x}.$$

Using undetermined coefficients we obtain

$$U(x,s) = c_1 e^{-sx} + c_2 e^{sx} - \frac{2s}{(s^2-1)^2} e^{-x} + \frac{s}{s^2-1} x e^{-x}.$$

The transformed boundary conditions $\lim_{x\to\infty} U(x,s) = 0$ and $U(0,s) = 0$ give, in turn, $c_2 = 0$ and $c_1 = 2s/(s^2-1)^2$. Therefore

$$U(x,s) = \frac{2s}{(s^2-1)^2} e^{-sx} - \frac{2s}{(s^2-1)^2} e^{-x} + \frac{s}{s^2-1} x e^{-x}.$$

From entries (13) and (26) in the table we obtain

$$u(x,t) = \mathcal{L}^{-1}\left\{ \frac{2s}{(s^2-1)^2} e^{-sx} - \frac{2s}{(s^2-1)^2} e^{-x} + \frac{s}{s^2-1} x e^{-x} \right\}$$

$$= 2(t-x)\sinh(t-x)\,\mathcal{U}(t-x) - te^{-x}\sinh t + xe^{-x}\cosh t.$$

**12.** Transforming the partial differential equation and using the initial condition gives

$$k\frac{d^2U}{dx^2} - sU = 0.$$

Since the domain of the variable $x$ is an infinite interval we write the general solution of this differential equation as

$$U(x,s) = c_1 e^{-\sqrt{s/k}\,x} + c_2 e^{-\sqrt{s/k}\,x}.$$

Transforming the boundary conditions gives $U'(0,s) = -A/s$ and $\lim_{x\to\infty} U(x,s) = 0$. Hence we find $c_2 = 0$ and $c_1 = A\sqrt{k}/s\sqrt{s}$. From

$$U(x,s) = A\sqrt{k}\,\frac{e^{-\sqrt{s/k}\,x}}{s\sqrt{s}}$$

we see that

$$u(x,t) = A\sqrt{k}\,\mathcal{L}^{-1}\left\{ \frac{e^{-\sqrt{s/k}\,x}}{s\sqrt{s}} \right\}.$$

With the identification $a = x/\sqrt{k}$ it follows from entry 47 of the table in Appendix II that

$$u(x,t) = A\sqrt{k}\left\{ 2\sqrt{\frac{t}{\pi}}\,e^{-x^2/4kt} - \frac{x}{\sqrt{k}}\,\mathrm{erfc}\left(x/2\sqrt{kt}\right) \right\}$$

$$= 2A\sqrt{\frac{kt}{\pi}}\,e^{-x^2/4kt} - Ax\,\mathrm{erfc}\left(x/2\sqrt{kt}\right).$$

**15.** We use

$$U(x,s) = c_1 e^{-\sqrt{s}\,x} + c_2 e^{\sqrt{s}\,x} + \frac{u_0}{s}.$$

The condition $\lim_{x \to \infty} u(x, t) = u_0$ implies $\lim_{x \to \infty} U(x, s) = u_0/s$, so we define $c_2 = 0$. Then

$$U(x, s) = c_1 e^{-\sqrt{s}\,x} + \frac{u_0}{s}.$$

The transform of the remaining boundary conditions gives

$$\left. \frac{dU}{dx} \right|_{x=0} = U(0, s).$$

This condition yields $c_1 = -u_0/s(\sqrt{s} + 1)$. Thus

$$U(x, s) = -u_0 \frac{e^{-\sqrt{s}\,x}}{s(\sqrt{s} + 1)} + \frac{u_0}{s}$$

and

$$u(x, t) = -u_0 \mathcal{L}^{-1} \left\{ \frac{e^{-x\sqrt{s}}}{s(\sqrt{s} + 1)} \right\} + u_0 \mathcal{L}^{-1} \left\{ \frac{1}{s} \right\}$$

$$= u_0 e^{x+t} \operatorname{erfc}\left( \sqrt{t} + \frac{x}{2\sqrt{t}} \right) - u_0 \operatorname{erfc}\left( \frac{x}{2\sqrt{t}} \right) + u_0 \qquad \boxed{\text{By (5) in the table in 14.1.}}$$

**18.** We use

$$U(x, s) = c_1 e^{-\sqrt{s}\,x} + c_2 e^{\sqrt{s}\,x}.$$

The condition $\lim_{x \to \infty} u(x, t) = 0$ implies $\lim_{x \to \infty} U(x, s) = 0$, so we define $c_2 = 0$. Then $U(x, s) = c_1 e^{-\sqrt{s}\,x}$. The transform of the remaining boundary condition gives

$$\left. \frac{dU}{dx} \right|_{x=0} = -F(s)$$

where $F(s) = \mathcal{L}\{f(t)\}$. This condition yields $c_1 = F(s)/\sqrt{s}$. Thus

$$U(x, s) = F(s) \frac{e^{-\sqrt{s}\,x}}{\sqrt{s}}.$$

Using entry (44) of the table and the convolution theorem we obtain

$$u(x, t) = \mathcal{L}^{-1} \left\{ F(s) \cdot \frac{e^{-\sqrt{s}\,x}}{\sqrt{s}} \right\} = \frac{1}{\sqrt{\pi}} \int_0^t f(\tau) \frac{e^{-x^2/4(t-\tau)}}{\sqrt{t - \tau}} \, d\tau.$$

**21.** Transforming the partial differential equation gives

$$\frac{d^2 U}{dx^2} - sU = 0$$

and so

$$U(x, s) = c_1 e^{-\sqrt{s}\,x} + c_2 e^{\sqrt{s}\,x}.$$

The condition $\lim_{x \to -\infty} u(x,t) = 0$ implies $\lim_{x \to -\infty} U(x,s) = 0$, so we define $c_1 = 0$. The transform of the remaining boundary condition gives

$$\frac{dU}{dx}\bigg|_{x=1} = \frac{100}{s} - U(1,s).$$

This condition yields

$$c_2 \sqrt{s}\, e^{\sqrt{s}} = \frac{100}{s} - c_2 e^{\sqrt{s}}$$

from which it follows that

$$c_2 = \frac{100}{s(\sqrt{s}+1)}\, e^{-\sqrt{s}}.$$

Thus

$$U(x,s) = 100\, \frac{e^{-(1-x)\sqrt{s}}}{s(\sqrt{s}+1)}.$$

Using entry (49) of the table we obtain

$$u(x,t) = 100\, \mathcal{L}^{-1}\left\{ \frac{e^{-(1-x)\sqrt{s}}}{s(\sqrt{s}+1)} \right\} = 100 \left[ -e^{1-x+t}\,\mathrm{erfc}\left( \sqrt{t} + \frac{1-x}{\sqrt{t}} \right) + \mathrm{erfc}\left( \frac{1-x}{2\sqrt{t}} \right) \right].$$

**24.** Transforming the partial differential equation gives

$$k\frac{d^2U}{dx^2} - sU = -\frac{r}{s}.$$

Using undetermined coefficients we obtian

$$U(x,s) = c_1 e^{-\sqrt{s/k}\,x} + c_2 e^{\sqrt{s/k}\,x} + \frac{r}{s^2}.$$

The condition $\lim_{x \to \infty} \frac{\partial u}{\partial x} = 0$ implies $\lim_{x \to \infty} \frac{dU}{dx} = 0$, so we define $c_2 = 0$. The transform of the remaining boundary condition gives $U(0,s) = 0$. This condition yields $c_1 = -r/s^2$. Thus

$$U(x,s) = r\left[ \frac{1}{s^2} - \frac{e^{-\sqrt{s/k}\,x}}{s^2} \right].$$

Using entries (3) and (46) of the table and the convolution theorem we obtain

$$u(x,t) = r\,\mathcal{L}^{-1}\left\{ \frac{1}{s^2} - \frac{1}{s}\cdot\frac{e^{-\sqrt{s/k}\,x}}{s} \right\} = rt - r\int_0^t \mathrm{erfc}\left( \frac{x}{2\sqrt{k\tau}} \right) d\tau.$$

**27.** The solution of

$$\frac{d^2U}{dx^2} - sU = -u_0 - u_0 \sin\frac{\pi}{L}x$$

is

$$U(x,s) = c_1 \cosh(\sqrt{s}\, x) + c_2 \sinh(\sqrt{s}\, x) + \frac{u_0}{s} + \frac{u_0}{s + \pi^2/L^2} \sin \frac{\pi}{L} x.$$

The transformed boundary conditions $U(0,s) = u_0/s$ and $U(L,s) = u_0/s$ give, in turn, $c_1 = 0$ and $c_2 = 0$. Therefore

$$U(x,s) = \frac{u_0}{s} + \frac{u_0}{s + \pi^2/L^2} \sin \frac{\pi}{L} x$$

and

$$u(x,t) = u_0 \mathscr{L}^{-1}\left\{\frac{1}{s}\right\} + u_0 \mathscr{L}^{-1}\left\{\frac{1}{s + \pi^2/L^2}\right\} \sin \frac{\pi}{L} x = u_0 + u_0 e^{-\pi^2 t/L^2} \sin \frac{\pi}{L} x.$$

**30.** The transform of the partial differential equation is

$$k\frac{d^2 U}{dx^2} - hU + h\frac{u_m}{s} = sU - u_0$$

or

$$k\frac{d^2 U}{dx^2} - (h + s)U = -h\frac{u_m}{s} - u_0.$$

By undetermined coefficients we find

$$U(x,s) = c_1 e^{\sqrt{(h+s)/k}\, x} + c_2 e^{-\sqrt{(h+s)/k}\, x} + \frac{h u_m + u_0 s}{s(s + h)}.$$

The transformed boundary conditions are $U'(0,s) = 0$ and $U'(L,s) = 0$. These conditions imply $c_1 = 0$ and $c_2 = 0$. By partial fractions we then get

$$U(x,s) = \frac{h u_m + u_0 s}{s(s + h)} = \frac{u_m}{s} - \frac{u_m}{s + h} + \frac{u_0}{s + h}.$$

Therefore,

$$u(x,t) = u_m \mathscr{L}^{-1}\left\{\frac{1}{s}\right\} - u_m \mathscr{L}^{-1}\left\{\frac{1}{s + h}\right\} + u_0 \mathscr{L}^{-1}\left\{\frac{1}{s + h}\right\} = u_m - u_m e^{-ht} + u_0 e^{-ht}.$$

**33.** We use

$$U(x,s) = c_1 e^{-\sqrt{RCs + RG}\, x} + c_2 e^{\sqrt{RCs + RG}} + \frac{C u_0}{Cs + G}.$$

The condition $\lim_{x \to \infty} \partial u/\partial x = 0$ implies $\lim_{x \to \infty} dU/dx = 0$, so we define $c_2 = 0$. Applying $U(0,s) = 0$ to

$$U(x,s) = c_1 e^{-\sqrt{RCsRG}\, x} + \frac{C u_0}{Cs + G}$$

gives $c_1 = -C u_0/(Cs + G)$. Therefore

$$U(x,s) = -C u_0 \frac{e^{-\sqrt{RCs + RG}\, x}}{Cs + G} + \frac{C u_0}{Cs + G}$$

and

$$u(x,t) = u_0 \mathscr{L}^{-1}\left\{\frac{1}{s+G/C}\right\} - u_0 \mathscr{L}^{-1}\left\{\frac{e^{-x\sqrt{RC}\sqrt{s+G/C}}}{s+G/C}\right\}$$

$$= u_0 e^{-Gt/C} - u_0 e^{-Gt/C}\operatorname{erfc}\left(\frac{x\sqrt{RC}}{2\sqrt{t}}\right)$$

$$= u_0 e^{-Gt/C}\left[1 - \operatorname{erfc}\left(\frac{x}{2}\sqrt{\frac{RC}{t}}\right)\right]$$

$$= u_0 e^{-Gt/C}\operatorname{erf}\left(\frac{x}{2}\sqrt{\frac{RC}{t}}\right)$$

**36. (a)** We use

$$U(x,s) = c_1 e^{-(s/a)x} + c_2 e^{(s/a)x} + \frac{v_0^2 F_0}{(a^2-v_0^2)s^2}e^{-(s/v_0)x}.$$

The condition $\lim_{x\to\infty} u(x,t) = 0$ implies $\lim_{x\to\infty} U(x,s) = 0$, so we must define $c_2 = 0$. Consequently

$$U(x,s) = c_1 e^{-(s/a)x} + \frac{v_0^2 F_0}{(a^2-v_0^2)s^2}e^{-(s/v_0)x}.$$

The remaining boundary condition transforms into $U(0,s) = 0$. From this we find

$$c_1 = -v_0^2 F_0/(a^2-v_0^2)s^2.$$

Therefore, by the second translation theorem

$$U(x,s) = -\frac{v_0^2 F_0}{(a^2-v_0^2)s^2}e^{-(s/a)x} + \frac{v_0^2 F_0}{(a^2-v_0^2)s^2}e^{-(s/v_0)x}$$

and

$$u(x,t) = \frac{v_0^2 F_0}{a^2-v_0^2}\left[\mathscr{L}^{-1}\left\{\frac{e^{-(x/v_0)s}}{s^2}\right\} - \mathscr{L}^{-1}\left\{\frac{e^{-(x/a)s}}{s^2}\right\}\right]$$

$$= \frac{v_0^2 F_0}{a^2-v_0^2}\left[\left(t-\frac{x}{v_0}\right)\mathscr{U}\left(t-\frac{x}{v_0}\right) - \left(t-\frac{x}{a}\right)\mathscr{U}\left(t-\frac{x}{a}\right)\right].$$

**(b)** In the case when $v_0 = a$ the solution of the transformed equation is

$$U(x,s) = c_1 e^{-(s/a)x} + c_2 e^{(s/a)x} - \frac{F_0}{2as}xe^{-(s/a)x}.$$

The usual analysis then leads to $c_1 = 0$ and $c_2 = 0$. Therefore

$$U(x,s) = -\frac{F_0}{2as}xe^{-(s/a)x}$$

and

$$u(x,t) = -\frac{xF_0}{2a} \mathscr{L}^{-1}\left\{\frac{e^{-(x/a)s}}{s}\right\} = -\frac{xF_0}{2a}\,\mathscr{U}\left(t - \frac{x}{a}\right).$$

## ──────── Exercises 14.3 ────────────

**3.** From formulas (5) and (6) in the text,

$$A(\alpha) = \int_0^3 x \cos \alpha x\, dx$$

$$= \frac{x \sin \alpha x}{\alpha}\Big|_0^3 - \frac{1}{\alpha}\int_0^3 \sin \alpha x\, dx$$

$$= \frac{3 \sin 3\alpha}{\alpha} + \frac{\cos \alpha x}{\alpha^2}\Big|_0^3$$

$$= \frac{3\alpha \sin 3\alpha + \cos 3\alpha - 1}{\alpha^2}$$

and

$$B(\alpha) = \int_0^3 x \sin \alpha x\, dx$$

$$= -\frac{x \cos \alpha x}{\alpha}\Big|_0^3 + \frac{1}{\alpha}\int_0^3 \cos \alpha x\, dx$$

$$= -\frac{3 \cos 3\alpha}{\alpha} + \frac{\sin \alpha x}{\alpha^2}\Big|_0^3$$

$$= \frac{\sin 3\alpha - 3\alpha \cos 3\alpha}{\alpha^2}.$$

Hence

$$f(x) = \frac{1}{\pi}\int_0^\infty \frac{(3\alpha \sin 3\alpha + \cos 3\alpha - 1)\cos \alpha x + (\sin 3\alpha - 3\alpha \cos 3\alpha)\sin \alpha x}{\alpha^2}\, d\alpha$$

$$= \frac{1}{\pi}\int_0^\infty \frac{3\alpha(\sin 3\alpha \cos \alpha x - \cos 3\alpha \sin \alpha x) + \cos 3\alpha \cos \alpha x + \sin 3\alpha \sin \alpha x - \cos \alpha x}{\alpha^2}\, d\alpha$$

$$= \frac{1}{\pi}\int_0^\infty \frac{3\alpha \sin \alpha(3 - x) + \cos \alpha(3 - x) - \cos \alpha x}{\alpha^2}\, d\alpha.$$

**215**

**6.** From formulas (5) and (6) in the text,

$$A(\alpha) = \int_{-1}^{1} e^x \cos \alpha x \, dx$$

$$= \frac{e(\cos \alpha + \alpha \sin \alpha) - e^{-1}(\cos \alpha - \alpha \sin \alpha)}{1 + \alpha^2}$$

$$= \frac{2(\sinh 1)\cos \alpha - 2\alpha(\cosh 1)\sin \alpha}{1 + \alpha^2}$$

and

$$B(\alpha) = \int_{-1}^{1} e^x \sin \alpha x \, dx$$

$$= \frac{e(\sin \alpha - \alpha \cos \alpha) - e^{-1}(-\sin \alpha - \alpha \cos \alpha)}{1 + \alpha^2}$$

$$= \frac{2(\cosh 1)\sin \alpha - 2\alpha(\sinh 1)\cos \alpha}{1 + \alpha^2} .$$

Hence

$$f(x) = \frac{1}{\pi} \int_0^{\infty} [A(\alpha) \cos \alpha x + B(\alpha) \sin \alpha x] \, d\alpha.$$

**9.** The function is even. Thus from formula (9) in the text

$$A(\alpha) = \int_0^{\pi} x \cos \alpha x \, dx = \frac{x \sin \alpha x}{\alpha} \Big|_0^{\pi} - \frac{1}{\alpha} \int_0^{\pi} \sin \alpha x \, dx$$

$$= \frac{\pi \sin \pi \alpha}{\alpha} + \frac{1}{\alpha^2} \cos \alpha x \Big|_0^{\pi} = \frac{\pi \alpha \sin \pi \alpha + \cos \pi \alpha - 1}{\alpha^2} .$$

Hence from formula (8) in the text

$$f(x) = \frac{2}{\pi} \int_0^{\infty} \frac{(\pi \alpha \sin \pi \alpha + \cos \pi \alpha - 1) \cos \alpha x}{\alpha^2} \, d\alpha.$$

**12.** The function is odd. Thus from formula (11) in the text

$$B(\alpha) = \int_0^{\infty} x e^{-x} \sin \alpha x \, dx.$$

Now recall

$$\mathscr{L}\{t \sin kt\} = -\frac{d}{ds} \mathscr{L}\{\sin kt\} = 2ks/(s^2 + k^2)^2.$$

If we set $s = 1$ and $k = \alpha$ we obtain

$$B(\alpha) = \frac{2\alpha}{(1 + \alpha^2)^2} .$$

Hence from formula (10) in the text

$$f(x) = \frac{4}{\pi} \int_0^{\infty} \frac{\alpha \sin \alpha x}{(1 + \alpha^2)^2} \, d\alpha.$$

**15.** For the cosine integral,

$$A(\alpha) = \int_0^\infty x e^{-2x} \cos \alpha x \, dx.$$

But we know

$$\mathcal{L}\{t \cos kt\} = -\frac{d}{ds}\frac{s}{(s^2+k^2)} = \frac{(s^2-k^2)}{(s^2+k^2)^2}.$$

If we set $s = 2$ and $k = \alpha$ we obtain

$$A(\alpha) = \frac{4-\alpha^2}{(4+\alpha^2)^2}.$$

Hence

$$f(x) = \frac{2}{\pi}\int_0^\infty \frac{(4-\alpha^2)\cos \alpha x}{(4+\alpha^2)^2}\, d\alpha.$$

For the sine integral,

$$B(\alpha) = \int_0^\infty x e^{-2x} \sin \alpha x \, dx.$$

From Problem 12, we know

$$\mathcal{L}\{t \sin kt\} = \frac{2ks}{(s^2+k^2)^2}.$$

If we set $s = 2$ and $k = \alpha$ we obtain

$$B(\alpha) = \frac{4\alpha}{(4+\alpha^2)^2}.$$

Hence

$$f(x) = \frac{8}{\pi}\int_0^\infty \frac{\alpha \sin \alpha x}{(4+\alpha^2)^2}\, d\alpha.$$

**18.** From the formula for sine integral of $f(x)$ we have

$$f(x) = \frac{2}{\pi}\int_0^\infty \left(\int_0^\infty f(x)\sin \alpha x \, dx\right) \sin \alpha x \, dx$$

$$= \frac{2}{\pi}\left[\int_0^1 1 \cdot \sin \alpha x \, d\alpha + \int_1^\infty 0 \cdot \sin \alpha x \, d\alpha\right]$$

$$= \frac{2}{\pi}\frac{(-\cos \alpha x)}{x}\Big|_0^1$$

$$= \frac{2}{\pi}\frac{1-\cos x}{x}.$$

For the boundary-value problems in this section it is sometimes useful to note that the identities

$$e^{i\alpha} = \cos\alpha + i\sin\alpha \quad \text{and} \quad e^{-i\alpha} = \cos\alpha - i\sin\alpha$$

imply

$$e^{i\alpha} + e^{-i\alpha} = 2\cos\alpha \quad \text{and} \quad e^{i\alpha} - e^{-i\alpha} = 2i\sin\alpha$$

**3.** Using the Fourier transform, the partial differential equation equation becomes

$$\frac{dU}{dt} + k\alpha^2 U = 0 \quad \text{and so} \quad U(\alpha, t) = ce^{-k\alpha^2 t}.$$

Now

$$\mathscr{F}\{u(x,0)\} = U(\alpha,0) = \sqrt{\pi}\, e^{-\alpha^2/4}$$

by the given result. This gives $c = \sqrt{\pi}\, e^{-\alpha^2/4}$ and so

$$U(\alpha, t) = \sqrt{\pi}\, e^{-(\frac{1}{4}+kt)\alpha^2}.$$

Using the given Fourier transform again we obtain

$$u(x,t) = \sqrt{\pi}\,\mathscr{F}^{-1}\{e^{-(\frac{1+4kt}{4})\alpha^2}\} = \frac{1}{\sqrt{1+4kt}}\, e^{-x^2/(1+4kt)}.$$

**6.** The solution of Problem 5 can be written

$$u(x,t) = \frac{2u_0}{\pi} \int_0^\infty \frac{\sin\alpha x}{\alpha}\, d\alpha - \frac{2u_0}{\pi} \int_0^\infty \frac{\sin\alpha x}{\alpha}\, e^{-k\alpha^2 t}\, d\alpha.$$

Using $\int_0^\infty \frac{\sin\alpha x}{\alpha}\, d\alpha = \pi/2$ the last line becomes

$$u(x,t) = u_0 - \frac{2u_0}{\pi} \int_0^\infty \frac{\sin\alpha x}{\alpha}\, e^{-k\alpha^2 t}\, d\alpha.$$

**9.** Using the Fourier cosine transform we find

$$U(\alpha, t) = ce^{-k\alpha^2 t}.$$

Now

$$\mathscr{F}_C\{u(x,0)\} = \int_0^1 \cos\alpha x\, dx = \frac{\sin\alpha}{\alpha} = U(\alpha,0).$$

From this we obtain $c = (\sin\alpha)/\alpha$ and so

$$U(\alpha, t) = \frac{\sin\alpha}{\alpha}\, e^{-k\alpha^2 t}$$

**218**

and

$$u(x,t) = \frac{2}{\pi} \int_0^\infty \frac{\sin \alpha}{\alpha} e^{-k\alpha^2 t} \cos \alpha x \, d\alpha.$$

**12.** Using the Fourier sine transform we obtain

$$U(\alpha, t) = c_1 \cos \alpha a t + c_2 \sin \alpha a t.$$

Now

$$\mathscr{F}_S\{u(x,0)\} = \mathscr{F}\{xe^{-x}\} = \int_0^\infty xe^{-x} \sin \alpha x \, dx = \frac{2\alpha}{(1+\alpha^2)^2} = U(\alpha, 0).$$

Also,

$$\mathscr{F}_S\{u_t(x,0)\} = \frac{dU}{dt}\bigg|_{t=0} = 0.$$

This last condition gives $c_2 = 0$. Then $U(\alpha, 0) = 2\alpha/(1+\alpha^2)^2$ yields $c_1 = 2\alpha/(1+\alpha^2)^2$. Therefore

$$U(\alpha, t) = \frac{2\alpha}{(1+\alpha^2)^2} \cos \alpha a t$$

and

$$u(x,t) = \frac{4}{\pi} \int_0^\infty \frac{\alpha \cos \alpha a t}{(1+\alpha^2)^2} \sin \alpha x \, d\alpha.$$

**15.** Using the Fourier cosine transform with respect to $x$ gives

$$U(\alpha, y) = c_1 e^{-\alpha y} + c_2 e^{\alpha y}.$$

Since we expect $u(x, y)$ to be bounded as $y \to \infty$ we define $c_2 = 0$. Thus

$$U(\alpha, y) = c_1 e^{-\alpha y}.$$

Now

$$\mathscr{F}_C\{u(x,0)\} = \int_0^1 50 \cos \alpha x \, dx = 50 \frac{\sin \alpha}{\alpha}$$

and so

$$U(\alpha, y) = 50 \frac{\sin \alpha}{\alpha} e^{-\alpha y}$$

and

$$u(x,y) = \frac{100}{\pi} \int_0^\infty \frac{\sin \alpha}{\alpha} e^{-\alpha y} \cos \alpha x \, d\alpha.$$

**18.** The domain of $y$ and the boundary condition at $y = 0$ suggest that we use a Fourier cosine transform. The transformed equation is

$$\frac{d^2 U}{dx^2} - \alpha^2 U - u_y(x,0) = 0 \quad \text{or} \quad \frac{d^2 U}{dx^2} - \alpha^2 U = 0.$$

Because the domain of the variable $x$ is a finite interval we choose to write the general solution of the latter equation as

$$U(x, \alpha) = c_1 \cosh \alpha x + c_2 \sinh \alpha x.$$

Now $U(0, \alpha) = F(\alpha)$, where $F(\alpha)$ is the Fourier cosine transform of $f(y)$, and $U'(\pi, \alpha) = 0$ imply $c_1 = F(\alpha)$ and $c_2 = -F(\alpha) \sinh \alpha \pi / \cosh \alpha \pi$. Thus

$$U(x, \alpha) = F(\alpha) \cosh \alpha x - F(\alpha) \frac{\sinh \alpha \pi}{\cosh \alpha \pi} \sinh \alpha x = F(\alpha) \frac{\cosh \alpha(\pi - x)}{\cosh \alpha \pi}.$$

Using the inverse transform we find that a solution to the problem is

$$u(x, y) = \frac{2}{\pi} \int_0^\infty F(\alpha) \frac{\cosh \alpha(\pi - x)}{\cosh \alpha \pi} \cos \alpha y \, d\alpha.$$

**21.** Using the Fourier transform with respect to $x$ gives

$$U(\alpha, y) = c_1 \cosh \alpha y + c_2 \sinh \alpha y.$$

The transform of the boundary condition $\left. \dfrac{\partial u}{\partial y} \right|_{y=0} = 0$ is $\left. \dfrac{dU}{dy} \right|_{y=0} = 0$. This condition gives $c_2 = 0$.

Hence

$$U(\alpha, y) = c_1 \cosh \alpha y.$$

Now by the given information the transform of the boundary condition $u(x, 1) = e^{-x^2}$ is $U(\alpha, 1) = \sqrt{\pi} \, e^{-\alpha^2/4}$. This condition then gives $c_1 = \sqrt{\pi} \, e^{-\alpha^2/4} \cosh \alpha$. Therefore

$$U(\alpha, y) = \sqrt{\pi} \, \frac{e^{-\alpha^2/4} \cosh \alpha y}{\cosh \alpha}$$

and

$$U(x, y) = \frac{1}{2\sqrt{\pi}} \int_{-\infty}^\infty \frac{e^{-\alpha^2/4} \cosh \alpha y}{\cosh \alpha} e^{-i\alpha x} \, d\alpha$$

$$= \frac{1}{2\sqrt{\pi}} \int_{-\infty}^\infty \frac{e^{-\alpha^2/4} \cosh \alpha y}{\cosh \alpha} \cos \alpha x \, d\alpha$$

$$= \frac{1}{\sqrt{\pi}} \int_0^\infty \frac{e^{-\alpha^2/4} \cosh \alpha y}{\cosh \alpha} \cos \alpha x \, d\alpha.$$

**24.** Using integration by parts,

$$\mathscr{F}_C\{f'(x)\} = \int_0^\infty f'(x) \cos \alpha x \, dx = f(x) \cos \alpha x \Big|_0^\infty + \alpha \int_0^\infty f(x) \sin \alpha x \, dx.$$

If we assume that $f(x) \cos \alpha x \to 0$ as $x \to \infty$ and that $f$ is bounded at $x = 0$, we obtain

$$\mathscr{F}_C\{f'(x)\} = -f(0) + \alpha \, \mathscr{F}_S\{f(x)\}.$$

## Chapter 14 Review Exercises

**3.** The Laplace transform gives

$$U(x,s) = c_1 e^{-\sqrt{s+h}\,x} + c_2 e^{\sqrt{s+h}\,x} + \frac{u_0}{s+h}.$$

The condition $\lim_{x\to\infty} \partial u/\partial x = 0$ implies $\lim_{x\to\infty} dU/dx = 0$ and so we define $c_2 = 0$. Thus

$$U(x,s) = c_1 e^{-\sqrt{s+h}\,x} + \frac{u_0}{s+h}.$$

The condition $U(0,s) = 0$ then gives $c_1 = -u_0/(s+h)$ and so

$$U(x,s) = \frac{u_0}{s+h} - u_0 \frac{e^{-\sqrt{s+h}\,x}}{s+h}.$$

With the help of the first translation theorem we then obtain

$$u(x,t) = u_0 \mathscr{L}^{-1}\left\{\frac{1}{s+h}\right\} - u_0 \mathscr{L}^{-1}\left\{\frac{e^{-\sqrt{s+h}\,x}}{s+h}\right\} = u_0 e^{-ht} - u_0 e^{-ht}\operatorname{erfc}\left(\frac{x}{2\sqrt{t}}\right)$$

$$= u_0 e^{-ht}\left[1 - \operatorname{erfc}\left(\frac{x}{2\sqrt{t}}\right)\right] = u_0 e^{-ht}\operatorname{erf}\left(\frac{x}{2\sqrt{t}}\right).$$

**6.** The Laplace transform and undetermined coefficients gives

$$U(x,s) = c_1 \cosh sx + c_2 \sinh sx + \frac{s-1}{s^2+\pi^2}\sin \pi x.$$

The conditions $U(0,s) = 0$ and $U(1,s) = 0$ give, in turn, $c_1 = 0$ and $c_2 = 0$. Thus

$$U(x,s) = \frac{s-1}{s^2+\pi^2}\sin \pi x$$

and

$$u(x,t) = \sin \pi x \,\mathscr{L}^{-1}\left\{\frac{s}{s^2+\pi^2}\right\} - \frac{1}{\pi}\sin \pi x \,\mathscr{L}^{-1}\left\{\frac{\pi}{s^2+\pi^2}\right\}$$

$$= (\sin \pi x)\cos \pi t - \frac{1}{\pi}(\sin \pi x)\sin \pi t.$$

**9.** We solve the two problems

$$\frac{\partial^2 u_1}{\partial x^2} + \frac{\partial^2 u_1}{\partial y^2} = 0, \quad x > 0, \quad y > 0,$$

$$u_1(0,y) = 0, \quad y > 0,$$

$$u_1(x,0) = \begin{cases} 100, & 0 < x < 1 \\ 0, & x > 1 \end{cases}$$

and

$$\frac{\partial^2 u_2}{\partial x^2} + \frac{\partial^2 u_2}{\partial y^2} = 0, \quad x > 0, \quad y > 0,$$

$$u_2(0, y) = \begin{cases} 50, & 0 < y < 1 \\ 0, & y > 1 \end{cases}$$

$$u_2(x, 0) = 0.$$

Using the Fourier sine transform with respect to $x$ we find

$$u_1(x, y) = \frac{200}{\pi} \int_0^\infty \left( \frac{1 - \cos \alpha}{\alpha} \right) e^{-\alpha y} \sin \alpha x \, d\alpha.$$

Using the Fourier sine transform with respect to $y$ we find

$$u_2(x, y) = \frac{100}{\pi} \int_0^\infty \left( \frac{1 - \cos \alpha}{\alpha} \right) e^{-\alpha x} \sin \alpha y \, d\alpha.$$

The solution of the problem is then

$$u(x, y) = u_1(x, y) + u_2(x, y).$$

**12.** Using the Fourier transform gives

$$U(\alpha, t) = c_1 e^{-k\alpha^2 t}.$$

Now

$$u(\alpha, 0) = \int_0^\infty e^{-x} e^{i\alpha x} \, dx = \frac{e^{(i\alpha - 1)x}}{i\alpha - 1} \Big|_0^\infty = 0 - \frac{1}{i\alpha - 1} = \frac{1}{1 - i\alpha} = c_1$$

so

$$U(\alpha, t) = \frac{1 + i\alpha}{1 + \alpha^2} e^{-k\alpha^2 t}$$

and

$$u(x, t) = \frac{1}{2\pi} \int_{-\infty}^\infty \frac{1 + i\alpha}{1 + \alpha^2} e^{-k\alpha^2 t} e^{-i\alpha x} \, d\alpha.$$

Since

$$\frac{1 + i\alpha}{1 + \alpha^2} (\cos \alpha x - i \sin \alpha x) = \frac{\cos \alpha x + \alpha \sin \alpha x}{1 + \alpha^2} + \frac{i(\alpha \cos \alpha x - \sin \alpha x)}{1 + \alpha^2}$$

and the integral of the product of the second term with $e^{-k\alpha^2 t}$ is 0 (it is an odd function), we have

$$u(x, t) = \frac{1}{2\pi} \int_{-\infty}^\infty \frac{\cos \alpha x + \alpha \sin \alpha x}{1 + \alpha^2} e^{-k\alpha^2 t} \, d\alpha.$$

# 15 Numerical Methods for Partial Differential Equations

―――――――― **Exercises 15.1** ――――――――

**3.** The figure shows the values of $u(x,y)$ along the boundary. We need to determine $u_{11}, u_{21}, u_{12}$, and $u_{22}$. By symmetry $u_{11} = u_{21}$ and $u_{12} = u_{22}$. The system is

$$u_{21} + u_{12} + 0 + 0 - 4u_{11} = 0$$

$$0 + u_{22} + u_{11} + 0 - 4u_{21} = 0 \qquad \qquad 3u_{11} + u_{12} = 0$$

$$\text{or}$$

$$u_{22} + \sqrt{3}/2 + 0 + u_{11} - 4u_{12} = 0 \qquad u_{11} - 3u_{12} = -\frac{\sqrt{3}}{2}.$$

$$0 + \sqrt{3}/2 + u_{12} + u_{21} - 4u_{22} = 0$$

Solving we obtain $u_{11} = u_{21} = \sqrt{3}/16$ and $u_{12} = u_{22} = 3\sqrt{3}/16$.

**6.** For Gauss-Seidel the coefficients of the unknowns $u_{11}, u_{21}, u_{31}, u_{12}, u_{22}, u_{32}, u_{13}, u_{23}, u_{33}$ are shown in the matrix

$$\begin{bmatrix}
0 & .25 & 0 & .25 & 0 & 0 & 0 & 0 & 0 \\
.25 & 0 & .25 & 0 & .25 & 0 & 0 & 0 & 0 \\
0 & .25 & 0 & 0 & 0 & .25 & 0 & 0 & 0 \\
.25 & 0 & 0 & 0 & .25 & 0 & .25 & 0 & 0 \\
0 & .25 & 0 & .25 & 0 & .25 & 0 & .25 & 0 \\
0 & 0 & .25 & 0 & .25 & 0 & 0 & 0 & .25 \\
0 & 0 & 0 & .25 & 0 & 0 & 0 & .25 & 0 \\
0 & 0 & 0 & 0 & .25 & 0 & .25 & 0 & .25 \\
0 & 0 & 0 & 0 & 0 & .25 & 0 & .25 & 0
\end{bmatrix}$$

The constant terms are 7.5, 5, 20, 10, 0, 15, 17.5, 5, 27.5. We use 32.5 as the initial guess for each variable. Then $u_{11} = 21.92$, $u_{21} = 28.30$, $u_{31} = 38.17$, $u_{12} = 29.38$, $u_{22} = 33.13$, $u_{32} = 44.38$, $u_{13} = 22.46$, $u_{23} = 30.45$, and $u_{33} = 46.21$.

**9.** Identifying $u_{ij} = u(x,t)$ the difference equation is given by

$$\frac{1}{h^2}(u_{i+1,j} - 2u_{ij} + u_{i-1,j}) = \frac{1}{k}(u_{i,j+1} - u_{ij})$$

or

$$u_{i,j+1} = \left(1 - \frac{2k}{h^2}\right)u_{ij} + \frac{k}{h^2}(u_{i+1,j} + u_{i-1,j}).$$

# Exercises 15.2

3. We identify $c = 1$, $a = 2$, $T = 1$, $n = 8$, and $m = 40$. Then $h = 2/8 = 0.25$, $k = 1/40 = 0.025$, and $\lambda = 2/5 = 0.4$.

| TIME | X=0.25 | X=0.50 | X=0.75 | X=1.00 | X=1.25 | X=1.50 | X=1.75 |
|------|--------|--------|--------|--------|--------|--------|--------|
| 0.000 | 1.0000 | 1.0000 | 1.0000 | 1.0000 | 0.0000 | 0.0000 | 0.0000 |
| 0.025 | 0.7074 | 0.9520 | 0.9566 | 0.7444 | 0.2545 | 0.0371 | 0.0053 |
| 0.050 | 0.5606 | 0.8499 | 0.8685 | 0.6633 | 0.3303 | 0.1034 | 0.0223 |
| 0.075 | 0.4684 | 0.7473 | 0.7836 | 0.6191 | 0.3614 | 0.1529 | 0.0462 |
| 0.100 | 0.4015 | 0.6577 | 0.7084 | 0.5837 | 0.3753 | 0.1871 | 0.0684 |
| 0.125 | 0.3492 | 0.5821 | 0.6428 | 0.5510 | 0.3797 | 0.2101 | 0.0861 |
| 0.150 | 0.3069 | 0.5187 | 0.5857 | 0.5199 | 0.3778 | 0.2247 | 0.0990 |
| 0.175 | 0.2721 | 0.4652 | 0.5359 | 0.4901 | 0.3716 | 0.2329 | 0.1078 |
| 0.200 | 0.2430 | 0.4198 | 0.4921 | 0.4617 | 0.3622 | 0.2362 | 0.1132 |
| 0.225 | 0.2186 | 0.3809 | 0.4533 | 0.4348 | 0.3507 | 0.2358 | 0.1160 |
| 0.250 | 0.1977 | 0.3473 | 0.4189 | 0.4093 | 0.3378 | 0.2327 | 0.1166 |
| 0.275 | 0.1798 | 0.3181 | 0.3881 | 0.3853 | 0.3240 | 0.2275 | 0.1157 |
| 0.300 | 0.1643 | 0.2924 | 0.3604 | 0.3626 | 0.3097 | 0.2208 | 0.1136 |
| 0.325 | 0.1507 | 0.2697 | 0.3353 | 0.3412 | 0.2953 | 0.2131 | 0.1107 |
| 0.350 | 0.1387 | 0.2495 | 0.3125 | 0.3211 | 0.2808 | 0.2047 | 0.1071 |
| 0.375 | 0.1281 | 0.2313 | 0.2916 | 0.3021 | 0.2666 | 0.1960 | 0.1032 |
| 0.400 | 0.1187 | 0.2150 | 0.2725 | 0.2843 | 0.2528 | 0.1871 | 0.0989 |
| 0.425 | 0.1102 | 0.2002 | 0.2549 | 0.2675 | 0.2393 | 0.1781 | 0.0946 |
| 0.450 | 0.1025 | 0.1867 | 0.2387 | 0.2517 | 0.2263 | 0.1692 | 0.0902 |
| 0.475 | 0.0955 | 0.1743 | 0.2236 | 0.2368 | 0.2139 | 0.1606 | 0.0858 |
| 0.500 | 0.0891 | 0.1630 | 0.2097 | 0.2228 | 0.2020 | 0.1521 | 0.0814 |
| 0.525 | 0.0833 | 0.1525 | 0.1967 | 0.2096 | 0.1906 | 0.1439 | 0.0772 |
| 0.550 | 0.0779 | 0.1429 | 0.1846 | 0.1973 | 0.1798 | 0.1361 | 0.0731 |
| 0.575 | 0.0729 | 0.1339 | 0.1734 | 0.1856 | 0.1696 | 0.1285 | 0.0691 |
| 0.600 | 0.0683 | 0.1256 | 0.1628 | 0.1746 | 0.1598 | 0.1214 | 0.0653 |
| 0.625 | 0.0641 | 0.1179 | 0.1530 | 0.1643 | 0.1506 | 0.1145 | 0.0617 |
| 0.650 | 0.0601 | 0.1106 | 0.1438 | 0.1546 | 0.1419 | 0.1080 | 0.0582 |
| 0.675 | 0.0564 | 0.1039 | 0.1351 | 0.1455 | 0.1336 | 0.1018 | 0.0549 |
| 0.700 | 0.0530 | 0.0976 | 0.1270 | 0.1369 | 0.1259 | 0.0959 | 0.0518 |
| 0.725 | 0.0497 | 0.0917 | 0.1194 | 0.1288 | 0.1185 | 0.0904 | 0.0488 |
| 0.750 | 0.0467 | 0.0862 | 0.1123 | 0.1212 | 0.1116 | 0.0852 | 0.0460 |
| 0.775 | 0.0439 | 0.0810 | 0.1056 | 0.1140 | 0.1050 | 0.0802 | 0.0433 |
| 0.800 | 0.0413 | 0.0762 | 0.0993 | 0.1073 | 0.0989 | 0.0755 | 0.0408 |
| 0.825 | 0.0388 | 0.0716 | 0.0934 | 0.1009 | 0.0931 | 0.0711 | 0.0384 |
| 0.850 | 0.0365 | 0.0674 | 0.0879 | 0.0950 | 0.0876 | 0.0669 | 0.0362 |
| 0.875 | 0.0343 | 0.0633 | 0.0827 | 0.0894 | 0.0824 | 0.0630 | 0.0341 |
| 0.900 | 0.0323 | 0.0596 | 0.0778 | 0.0841 | 0.0776 | 0.0593 | 0.0321 |
| 0.925 | 0.0303 | 0.0560 | 0.0732 | 0.0791 | 0.0730 | 0.0558 | 0.0302 |
| 0.950 | 0.0285 | 0.0527 | 0.0688 | 0.0744 | 0.0687 | 0.0526 | 0.0284 |
| 0.975 | 0.0268 | 0.0496 | 0.0647 | 0.0700 | 0.0647 | 0.0495 | 0.0268 |
| 1.000 | 0.0253 | 0.0466 | 0.0609 | 0.0659 | 0.0608 | 0.0465 | 0.0252 |

| (x,y) | exact | approx | abs error |
|-------|-------|--------|-----------|
| (0.25,0.1) | 0.3794 | 0.4015 | 0.0221 |
| (1,0.5) | 0.1854 | 0.2228 | 0.0374 |
| (1.5,0.8) | 0.0623 | 0.0755 | 0.0132 |

**6. (a)** We identify $c = 15/88 \approx 0.1705$, $a = 20$, $T = 10$, $n = 10$, and $m = 10$. Then $h = 2$, $k = 1$, and $\lambda = 15/352 \approx 0.0426$.

| TIME | X=2 | X=4 | X=6 | X=8 | X=10 | X=12 | X=14 | X=16 | X=18 |
|---|---|---|---|---|---|---|---|---|---|
| 0 | 30.0000 | 30.0000 | 30.0000 | 30.0000 | 30.0000 | 30.0000 | 30.0000 | 30.0000 | 30.0000 |
| 1 | 28.7216 | 30.0000 | 30.0000 | 30.0000 | 30.0000 | 30.0000 | 30.0000 | 30.0000 | 28.7216 |
| 2 | 27.5521 | 29.9455 | 30.0000 | 30.0000 | 30.0000 | 30.0000 | 30.0000 | 29.9455 | 27.5521 |
| 3 | 26.4800 | 29.8459 | 29.9977 | 30.0000 | 30.0000 | 30.0000 | 29.9977 | 29.8459 | 26.4800 |
| 4 | 25.4951 | 29.7089 | 29.9913 | 29.9999 | 30.0000 | 29.9999 | 29.9913 | 29.7089 | 25.4951 |
| 5 | 24.5882 | 29.5414 | 29.9796 | 29.9995 | 30.0000 | 29.9995 | 29.9796 | 29.5414 | 24.5882 |
| 6 | 23.7515 | 29.3490 | 29.9618 | 29.9987 | 30.0000 | 29.9987 | 29.9618 | 29.3490 | 23.7515 |
| 7 | 22.9779 | 29.1365 | 29.9373 | 29.9972 | 29.9998 | 29.9972 | 29.9373 | 29.1365 | 22.9779 |
| 8 | 22.2611 | 28.9082 | 29.9057 | 29.9948 | 29.9996 | 29.9948 | 29.9057 | 28.9082 | 22.2611 |
| 9 | 21.5958 | 28.6675 | 29.8670 | 29.9912 | 29.9992 | 29.9912 | 29.8670 | 28.6675 | 21.5958 |
| 10 | 20.9768 | 28.4172 | 29.8212 | 29.9862 | 29.9985 | 29.9862 | 29.8212 | 28.4172 | 20.9768 |

**(b)** We identify $c = 15/88 \approx 0.1705$, $a = 50$, $T = 10$, $n = 10$, and $m = 10$. Then $h = 5$, $k = 1$, and $\lambda = 3/440 \approx 0.0068$.

| TIME | X=5 | X=10 | X=15 | X=20 | X=25 | X=30 | X=35 | X=40 | X=45 |
|---|---|---|---|---|---|---|---|---|---|
| 0 | 30.0000 | 30.0000 | 30.0000 | 30.0000 | 30.0000 | 30.0000 | 30.0000 | 30.0000 | 30.0000 |
| 1 | 29.7955 | 30.0000 | 30.0000 | 30.0000 | 30.0000 | 30.0000 | 30.0000 | 30.0000 | 29.7955 |
| 2 | 29.5937 | 29.9986 | 30.0000 | 30.0000 | 30.0000 | 30.0000 | 30.0000 | 29.9986 | 29.5937 |
| 3 | 29.3947 | 29.9959 | 30.0000 | 30.0000 | 30.0000 | 30.0000 | 30.0000 | 29.9959 | 29.3947 |
| 4 | 29.1984 | 29.9918 | 30.0000 | 30.0000 | 30.0000 | 30.0000 | 30.0000 | 29.9918 | 29.1984 |
| 5 | 29.0047 | 29.9864 | 29.9999 | 30.0000 | 30.0000 | 30.0000 | 29.9999 | 29.9864 | 29.0047 |
| 6 | 28.8136 | 29.9798 | 29.9998 | 30.0000 | 30.0000 | 30.0000 | 29.9998 | 29.9798 | 28.8136 |
| 7 | 28.6251 | 29.9720 | 29.9997 | 30.0000 | 30.0000 | 30.0000 | 29.9997 | 29.9720 | 28.6251 |
| 8 | 28.4391 | 29.9630 | 29.9995 | 30.0000 | 30.0000 | 30.0000 | 29.9995 | 29.9630 | 28.4391 |
| 9 | 28.2556 | 29.9529 | 29.9992 | 30.0000 | 30.0000 | 30.0000 | 29.9992 | 29.9529 | 28.2556 |
| 10 | 28.0745 | 29.9416 | 29.9989 | 30.0000 | 30.0000 | 30.0000 | 29.9989 | 29.9416 | 28.0745 |

**(c)** We identify $c = 50/27 \approx 1.8519$, $a = 20$, $T = 10$, $n = 10$, and $m = 10$. Then $h = 2$, $k = 1$, and $\lambda = 25/54 \approx 0.4630$.

| TIME | X=2 | X=4 | X=6 | X=8 | X=10 | X=12 | X=14 | X=16 | X=18 |
|---|---|---|---|---|---|---|---|---|---|
| 0 | 18.0000 | 32.0000 | 42.0000 | 48.0000 | 50.0000 | 48.0000 | 42.0000 | 32.0000 | 18.0000 |
| 1 | 16.1481 | 30.1481 | 40.1481 | 46.1481 | 48.1481 | 46.1481 | 40.1481 | 30.1481 | 16.1481 |
| 2 | 15.1536 | 28.2963 | 38.2963 | 44.2963 | 46.2963 | 44.2963 | 38.2963 | 28.2963 | 15.1536 |
| 3 | 14.2226 | 26.8414 | 36.4444 | 42.4444 | 44.4444 | 42.4444 | 36.4444 | 26.8414 | 14.2226 |
| 4 | 13.4801 | 25.4452 | 34.7764 | 40.5926 | 42.5926 | 40.5926 | 34.7764 | 25.4452 | 13.4801 |
| 5 | 12.7787 | 24.2258 | 33.1491 | 38.8258 | 40.7407 | 38.8258 | 33.1491 | 24.2258 | 12.7787 |
| 6 | 12.1622 | 23.0574 | 31.6460 | 37.0842 | 38.9677 | 37.0842 | 31.6460 | 23.0574 | 12.1622 |
| 7 | 11.5756 | 21.9895 | 30.1875 | 35.4385 | 37.2238 | 35.4385 | 30.1875 | 21.9895 | 11.5756 |
| 8 | 11.0378 | 20.9636 | 28.8232 | 33.8340 | 35.5707 | 33.8340 | 28.8232 | 20.9636 | 11.0378 |
| 9 | 10.5230 | 20.0070 | 27.5043 | 32.3182 | 33.9626 | 32.3182 | 27.5043 | 20.0070 | 10.5230 |
| 10 | 10.0420 | 19.0872 | 26.2620 | 30.8509 | 32.4400 | 30.8509 | 26.2620 | 19.0872 | 10.0420 |

**(d)** We identify $c = 260/159 \approx 1.6352$, $a = 100$, $T = 10$, $n = 10$, and $m = 10$. Then $h = 10$, $k = 1$, and $\lambda = 13/795 \approx 00164$.

| TIME | X=10 | X=20 | X=30 | X=40 | X=50 | X=60 | X=70 | X=80 | X=90 |
|---|---|---|---|---|---|---|---|---|---|
| 0 | 8.0000 | 16.0000 | 24.0000 | 32.0000 | 40.0000 | 32.0000 | 24.0000 | 16.0000 | 8.0000 |
| 1 | 8.0000 | 16.0000 | 24.0000 | 32.0000 | 39.7384 | 32.0000 | 24.0000 | 16.0000 | 8.0000 |
| 2 | 8.0000 | 16.0000 | 24.0000 | 31.9957 | 39.4853 | 31.9957 | 24.0000 | 16.0000 | 8.0000 |
| 3 | 8.0000 | 16.0000 | 23.9999 | 31.9874 | 39.2403 | 31.9874 | 23.9999 | 16.0000 | 8.0000 |
| 4 | 8.0000 | 16.0000 | 23.9997 | 31.9754 | 39.0031 | 31.9754 | 23.9997 | 16.0000 | 8.0000 |
| 5 | 8.0000 | 16.0000 | 23.9993 | 31.9599 | 38.7733 | 31.9599 | 23.9993 | 16.0000 | 8.0000 |
| 6 | 8.0000 | 16.0000 | 23.9987 | 31.9412 | 38.5505 | 31.9412 | 23.9987 | 16.0000 | 8.0000 |
| 7 | 8.0000 | 16.0000 | 23.9978 | 31.9194 | 38.3343 | 31.9194 | 23.9978 | 16.0000 | 8.0000 |
| 8 | 8.0000 | 15.9999 | 23.9965 | 31.8947 | 38.1245 | 31.8947 | 23.9965 | 15.9999 | 8.0000 |
| 9 | 8.0000 | 15.9999 | 23.9949 | 31.8675 | 37.9208 | 31.8675 | 23.9949 | 15.9999 | 8.0000 |
| 10 | 8.0000 | 15.9998 | 23.9929 | 31.8377 | 37.7228 | 31.8377 | 23.9929 | 15.9998 | 8.0000 |

**9. (a)** We identify $c = 15/88 \approx 0.1705$, $a = 20$, $T = 10$, $n = 10$, and $m = 10$. Then $h = 2$, $k = 1$, and $\lambda = 15/352 \approx 0.0426$.

| TIME | X=2.00 | X=4.00 | X=6.00 | X=8.00 | X=10.00 | X=12.00 | X=14.00 | X=16.00 | X=18.00 |
|---|---|---|---|---|---|---|---|---|---|
| 0.00 | 30.0000 | 30.0000 | 30.0000 | 30.0000 | 30.0000 | 30.0000 | 30.0000 | 30.0000 | 30.0000 |
| 1.00 | 28.7733 | 29.9749 | 29.9995 | 30.0000 | 30.0000 | 30.0000 | 29.9998 | 29.9916 | 29.5911 |
| 2.00 | 27.6450 | 29.9037 | 29.9970 | 29.9999 | 30.0000 | 30.0000 | 29.9990 | 29.9679 | 29.2150 |
| 3.00 | 26.6051 | 29.7938 | 29.9911 | 29.9997 | 30.0000 | 29.9999 | 29.9970 | 29.9313 | 28.8684 |
| 4.00 | 25.6452 | 29.6517 | 29.9805 | 29.9991 | 30.0000 | 29.9997 | 29.9935 | 29.8839 | 28.5484 |
| 5.00 | 24.7573 | 29.4829 | 29.9643 | 29.9981 | 29.9999 | 29.9994 | 29.9881 | 29.8276 | 28.2524 |
| 6.00 | 23.9347 | 29.2922 | 29.9421 | 29.9963 | 29.9997 | 29.9988 | 29.9807 | 29.7641 | 27.9782 |
| 7.00 | 23.1711 | 29.0836 | 29.9134 | 29.9936 | 29.9995 | 29.9979 | 29.9711 | 29.6945 | 27.7237 |
| 8.00 | 22.4612 | 28.8606 | 29.8782 | 29.9899 | 29.9991 | 29.9966 | 29.9594 | 29.6202 | 27.4870 |
| 9.00 | 21.7999 | 28.6263 | 29.8362 | 29.9848 | 29.9985 | 29.9949 | 29.9454 | 29.5421 | 27.2666 |
| 10.00 | 21.1829 | 28.3831 | 29.7878 | 29.9783 | 29.9976 | 29.9927 | 29.9293 | 29.4610 | 27.0610 |

**(b)** We identify $c = 15/88 \approx 0.1705$, $a = 50$, $T = 10$, $n = 10$, and $m = 10$. Then $h = 5$, $k = 1$, and $\lambda = 3/440 \approx 0.0068$.

| TIME | X=5.00 | X=10.00 | X=15.00 | X=20.00 | X=25.00 | X=30.00 | X=35.00 | X=40.00 | X=45.00 |
|---|---|---|---|---|---|---|---|---|---|
| 0.00 | 30.0000 | 30.0000 | 30.0000 | 30.0000 | 30.0000 | 30.0000 | 30.0000 | 30.0000 | 30.0000 |
| 1.00 | 29.7968 | 29.9993 | 30.0000 | 30.0000 | 30.0000 | 30.0000 | 30.0000 | 29.9998 | 29.9323 |
| 2.00 | 29.5964 | 29.9973 | 30.0000 | 30.0000 | 30.0000 | 30.0000 | 30.0000 | 29.9991 | 29.8655 |
| 3.00 | 29.3987 | 29.9939 | 30.0000 | 30.0000 | 30.0000 | 30.0000 | 30.0000 | 29.9980 | 29.7996 |
| 4.00 | 29.2036 | 29.9893 | 29.9999 | 30.0000 | 30.0000 | 30.0000 | 30.0000 | 29.9964 | 29.7345 |
| 5.00 | 29.0112 | 29.9834 | 29.9998 | 30.0000 | 30.0000 | 30.0000 | 29.9999 | 29.9945 | 29.6704 |
| 6.00 | 28.8212 | 29.9762 | 29.9997 | 30.0000 | 30.0000 | 30.0000 | 29.9999 | 29.9921 | 29.6071 |
| 7.00 | 28.6339 | 29.9679 | 29.9995 | 30.0000 | 30.0000 | 30.0000 | 29.9998 | 29.9893 | 29.5446 |
| 8.00 | 28.4490 | 29.9585 | 29.9992 | 30.0000 | 30.0000 | 30.0000 | 29.9997 | 29.9862 | 29.4830 |
| 9.00 | 28.2665 | 29.9479 | 29.9989 | 30.0000 | 30.0000 | 30.0000 | 29.9996 | 29.9827 | 29.4222 |
| 10.00 | 28.0864 | 29.9363 | 29.9986 | 30.0000 | 30.0000 | 30.0000 | 29.9995 | 29.9788 | 29.3621 |

(c) We identify $c = 50/27 \approx 1.8519$, $a = 20$, $T = 10$, $n = 10$, and $m = 10$. Then $h = 2$, $k = 1$, and $\lambda = 25/54 \approx 0.4630$.

| TIME | X=2.00 | X=4.00 | X=6.00 | X=8.00 | X=10.00 | X=12.00 | X=14.00 | X=16.00 | X=18.00 |
|------|--------|--------|--------|--------|---------|---------|---------|---------|---------|
| 0.00 | 18.0000 | 32.0000 | 42.0000 | 48.0000 | 50.0000 | 48.0000 | 42.0000 | 32.0000 | 18.0000 |
| 1.00 | 16.4489 | 30.1970 | 40.1562 | 46.1502 | 48.1531 | 46.1773 | 40.3274 | 31.2520 | 22.9449 |
| 2.00 | 15.3312 | 28.5350 | 38.3477 | 44.3130 | 46.3327 | 44.4671 | 39.0872 | 31.5755 | 24.6930 |
| 3.00 | 14.4219 | 27.0429 | 36.6090 | 42.5113 | 44.5759 | 42.9362 | 38.1976 | 31.7478 | 25.4131 |
| 4.00 | 13.6381 | 25.6913 | 34.9606 | 40.7728 | 42.9127 | 41.5716 | 37.4340 | 31.7086 | 25.6986 |
| 5.00 | 12.9409 | 24.4545 | 33.4091 | 39.1182 | 41.3519 | 40.3240 | 36.7033 | 31.5136 | 25.7663 |
| 6.00 | 12.3088 | 23.3146 | 31.9546 | 37.5566 | 39.8880 | 39.1565 | 35.9745 | 31.2134 | 25.7128 |
| 7.00 | 11.7294 | 22.2589 | 30.5939 | 36.0884 | 38.5109 | 38.0470 | 35.2407 | 30.8434 | 25.5871 |
| 8.00 | 11.1946 | 21.2785 | 29.3217 | 34.7092 | 37.2109 | 36.9834 | 34.5032 | 30.4279 | 25.4167 |
| 9.00 | 10.6987 | 20.3660 | 28.1318 | 33.4130 | 35.9801 | 35.9591 | 33.7660 | 29.9836 | 25.2181 |
| 10.00 | 10.2377 | 19.5150 | 27.0178 | 32.1929 | 34.8117 | 34.9710 | 33.0338 | 29.5224 | 25.0019 |

(d) We identify $c = 260/159 \approx 1.6352$, $a = 100$, $T = 10$, $n = 10$, and $m = 10$. Then $h = 10$, $k = 1$, and $\lambda = 13/795 \approx 00164$.

| TIME | X=10.00 | X=20.00 | X=30.00 | X=40.00 | X=50.00 | X=60.00 | X=70.00 | X=80.00 | X=90.00 |
|------|---------|---------|---------|---------|---------|---------|---------|---------|---------|
| 0.00 | 8.0000 | 16.0000 | 24.0000 | 32.0000 | 40.0000 | 32.0000 | 24.0000 | 16.0000 | 8.0000 |
| 1.00 | 8.0000 | 16.0000 | 24.0000 | 31.9979 | 39.7425 | 31.9979 | 24.0000 | 16.0026 | 8.3218 |
| 2.00 | 8.0000 | 16.0000 | 23.9999 | 31.9918 | 39.4932 | 31.9918 | 24.0000 | 16.0102 | 8.6333 |
| 3.00 | 8.0000 | 16.0000 | 23.9997 | 31.9820 | 39.2517 | 31.9820 | 24.0001 | 16.0225 | 8.9350 |
| 4.00 | 8.0000 | 16.0000 | 23.9993 | 31.9686 | 39.0175 | 31.9687 | 24.0002 | 16.0391 | 9.2272 |
| 5.00 | 8.0000 | 16.0000 | 23.9987 | 31.9520 | 38.7905 | 31.9520 | 24.0003 | 16.0599 | 9.5103 |
| 6.00 | 8.0000 | 15.9999 | 23.9978 | 31.9323 | 38.5701 | 31.9324 | 24.0005 | 16.0845 | 9.7846 |
| 7.00 | 8.0000 | 15.9999 | 23.9966 | 31.9097 | 38.3561 | 31.9098 | 24.0008 | 16.1126 | 10.0506 |
| 8.00 | 8.0000 | 15.9998 | 23.9950 | 31.8844 | 38.1483 | 31.8846 | 24.0012 | 16.1441 | 10.3084 |
| 9.00 | 8.0000 | 15.9997 | 23.9931 | 31.8566 | 37.9463 | 31.8569 | 24.0017 | 16.1786 | 10.5585 |
| 10.00 | 8.0000 | 15.9996 | 23.9908 | 31.8265 | 37.7499 | 31.8269 | 24.0023 | 16.2160 | 10.8012 |

**227**

**3. (a)** Identifying $h = 1/5$ and $k = 0.5/10 = 0.05$ we see that $\lambda = 0.25$.

| TIME | X=0.2 | X=0.4 | X=0.6 | X=0.8 |
|------|-------|-------|-------|-------|
| 0.00 | 0.5878 | 0.9511 | 0.9511 | 0.5878 |
| 0.05 | 0.5808 | 0.9397 | 0.9397 | 0.5808 |
| 0.10 | 0.5599 | 0.9059 | 0.9059 | 0.5599 |
| 0.15 | 0.5256 | 0.8505 | 0.8505 | 0.5256 |
| 0.20 | 0.4788 | 0.7748 | 0.7748 | 0.4788 |
| 0.25 | 0.4206 | 0.6806 | 0.6806 | 0.4206 |
| 0.30 | 0.3524 | 0.5701 | 0.5701 | 0.3524 |
| 0.35 | 0.2757 | 0.4460 | 0.4460 | 0.2757 |
| 0.40 | 0.1924 | 0.3113 | 0.3113 | 0.1924 |
| 0.45 | 0.1046 | 0.1692 | 0.1692 | 0.1046 |
| 0.50 | 0.0142 | 0.0230 | 0.0230 | 0.0142 |

**(b)** Identifying $h = 1/5$ and $k = 0.5/20 = 0.025$ we see that $\lambda = 0.125$.

| TIME | X=0.2 | X=0.4 | X=0.6 | X=0.8 |
|------|-------|-------|-------|-------|
| 0.00 | 0.5878 | 0.9511 | 0.9511 | 0.5878 |
| 0.03 | 0.5860 | 0.9482 | 0.9482 | 0.5860 |
| 0.05 | 0.5808 | 0.9397 | 0.9397 | 0.5808 |
| 0.08 | 0.5721 | 0.9256 | 0.9256 | 0.5721 |
| 0.10 | 0.5599 | 0.9060 | 0.9060 | 0.5599 |
| 0.13 | 0.5445 | 0.8809 | 0.8809 | 0.5445 |
| 0.15 | 0.5257 | 0.8507 | 0.8507 | 0.5257 |
| 0.18 | 0.5039 | 0.8153 | 0.8153 | 0.5039 |
| 0.20 | 0.4790 | 0.7750 | 0.7750 | 0.4790 |
| 0.23 | 0.4513 | 0.7302 | 0.7302 | 0.4513 |
| 0.25 | 0.4209 | 0.6810 | 0.6810 | 0.4209 |
| 0.28 | 0.3879 | 0.6277 | 0.6277 | 0.3879 |
| 0.30 | 0.3527 | 0.5706 | 0.5706 | 0.3527 |
| 0.33 | 0.3153 | 0.5102 | 0.5102 | 0.3153 |
| 0.35 | 0.2761 | 0.4467 | 0.4467 | 0.2761 |
| 0.38 | 0.2352 | 0.3806 | 0.3806 | 0.2352 |
| 0.40 | 0.1929 | 0.3122 | 0.3122 | 0.1929 |
| 0.43 | 0.1495 | 0.2419 | 0.2419 | 0.1495 |
| 0.45 | 0.1052 | 0.1701 | 0.1701 | 0.1052 |
| 0.48 | 0.0602 | 0.0974 | 0.0974 | 0.0602 |
| 0.50 | 0.0149 | 0.0241 | 0.0241 | 0.0149 |

**6.** We identify $c = 24944.4$, $k = 0.00010022$ seconds $= 0.10022$ milliseconds, and $\lambda = 0.25$. Time in the table is expressed in milliseconds.

| TIME | X=10 | X=20 | X=30 | X=40 | X=50 |
|---|---|---|---|---|---|
| 0.00000 | 0.2000 | 0.2667 | 0.2000 | 0.1333 | 0.0667 |
| 0.10022 | 0.1958 | 0.2625 | 0.2000 | 0.1333 | 0.0667 |
| 0.20045 | 0.1836 | 0.2503 | 0.1997 | 0.1333 | 0.0667 |
| 0.30067 | 0.1640 | 0.2307 | 0.1985 | 0.1333 | 0.0667 |
| 0.40089 | 0.1384 | 0.2050 | 0.1952 | 0.1332 | 0.0667 |
| 0.50111 | 0.1083 | 0.1744 | 0.1886 | 0.1328 | 0.0667 |
| 0.60134 | 0.0755 | 0.1407 | 0.1777 | 0.1318 | 0.0666 |
| 0.70156 | 0.0421 | 0.1052 | 0.1615 | 0.1295 | 0.0665 |
| 0.80178 | 0.0100 | 0.0692 | 0.1399 | 0.1253 | 0.0661 |
| 0.90201 | -0.0190 | 0.0340 | 0.1129 | 0.1184 | 0.0654 |
| 1.00223 | -0.0435 | 0.0004 | 0.0813 | 0.1077 | 0.0638 |
| 1.10245 | -0.0626 | -0.0309 | 0.0464 | 0.0927 | 0.0610 |
| 1.20268 | -0.0758 | -0.0593 | 0.0095 | 0.0728 | 0.0564 |
| 1.30290 | -0.0832 | -0.0845 | -0.0278 | 0.0479 | 0.0493 |
| 1.40312 | -0.0855 | -0.1060 | -0.0639 | 0.0184 | 0.0390 |
| 1.50334 | -0.0837 | -0.1237 | -0.0974 | -0.0150 | 0.0250 |
| 1.60357 | -0.0792 | -0.1371 | -0.1275 | -0.0511 | 0.0069 |
| 1.70379 | -0.0734 | -0.1464 | -0.1533 | -0.0882 | -0.0152 |
| 1.80401 | -0.0675 | -0.1515 | -0.1747 | -0.1249 | -0.0410 |
| 1.90424 | -0.0627 | -0.1528 | -0.1915 | -0.1595 | -0.0694 |
| 2.00446 | -0.0596 | -0.1509 | -0.2039 | -0.1904 | -0.0991 |
| 2.10468 | -0.0585 | -0.1467 | -0.2122 | -0.2165 | -0.1283 |
| 2.20491 | -0.0592 | -0.1410 | -0.2166 | -0.2368 | -0.1551 |
| 2.30513 | -0.0614 | -0.1349 | -0.2175 | -0.2507 | -0.1772 |
| 2.40535 | -0.0643 | -0.1294 | -0.2154 | -0.2579 | -0.1929 |
| 2.50557 | -0.0672 | -0.1251 | -0.2105 | -0.2585 | -0.2005 |
| 2.60580 | -0.0696 | -0.1227 | -0.2033 | -0.2524 | -0.1993 |
| 2.70602 | -0.0709 | -0.1219 | -0.1942 | -0.2399 | -0.1889 |
| 2.80624 | -0.0710 | -0.1225 | -0.1833 | -0.2214 | -0.1699 |
| 2.90647 | -0.0699 | -0.1236 | -0.1711 | -0.1972 | -0.1435 |
| 3.00669 | -0.0678 | -0.1244 | -0.1575 | -0.1681 | -0.1115 |
| 3.10691 | -0.0649 | -0.1237 | -0.1425 | -0.1348 | -0.0761 |
| 3.20713 | -0.0617 | -0.1205 | -0.1258 | -0.0983 | -0.0395 |
| 3.30736 | -0.0583 | -0.1139 | -0.1071 | -0.0598 | -0.0042 |
| 3.40758 | -0.0547 | -0.1035 | -0.0859 | -0.0209 | 0.0279 |
| 3.50780 | -0.0508 | -0.0889 | -0.0617 | 0.0171 | 0.0552 |
| 3.60803 | -0.0460 | -0.0702 | -0.0343 | 0.0525 | 0.0767 |
| 3.70825 | -0.0399 | -0.0478 | -0.0037 | 0.0840 | 0.0919 |
| 3.80847 | -0.0318 | -0.0221 | 0.0297 | 0.1106 | 0.1008 |
| 3.90870 | -0.0211 | 0.0062 | 0.0648 | 0.1314 | 0.1041 |
| 4.00892 | -0.0074 | 0.0365 | 0.1005 | 0.1464 | 0.1025 |
| 4.10914 | 0.0095 | 0.0680 | 0.1350 | 0.1558 | 0.0973 |
| 4.20936 | 0.0295 | 0.1000 | 0.1666 | 0.1602 | 0.0897 |
| 4.30959 | 0.0521 | 0.1318 | 0.1937 | 0.1606 | 0.0808 |
| 4.40981 | 0.0764 | 0.1625 | 0.2148 | 0.1581 | 0.0719 |
| 4.51003 | 0.1013 | 0.1911 | 0.2291 | 0.1538 | 0.0639 |
| 4.61026 | 0.1254 | 0.2164 | 0.2364 | 0.1485 | 0.0575 |
| 4.71048 | 0.1475 | 0.2373 | 0.2369 | 0.1431 | 0.0532 |
| 4.81070 | 0.1659 | 0.2526 | 0.2315 | 0.1379 | 0.0512 |
| 4.91093 | 0.1794 | 0.2611 | 0.2217 | 0.1331 | 0.0514 |
| 5.01115 | 0.1867 | 0.2620 | 0.2087 | 0.1288 | 0.0535 |

# Chapter 15 Review Exercises

**3. (a)**

| TIME | X=0.0 | X=0.2 | X=0.4 | X=0.6 | X=0.8 | X=1.0 |
|------|-------|-------|-------|-------|-------|-------|
| 0.00 | 0.0000 | 0.2000 | 0.4000 | 0.6000 | 0.8000 | 0.0000 |
| 0.01 | 0.0000 | 0.2000 | 0.4000 | 0.6000 | 0.5500 | 0.0000 |
| 0.02 | 0.0000 | 0.2000 | 0.4000 | 0.5375 | 0.4250 | 0.0000 |
| 0.03 | 0.0000 | 0.2000 | 0.3844 | 0.4750 | 0.3469 | 0.0000 |
| 0.04 | 0.0000 | 0.1961 | 0.3609 | 0.4203 | 0.2922 | 0.0000 |
| 0.05 | 0.0000 | 0.1883 | 0.3346 | 0.3734 | 0.2512 | 0.0000 |

**(b)**

| TIME | X=0.0 | X=0.2 | X=0.4 | X=0.6 | X=0.8 | X=1.0 |
|------|-------|-------|-------|-------|-------|-------|
| 0.00 | 0.0000 | 0.2000 | 0.4000 | 0.6000 | 0.8000 | 0.0000 |
| 0.01 | 0.0000 | 0.2000 | 0.4000 | 0.6000 | 0.8000 | 0.0000 |
| 0.02 | 0.0000 | 0.2000 | 0.4000 | 0.6000 | 0.5500 | 0.0000 |
| 0.03 | 0.0000 | 0.2000 | 0.4000 | 0.5375 | 0.4250 | 0.0000 |
| 0.04 | 0.0000 | 0.2000 | 0.3844 | 0.4750 | 0.3469 | 0.0000 |
| 0.05 | 0.0000 | 0.1961 | 0.3609 | 0.4203 | 0.2922 | 0.0000 |

**(c)** The table in part (b) is the same as the table in part (a) shifted downward one row.

# Appendix

## ———— Appendix I ————

**3.** If $t = x^3$, then $dt = 3x^2 \, dx$ and $x^4 \, dx = \frac{1}{3} t^{2/3} \, dt$. Now

$$\int_0^\infty x^4 e^{-x^3} \, dx = \int_0^\infty \frac{1}{3} t^{2/3} e^{-t} \, dt = \frac{1}{3} \int_0^\infty t^{2/3} e^{-t} \, dt$$

$$= \frac{1}{3} \Gamma\left(\frac{5}{3}\right) = \frac{1}{3}(0.89) \approx 0.297.$$

**6.** For $x > 0$

$$\Gamma(x + 1) = \int_0^\infty t^x e^{-t} dt$$

| | |
|---|---|
| $u = t^x$ | $dv = e^{-t} \, dt$ |
| $du = xt^{x-1} \, dt$ | $v = -e^{-t}$ |

$$= -t^x e^{-t} \Big|_0^\infty - \int_0^\infty xt^{x-1}(-e^{-t}) \, dt$$

$$= x \int_0^\infty t^{x-1} e^{-t} dt = x\Gamma(x).$$

## ———— Appendix III ————

**3.** Expanding by the first row gives

$$\begin{vmatrix} 2 & 0 & 5 \\ 0 & 7 & 9 \\ -6 & 1 & 4 \end{vmatrix} = 2 \begin{vmatrix} 7 & 9 \\ 1 & 4 \end{vmatrix} + 5 \begin{vmatrix} 0 & 7 \\ -6 & 1 \end{vmatrix} = 2(28 - 9) + 5(0 + 42) = 248.$$

**6.** Expanding by the third row gives

$$\begin{vmatrix} 1 & 0 & 9 & 0 & 3 \\ 2 & 1 & 7 & 0 & 0 \\ 0 & 0 & 2 & 0 & 0 \\ -1 & 1 & 5 & 2 & 2 \\ 2 & 2 & 8 & 1 & 1 \end{vmatrix} = 2 \begin{vmatrix} 1 & 0 & 0 & 3 \\ 2 & 1 & 0 & 0 \\ -1 & 1 & 2 & 2 \\ 2 & 2 & 1 & 1 \end{vmatrix} = 2 \left( 1 \begin{vmatrix} 1 & 0 & 0 \\ 1 & 2 & 2 \\ 2 & 1 & 1 \end{vmatrix} - 3 \begin{vmatrix} 2 & 1 & 0 \\ -1 & 1 & 2 \\ 2 & 2 & 1 \end{vmatrix} \right)$$

$$= 2(1) \begin{vmatrix} 2 & 2 \\ 1 & 1 \end{vmatrix} + 2(-3) \left( 2 \begin{vmatrix} 1 & 2 \\ 2 & 1 \end{vmatrix} - 1 \begin{vmatrix} -1 & 2 \\ 2 & 1 \end{vmatrix} \right)$$

$$= 2(1)(2 - 2) + 2(-3)[2(1 - 4) - (-1 - 4)] = 6.$$

**9.** We first compute

$$\begin{vmatrix} 2 & 1 \\ 3 & 2 \end{vmatrix} = 1.$$

Then

$$x = \frac{1}{1}\begin{vmatrix} 1 & 1 \\ -2 & 2 \end{vmatrix} = 4 \quad \text{and} \quad y = \frac{1}{1}\begin{vmatrix} 2 & 1 \\ 3 & -2 \end{vmatrix} = -7.$$

**12.** We first compute

$$\begin{vmatrix} 4 & 3 & 2 \\ -1 & 0 & 2 \\ 3 & 2 & 1 \end{vmatrix} = 1.$$

Then

$$x = \begin{vmatrix} 8 & 3 & 2 \\ 12 & 0 & 2 \\ 3 & 2 & 1 \end{vmatrix} = -2, \quad y = \begin{vmatrix} 4 & 8 & 2 \\ -1 & 12 & 2 \\ 3 & 3 & 1 \end{vmatrix} = 2, \quad \text{and} \quad z = \begin{vmatrix} 4 & 3 & 8 \\ -1 & 0 & 12 \\ 3 & 2 & 3 \end{vmatrix} = 5.$$

# Appendix IV

**3.** $2z_1 - 3z_2 = 2(2 - i) - 3(5 + 3i) = -11 - 11i$

**6.** $\bar{z}_1(i + z_2) = (2 + i)(5 + 4i) = 10 + 13i + 4i^2 = 6 + 13i$

**9.** $\dfrac{1}{z_2} = \dfrac{1}{5 + 3i}\dfrac{5 - 3i}{5 - 3i} = \dfrac{5 - 3i}{25 + 9} = \dfrac{5}{34} - \dfrac{3}{34}i$

**12.** The modulus is $r = \sqrt{0^2 + 4^2} = 4$ and $\theta = -\pi/2$, so

$$z = 4e^{-i\pi/2}.$$

**15.** The modulus is $r = \sqrt{2^2 + 2^2} = 2\sqrt{2}$ and $\theta = \pi/4$, so

$$z = 2\sqrt{2}e^{i\pi/4}.$$

**18.** The modulus is $r = \sqrt{10^2(3) + 10^2} = 20$ and $\tan\theta = 10/(-10\sqrt{3}) = -\sqrt{3}/3$, where $\theta$ is in the second quadrant, so $\theta = 5\pi/6$ and

$$z = 20e^{5i\pi/6}.$$

**21.** $z = 8e^{-i\pi} = 8[\cos(-\pi) + i\sin(-\pi)] = -8$

**24.** We first express $1 + i$ in polar form as

$$1 + i = \sqrt{2}\left(\cos\frac{\pi}{4} + i\sin\frac{\pi}{4}\right).$$

Then

$$(1 + i)^{10} = (\sqrt{2})^{10}\left(\cos\frac{10\pi}{4} + i\sin\frac{10\pi}{4}\right) = 32\left(\cos\frac{5\pi}{2} + i\sin\frac{5\pi}{2}\right) = 32(0 + i) = 32i.$$